Intermediate
Statistical
Mechanics

Intermediate Statistical Mechanics

Jayanta Bhattacharjee
Harish-Chandra Research Institute, Allahabad, India

Dhruba Banerjee
Jadavpur University, India

HINDUSTAN
BOOK AGENCY

NEW JERSEY · LONDON · SINGAPORE · BEIJING · SHANGHAI · HONG KONG · TAIPEI · CHENNAI · TOKYO

Published by

World Scientific Publishing Co. Pte. Ltd.

5 Toh Tuck Link, Singapore 596224

USA office: 27 Warren Street, Suite 401-402, Hackensack, NJ 07601

UK office: 57 Shelton Street, Covent Garden, London WC2H 9HE

British Library Cataloguing-in-Publication Data
A catalogue record for this book is available from the British Library.

ISBN 978-981-3201-14-9

Printed in India, bookbinding made in Singapore

Preface

Why another book on statistical physics? A large number of excellent texts exist. The amazing thing about statistical physics texts is that virtually no two authors (or set of authors) have the same point of view. The differences in points of view can occur at the level of introducing the subject and extend to deciding what applications should be included. After thirty years of teaching statistical physics courses at different institutions in India, namely Indian Institute of Technology (Kanpur), Indian Association for the Cultivation of Science (Kolkata), S.N.Bose National Centre for Sciences (Kolkata), Jadavpur University (Kolkata) and Harish-Chandra Research Institute (Allahabad), we feel that we have a reasonably balanced point of view about what should appear in a textbook designed for the senior undergraduate and first year graduate student. Hence this effort of putting our thoughts in print.

Statistical physics deals with macroscopic (or at best mesoscopic) systems – systems with a large number of particles whose behaviour can only be followed on an average from a practical point of view. The dynamics of individual particles follow Newton's equations or Schrodinger equation, as the case may be. The passage from a microscopic description of individual particles to a statistical description of a large number of them lays the foundation of statistical mechanics and we decided to follow this passage carefully, in the process introducing the notions of dynamical systems and chaotic dynamics. The microcanonical ensemble is a direct outcome and hence, contrary to usual practice, we have discussed a large number of non-interacting systems, both classical and quantum, within the ambit of microcanonical ensemble.

Close attention has been paid to the connection between thermodynamics and the canonical and the grand canonical distributions. The thermodynamic properties of a variety of non-interacting and interacting systems have been obtained as the distribution functions of the different ensembles. The study of correlation functions has been introduced to give a flavour of the techniques used to describe liquids. For quantum systems, a few non-standard topics have been included like correlation functions in a free Fermi gas, Peierl's instability, the quantum Hall effects, Bose-Einstein condensation and effects of interactions in a trapped gas, Casimir effect etc. After treating fermions and bosons separately, a qualitative analysis of superconductivity where two fermions form a lightly bound boson has been discussed in detail. Critical phenomena have been studied by introducing the renormalization group in both real and momentum spaces. As an application of these ideas, disordered systems, with emphasis on models of quenched disorder and electron localization, have also been included.

The last one-third of the text has been devoted to the dynamics of evolution of the probability distribution and the various moments. In this context Boltzmann equation, Langevin equation and Fokker-Planck equations have been extensively studied for both classical and quantum particles. Examples have been included that are not usually included in traditional texts. We hope that instructors and students will find the text useful.

Acknowledgments

This book has been a long time in the writing, and also in the polishing-up to get it into final shape. Any publishers other than the Hindustan Book Agency would have given up on us long ago. They, as well as the Managing Editors of TRiPS, the series under which this book will appear, have shown the patience of Job. In the Old Testament the patience of Job eventually paid off; we can only hope that in this case the persistence of our Editors and Publishers have not been in vain. Our sincerest thanks to these long suffering managers of this virtually interminable project.

It is a pleasure to thank Ms. Debapriya Das for her technical help in putting the manuscript together. One of us (JKB) is deeply indebted to Prof Sagar Chakraborty of IIT Kanpur for a thorough reading of an older version of the text and numerous suggestions that have helped to improve it and to make it more accurate. He would also like to acknowledge several helpful conversations and correspondence with Mr D. K. Jain and Prof. Ram Ramaswamy. And finally, the visions and revisions of the last few years have been made possible by the amazing academic environment provided by the Harishchandra Research Institute, Allahabad which is impossible to describe unless one has actually experienced it.

Jayanta Bhattacharjee

Dhruba Banerjee

Contents

1

Statistical Mechanics: The Basics

1.1   From Micro to Macro

Thermodynamics describes the state of a system in terms of macroscopic variables like pressure, volume, temperature, energy, entropy etc. The laws of thermodynamics relate changes in these macroscopic variables. In the process thermodynamics requires some inputs from elsewhere. The best known example is that of an equation of state (i.e., relation between macroscopic variables) for an ideal gas, written as $PV = nRT$, where P is the pressure and V the volume of n moles of the gas. The temperature of the gas is T and R is the gas constant. Similarly for a non ideal gas the pressure is given by Van der Waals equation of state. One also needs to know that the molar specific heat of an ideal monatomic gas at constant volume is $(3/2)R$. Thermodynamics cannot calculate the pressure or the specific heat. This is because these calculations entail a microscopic picture that depends on the details of inter-molecular interactions. To calculate the pressure one needs to know what is the force exerted by the gas molecules on the walls of the container. This question has to do with knowledge of the dynamics of each of the particles. In principle this dynamics can be specified if one knows how a given molecule interacts with all the other molecules. All the other molecules mean an Avogadro number of them i.e. of the order of 10^{23}. In practice then, one would not dream of doing this calculation. How does one arrive at an answer?

Let us consider the wall which is parallel to the yz plane as shown in Fig.1.1. The container is rectangular and the surface shown is the right hand edge. The molecule which strikes it must have a positive x-component of velocity ($v_x > 0$). Collisions with the wall are elastic and hence the collision with the wall shown in Fig.(1.1) will result in a change of momentum of $2mv_x$ (m is the mass of the molecule) in the negative x-direction. This change of momentum

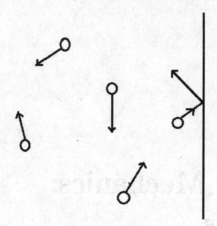

Figure 1.1: Molecules colliding with a wall.

of the molecule caused by the wall will result in an equal and opposite effect that produces the pressure. Since the gas is taken to be ideal (i.e., non interacting) the molecule will not change its momentum on its way to the wall. But constancy of velocity of a given molecule does not imply that velocities of all molecules are equal. In fact the gas will contain molecules with all possible velocities with the number of molecules in a given velocity range given by a definite distribution function. Interestingly enough, this distribution will be seen to be the outcome of interactions (however weak) between the gas molecules! Our ignorance of the exact dynamics of the gas molecules forces us to take recourse to the existence of this distribution function. This is the statistical description that we take recourse to whenever an exact description becomes impossible in practice (not in principle) and that sets up the subject of statistical mechanics.

To calculate the force on the area δA of the wall, we need to how much momentum transfer occurred in unit time. This means we need to know how many collisions occurred in unit time. For the molecules with v_x as the x-component of the velocity, the number of collisions per unit time is the number of molecules with x-velocity of v_x that lie in a cylinder of cross-section δA and height v_x. This number is $\frac{1}{2}n(v_x)v_x\delta A$ where $n(v_x)$ is the density of molecules with a x-velocity of v_x and the factor of $\frac{1}{2}$ comes from the fact that we need only the right moving molecules. The momentum transferred per unit time is $mv_x{}^2n(v_x)\delta A$ and the force δF is obtained by summing over all v_x i.e.,

$$
\begin{aligned}
\delta F &= m\sum v_x^2 n(v_x)\delta A \\
&= m.\bar{n}\frac{1}{n}\sum v_x^2 n(v_x)\delta A \\
&= m.\bar{n}\langle v_x^2\rangle\delta A \\
&= \frac{1}{3}m.\bar{n}\langle v^2\rangle\delta A.
\end{aligned}
\tag{1.1.1}
$$

In the above $\bar{n}$ is the total number density and we have made the assumption that the three components v_x, v_y and v_z are statistically independent of each other, which is reasonable because the molecules are totally independent and there is no bias in any direction. Writing $\bar{n} = N/V$ the pressure is found as

$$P = \frac{\delta F}{\delta A} = \frac{1}{3} \frac{N}{V} . m \langle v^2 \rangle = \frac{2}{3} \frac{N}{V} \bar{E} \qquad (1.1.2)$$

where $\bar{E}$ is the average kinetic energy per particle for a monatomic gas (only translational kinetic energy). We can write Eq.(1.1.2) as

$$PV = \frac{2}{3} E \qquad (1.1.3)$$

where E is the total energy of the gas. We have thus gone from a microscopic to a macroscopic description.

To make further progress, we need to bring in the concept of temperature. For the moment if we accept as known that the constant volume specific heat of an ideal monatomic gas is $\frac{3}{2}R$ for one mole of gas, then thermodynamics would give $E = \frac{3}{2}RT$ for one mole and the ideal gas law would follow as $PV = nRT$, where n is the number of moles. The above considerations make it clear that

1. Pressure can only be calculated from the dynamics of the gas molecules.

2. It is a practical impossibility to follow the dynamics of individual molecules, thus necessitating a statistical picture.

3. The concept of temperature needs to be introduced separately.

It should be clear that the above methodology would fail if we introduced interactions between gas molecules (non ideal gas). Hence the bridge between the microscopic and the macroscopic picture has to be built with greater care.

We begin by describing the microscopic state of a system of N particles. In a three dimensional space, each particle requires three coordinates and three momenta to specify its state. The N particle system requires the specification of $6N$ variables. It is customary to write the coordinates as q_i ($i = 1, 2, \ldots 3N$) and the momenta as p_i ($i = 1, 2, \ldots 3N$). The microscopic state of a system corresponds to specific values of each of the $6N$ variables. The state is generally exhibited as a point in a $6N$ dimensional space called the phase space of the system (Fig.1.2). The macroscopic state on the other hand, is specified by macroscopic variables like the total energy, the volume etc. For a given total energy of a set of free particles the individual momenta could be different from each other. The coordinates, of course, can be totally arbitrary without changing the total energy. Thus corresponding to a given macrostate, there can be tremendously large number of microstates. Hence, a macrostate occupies a volume in phase space (Fig.1.2).

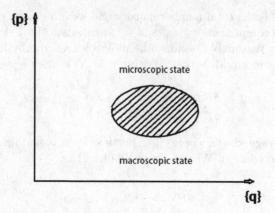

Figure 1.2: Phase space for N particles. {**p**} and {**q**} schematically denote the complete sets of momenta and coordinates respectively.

As time evolves, the microscopic state of the system changes, the evolution occurring according to the usual Hamiltonian prescription.

$$\dot{q}_i \;=\; \frac{\partial H}{\partial p_i}$$
$$\dot{p}_i \;=\; \frac{-\partial H}{\partial q_i} \tag{1.1.4}$$

Here $H = H(q_i, p_i)$ is the Hamiltonian of the system which exists regardless of whether we can write it down or not and can be safely considered time-independent for an isolated system. The dynamics is captured as a trajectory in phase space(Fig.1.3).

A macroscopic quantity, like energy or pressure, has a microscopic expression in terms of the microscopic variables $\{q_i\}$, $\{p_i\}$ i.e., can be written as

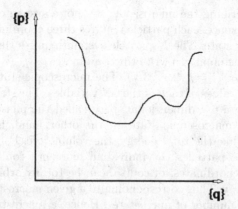

Figure 1.3: Evolution of a trajectory in phase space.

$f(\{q_i\}, \{p_i\})$. Now, the $\{q_i\}$ and $\{p_i\}$ change with time but the macroscopic quantities that we are familiar with certainly do not if things are in equilibrium. Equilibrium means that there is no time variation, at least on a long time scale. This happens because a measurement time scale is typically microseconds, while the dynamics changes the state of a system at the order of nanoseconds and hence the measured quantity is really a time average which we can write as

$$\bar{f} = \lim_{T \to \infty} \frac{1}{T} \int\limits_0^T f(\{q_i(t)\}, \{p_i(t)\}) dt. \qquad (1.1.5)$$

In principle if one knew $\{q_i(t)\}$ and $\{p_i(t_i)\}$ then the macroscopic quantity $\bar{f}$ could be calculated and one would have the passage from Newtonian dynamics to thermodynamics. In practice, this fails because of the large number of particles involved, the impossibility of ascertaining the Hamiltonian etc. Thus due to one's ignorance, one has to resort to a statistical description.

In a statistical description, one does not need to specify the trajectory any more. Instead one talks about the probability of the trajectory being in the vicinity of a particular point (q_i, p_i) $[i = 1, 2, \ldots, 3N]$ of phase space. To quantify this, we need to introduce a volume element of phase space – a volume of magnitude $d^{3N}q_i d^{3N}p_i$ around the point (q_i, p_i) in question. We ask, what is the probability, dw, that the system was in that volume of phase space during the evolution? If we track the dynamics for a period T and of that a time, dt, was spent in the volume discussed (i.e., in that time, dt, the system was very close to the state (q_i, p_i), then

$$dw = \lim_{T \to \infty} \frac{dt}{T}. \qquad (1.1.6)$$

Clearly, dw will be proportional to the volume $d\Gamma = d^{3N}q d^{3N}p$ of the phase space concerned and we can write

$$dw = \rho(\{q_i\}, \{p_i\}) d^{3N}q d^{3N}p \qquad (1.1.7)$$

where $\rho(\{q_i\}, \{p_i\})$ is a constant of proportionality which is a local quantity and is like a density.

The question is can we define a density? What constraint does this put on the trajectory of the system? This can be understood from Figs.(1.4a) and Figs.(1.4b).

The combinations of Eqs.(1.1.6) and (1.1.7) means that the time spent by the trajectory in $d\Gamma$ can be measured by counting how many distinct points $d\bar{N}$ of $d\Gamma$ happen to be part of the trajectory. Only if the trajectory happens to be space filling as in Fig.(1.4b), can we expect a relation like Eq.(1.1.7). For the periodic trajectory of Fig.1.4a, the measure of the points in $d\Gamma$ is zero regardless of the size of $d\Gamma$ and hence Eq.(1.1.7) cannot hold.

We have tacitly brought in a question of equilibrium. The proportionality constant $\rho(q_i, p_i)$ is not a function of time and hence a picture as in Fig.1.4b

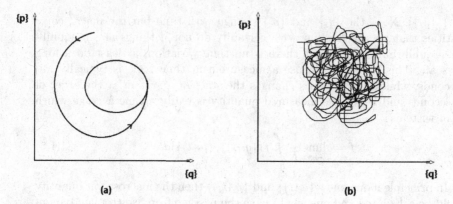

Figure 1.4: (a) Periodic trajectory in phase space. (b) Space-filling trajectory in phase space.

implies that while the trajectory would keep on adding more points as time increases, the density ρ would not change if equilibrium is established. This space filling nature of the trajectory – wandering over all the allowed region of phase space – corresponds to ergodicity of the dynamics and we find that the average value of Eq.(1.1.5) can now be written as

$$\bar{f} = \lim_{T \to \infty} \frac{1}{T} \int\limits_0^T f(\{q_i(t)\}, \{p_i(t)\}) dt$$

$$= \int f(q_i, p_i) \rho(q_i, p_i) d^{3N} q_i d^{3N} p_i. \tag{1.1.8}$$

We can assign a picture to the second form of the average in Eq.(1.1.8). We can imagine we have $\bar{N}$ identical copies of our system and at any given time we can ask, how many of those $\bar{N}$ copies are in a state which happens to be within a volume $d\Gamma$ round the point (q_i, p_i) the number $d\bar{N} = \rho(p_i, q_i) d\Gamma$ and the average of any function $f(q_i, p_i)$ would be as given by the second from of Eq.(1.1.8). This form is consequently called an ensemble average and Eq.(1.1.8) expresses what is called the ergodic hypothesis – the time averages and ensemble averages are equal. Thus, we have reached the central point of building a bridge between the microscopic and the macroscopic world – we need to calculate the density – the probability density $\rho(q_i, p_i)$. It is $\rho(q_i, p_i)$ which will take us from the microworld to the everyday world.

This however is like begging the question. From what we have said so far $\rho(q_i, p_i)$ can be constructed only if we know the trajectory and trajectory is what we cannot calculate. Consequently we need to make another hypothesis which will essentially tell us what $\rho(q_i, p_i)$ is. For this, we go back to Fig.1.2. We note that a given macrostate consists of a large number of microstates. We postulate that for an isolated system i.e., a system characterized by a fixed (almost) energy E, fixed number of particles N and fixed volume V, all the microstates corresponding to this macrostate are equally probable. This is called the equal a-priori probability hypothesis.

Now that we know how to calculate ρ, we are in a position to calculate thermodynamic quantities. This is where our experience with the calculation of the equation of state carried out earlier in the section comes in handy. We could arrive at Eq.(1.1.3) by postulating the existence of a distribution function but we could not introduce the temperature. That needed another input. It is a similar situation here. Form what we have said so far, the concept of temperature does not follow. Temperature comes from the laws of thermodynamics as

$$\left(\frac{\partial S}{\partial E}\right)_V = \frac{1}{T} \tag{1.1.9}$$

where S is the entropy of the system. We now need to know S and that is where the next postulate comes in – Boltzmann's hypothesis, which says that

$$S = k_B \ln \Gamma(E, V, N) \tag{1.1.10}$$

where $\Gamma(E, V, N)$ is the total phase space available to the particle. It needs to be a dimensionless quantity and hence is related to our previously introduced $d\Gamma$ as

$$\Gamma(E, N, V) = \int\int \frac{d^{3N}q \, d^{3N}p}{h^{3N}} \quad \text{(over the allowed region)}. \tag{1.1.11}$$

The elementary volume $d\Gamma$ introduced before has been made dimensionless by the factor h^{3N}. We are now in a position to start from a microscopic world and go to a macroscopic description. The information of the microscopic world is contained in the form of the Hamiltonian and in the $\Gamma(E, V, N)$ above. Success of statistical mechanics depends on three postulates:

1. Ergodic hypothesis which relates the time average to an ensemble average and thus allows for the existence of $\rho(q_i, p_i)$.

2. Equal a-priori probability hypothesis which tells us how to calculate ρ by saying that $\rho(q_i, p_i) = $ constant, independent of the microstate for an isolated system.

3. Boltzmann's hypothesis which allows us to introduce temperature into the formalism.

A basic question that can be asked is whether there is an unifying theme behind the three postulates stated above. It is believed that chaotic dynamics which leads to ergodicity, mixing and loss of correlations is the single factor that provides the foundation of statistical mechanics. We will describe the features associated with this dynamics in Secs. 1.4 and 1.5. Before that we provide a picture of the equilibrium distribution and a motivation behind Boltzmann's hypothesis.

1.2   Equilibrium Distribution

Our picture of the equilibrium distribution is that as time goes on and the trajectory develops, the density $\rho(q,p)$ doesn't change. We want to find out what this statement implies.

To clarify the implications of a time-independent $\rho(\{q_i\}, \{p_i\})$, we need to follow the time development of $\rho(\{q_i\}, \{p_i\}, t)$. The relevant principle for determining the dynamics of ρ is a conservation law-conservation of the number of ensembles. In a given elementary volume $d^{3N}q\, d^{3N}p$, the number of ensemble is $\rho d^{3N}q\, d^{3N}p$ and this number can change only if in the course of the evolution some ensemble points enter or leave the volume across the surface which bounds it. The result is a local conservation law

$$\frac{\partial \rho}{\partial t} + \vec{\nabla}.(\rho \vec{v}) = 0 \tag{1.2.1}$$

where $\vec{v}$ is the velocity $(\{\dot{q}_i\}, \{\dot{p}_i\})$, in the $6N$-dimensional space. Writing out Eq.(1.2.1) explicitly.

$$\frac{\partial \rho}{\partial t} + \sum_i \frac{\partial}{\partial q_i}(\rho \dot{q}_i) + \sum_i \frac{\partial}{\partial p_i}(\rho \dot{p}_i) = 0 \tag{1.2.2}$$

or

$$\frac{\partial \rho}{\partial t} + \sum_i \left[\frac{\partial \rho}{\partial q_i}\dot{q}_i + \frac{\partial \rho}{\partial p_i}\dot{p}_i \right] + \sum_i \rho \left[\frac{\partial \dot{q}_i}{\partial q_i} + \frac{\partial \dot{p}_i}{\partial p_i} \right] = 0$$

But,

$$\frac{\partial \dot{q}_i}{\partial q_i} + \frac{\partial \dot{p}_i}{\partial p_i} = \frac{\partial}{\partial q_i}\left[\frac{\partial H}{\partial p_i} \right] + \frac{\partial}{\partial p_i}\left[-\frac{\partial H}{\partial q_i} \right] = 0$$

where we have used Eq.(1.1.4)and thus Eq.(1.2.2) reduces to

$$\frac{d\rho}{dt} = \frac{\partial \rho}{\partial t} + \sum_i \left[\frac{\partial \rho}{\partial q_i}\dot{q}_i + \frac{\partial \rho}{\partial p_i}\dot{p}_i \right] = 0. \tag{1.2.3}$$

An alternative proof uses the fact that the volume element $d^{3N}q\, d^{3N}p$ is preserved by Hamilton's equations and hence number conservation implies $\frac{d\rho}{dt} = 0$ (see any classical mechanics text).

To have an equilibrium distribution, we require an even stronger condition on ρ, namely $\frac{\partial \rho}{\partial t} = 0$. This implies that

$$\frac{\partial \rho}{\partial q_i}\dot{q}_i + \frac{\partial \rho}{\partial p_i}\dot{p}_i = 0$$

or

$$\frac{\partial \rho}{\partial q_i}\frac{\partial H}{\partial p_i} - \frac{\partial \rho}{\partial p_i}\frac{\partial H}{\partial q_i} = 0 \tag{1.2.4}$$

a condition which can be easily satisfied if $\rho = \rho(H(\{q_i\}, \{p_i\}))$ since then

$$\frac{\partial \rho}{\partial q_i} = \rho'\frac{\partial H}{\partial q_i} \tag{1.2.5}$$

and

$$\frac{\partial \rho}{\partial p_i} = \rho' \frac{\partial H}{\partial p_i} \tag{1.2.6}$$

and Eq.(1.2.4) is clearly satisfied. In the above the prime denotes differentiation with respect to H. Hence the equilibrium distribution can be chosen to have the form

$$\rho(\{q_i\}, \{p_i\}) = \rho(H(\{q_i\}, \{p_i\})). \tag{1.2.7}$$

Thus there is a constraint on the equilibrium distribution. The information on microscopic dynamics thus enters the bridging formula through the Hamiltonian.

1.3 Boltzmann's Hypothesis

To begin with let us use the laws of thermodynamics to find the entropy of a perfect gas. The starting point is

$$TdS = dE + PdV. \tag{1.3.1}$$

Now, for a perfect monatomic gas at a temperature T, the energy is $E = (3/2)Nk_BT$ (we are considering N molecules) and the equation of state is $PV = Nk_BT$, so that Eq.(1.3.1) becomes

$$TdS = \frac{3}{2}Nk_BdT + \frac{Nk_BT}{V}dV$$

or

$$dS = \frac{3}{2}Nk_B\frac{dT}{T} + \frac{Nk_B}{V}dV. \tag{1.3.2}$$

Integrating,

$$
\begin{aligned}
S &= \frac{3}{2}Nk_B \ln T + Nk_B \ln V + S_0 \\
&= k_B \ln (VT^{3/2})^N + S_0 \\
&= k_B \ln \left[V\left(\frac{2E}{3Nk_B}\right)^{3/2} \right]^N + S_0. \tag{1.3.3}
\end{aligned}
$$

We now need to look at the volume occupied by the trajectories in phase space. We consider an extremely small spread Δ in the total energy E, i.e., the total energy lies between E and $E + \Delta$, $\Delta \ll E$ and in fact as $N \to \infty$, $\Delta/E \to 0$. The volume in phase space is defined by the restriction

$$E < \sum_{i=1}^{3N} \frac{p_i^2}{2m} < E + \Delta. \tag{1.3.4}$$

To calculate the volume, we proceed to find the volume in phase space occupied by all trajectories corresponding to energies $\leq E$. This volume $\Gamma(E)$ is given by (in non-dimensional form)

$$\Gamma(E) = \frac{1}{h^{3N}} \prod_i \int_V d^{3N} q_i \int_{0 < \frac{p^2}{2m} < E} d^{3N} p_i \qquad (1.3.5)$$

where $p^2 = \sum_{i=1}^{3N} p_i^2$. To appreciate why h enters a classical formula, see Problem(1.1). Clearly,

$$\Gamma(E) = h^{-3N} V^N \, (\text{Area of } 3N-\text{dimensional sphere of radius } (2mE)^{1/2}) \qquad (1.3.6)$$

where V is the volume of the container. Now recall (see Problem 1.2) that the volume of a sphere of radius R in a D-dimensional space is $2\pi^{D/2} R^{D-1} / \bar{\Gamma}(D/2)$ and hence (note that $\bar{\Gamma}$ is the usual gamma function)

$$\Gamma(E) = \frac{V^N 2\pi^{3N/2} (2mE)^{3N/2}}{3N \bar{\Gamma}(3N/2) (2\pi\hbar)^{3N}}. \qquad (1.3.7)$$

The volume $\Gamma(E, \Delta)$ which is the volume of the annulus lying between spheres of radii E and $E + \Delta$ is obtained as (for convenience we will call the gamma function Γ from now on, it being clear that whenever Γ is a function of N alone, it is the gamma function.

$$\Gamma(E, \Delta) \approx \frac{\partial \Gamma(E)}{\partial E} \Delta = \frac{V^N \pi^{3N/2} (2mE)^{3N/2} \Delta}{(2\pi\hbar)^{3N/2} \Gamma(3N/2) E}. \qquad (1.3.8)$$

Thus,

$$
\begin{aligned}
\ln \Gamma(E, \Delta) &= N \ln V + \frac{3N}{2} \ln 2mE + \frac{3N}{2} \ln 2\pi \\
&\quad + \ln \frac{\Delta}{2E} - \frac{3N}{2} \left(\ln \frac{3N}{2} - 1 \right) - 3N \ln (2\pi\hbar) \\
&= N \ln V + \frac{3N}{2} \ln \left(\frac{mE}{3N/2} \right) + \frac{3N}{2} (1 - 2 \ln h) + \ln \frac{\Delta}{E} \\
&\approx N \ln V + \frac{3N}{2} \ln \frac{E}{N} + \frac{3N}{2} \left(\ln \frac{m}{h^2} + 1 \right) \qquad (1.3.9)
\end{aligned}
$$

where we have made the approximation of N being very large and dropped terms of $O(1)$. Comparing with Eq.(1.3.3), we note that the Boltzmann's hypothesis $S = k_B \ln \Gamma(E, \Delta)$ is indeed satisfied if S_0 the undetermined constant in thermodynamics, is fixed from Eq.(1.3.9). It should be pointed out that on taking the logarithm, there is no difference between $\Gamma(E, \Delta)$, $\frac{\partial \Gamma}{\partial E} \Delta, \ldots$ etc and the same notation will be used in future for either. The meaning will be clear from the context.

At this point, we make a small digression to point out a difficulty with Eq.(1.3.3). Let us imagine a container of volume $2V$ divided into equal volumes carrying N molecules each of the same gas. The entropy of the total system is

$$S_{tot} = 2Nk_B \ln V + 3Nk_B \ln (E/N) + S_0. \qquad (1.3.10)$$

If the partition is removed and the gases allowed to mix, then the final entropy is that of $2N$ molecules occupying a volume $2V$ and thus

$$S_{Final} = 2Nk_B \ln 2V + 3Nk_B \ln \frac{E}{N} + S_0 \qquad (1.3.11)$$

leading to an entropy change $S = S_{Final} - S_{tot} = 2N \ln 2$. This cannot be correct as the mixing process described is clearly reversible and there can be no change in entropy. This is known as Gibbs paradox. The error lies in a wrong enumeration of the number of available states for N totally indistinguishable particles. In this case one has to recognize the fact that $N!$ permutations of the particles among themselves cannot lead to new states and hence the number of available states has to be changed from Eq.(1.3.5) to

$$\Gamma(E) = \frac{1}{N!} \left(\frac{1}{2\pi\hbar} \right)^{3N} \int_V d^{3N}q_i \int_{0 < \sum_i \frac{p_i^2}{2m} < E} d^{3N}p_i \qquad (1.3.12)$$

with the corresponding change in $\Gamma(E, \Delta)$. With this change in counting, Eq.(1.3.3) is altered to

$$S = Nk_B \ln \frac{V}{N} + \frac{3}{2} Nk_B \ln \frac{E}{N} + S_0 \qquad (1.3.13)$$

and Gibbs paradox is removed as ΔS now does turn out to be zero.

After this digression, we return to a consideration of the Boltzmann relation $S = k_B \ln \Gamma(E)$ (note that for large N, it does not matter whether we consider $\ln \Gamma(E)$ or $\ln \Gamma(E, \Delta)$. The thing to note is that entropy being extensive is $O(N)$ for a N-particle system and thus the region of phase space $\Gamma(E)$ covered by a trajectory is $O(E^N)$, which is the same as $O(V^N)$. This fact is significant because it hints at the possibility of any particular trajectory being totally uncorrelated with any other trajectory and can wander over the allowed phase space independent of the other trajectories. Once again this loss of correlation (in different trajectories independent of each other) is a characteristic of chaotic dynamics, so that the justification behind the Boltzmann's hypothesis lies in the chaotic nature of the microscopic dynamics.

Finally, we note that ergodicity (ability of a trajectory to cover all of phase space) does not imply loss of correlation. This can be seen most easily by considering the mapping $\theta_{n+1} = \theta_n + \Omega(\mathrm{mod}\ 2\pi)$, where Ω is irrational. Clearly the successive iterates of an arbitrary initial θ_0 cover the entire circumference and thus assures ergodicity. A moment's thought, however, convinces one that there can be no loss of correlation and thus we arrive at the conclusion that ergodicity does not necessarily lead to loss of correlation. However, if we have a loss of correlation, ergodicity will be assured.

1.4   Dynamics, Dynamical Systems and Chaos

As we have repeatedly emphasized, the ultimate justification of the postulates of statistical mechanics lies in the microscopic dynamics exhibiting chaotic trajectories. This leads to loss of correlation due to the sensitive dependence on initial conditions. This particular feature of dynamics is what we intend to explore in this Section. To do so, we do not need to worry about an N-particle system – only a few degrees of freedom are necessary to establish these concepts. Consequently, in this Section we will not talk about large N, rather we will discuss simple systems like oscillators, forced oscillators or two coupled oscillators. We first need to establish certain terminologies. The equations of motion in each case will obviously be Newton's law which has the general form

$$\dot{q} = f(q, p, t) \tag{1.4.1}$$

where q is the position coordinate and p is the momentum (we take the mass to be unity for convenience) and the force on the right hand side of Eq.(1.4.1) is the most general possible including, if need be, an explicit dependence on time. It is clear that this second order differential equation can be written as two first order differential equations by writing

$$\dot{q} = p \tag{1.4.2}$$
$$\dot{p} = f(q, p, t). \tag{1.4.3}$$

We consider a few examples:

Harmonic oscillator:

$$\ddot{X} + \omega^2 X = 0.$$

This can be written as

$$\dot{X} = Y \tag{1.4.4}$$
$$\dot{Y} = -\omega^2 X. \tag{1.4.5}$$

Damped harmonic oscillator:

$$\ddot{X} + 2k\dot{X} + \omega^2 X = 0.$$

This can be written as

$$\dot{X} = Y \tag{1.4.6}$$
$$\dot{Y} = -2kY - \omega^2 X. \tag{1.4.7}$$

Non-Linear damped oscillator:

$$\ddot{X} + 2k\dot{X} + \omega^2 X + \lambda X^3 = 0$$

This can be cast as

$$\dot{X} = Y \tag{1.4.8}$$
$$\dot{Y} = -2kY - \omega^2 X - \lambda X^3. \tag{1.4.9}$$

Forced damped harmonic oscillator:

$$\ddot{X} + 2k\dot{X} + \omega^2 X = f_0 \cos \Omega t.$$

This can be recast as

$$\dot{X} = Y \tag{1.4.10}$$
$$\dot{Y} = -2kY - \omega^2 X + f_0 \cos \Omega t. \tag{1.4.11}$$

Yet another way of rewriting it yields

$$\dot{X} = Y \tag{1.4.12}$$
$$\dot{Y} = -2kY - \omega^2 X + f_0 Z \tag{1.4.13}$$
$$\dot{Z} = -\Omega\sqrt{1 - Z^2}. \tag{1.4.14}$$

Forced damped non-linear oscillator:

$$\ddot{X} + 2k\dot{X} + \omega^2 X + \lambda X^3 = f_0 \cos \Omega t \tag{1.4.15}$$

which can be put in the form

$$\dot{X} = Y \tag{1.4.16}$$
$$\dot{Y} = -2kY - \omega^2 X - \lambda X^3 + f_0 \cos \Omega t. \tag{1.4.17}$$

From the above it is clear that the equation of motion can always be written as a set of first order differential equations. A set of first-order differential equations is called a dynamical system. The dynamical system can be linear [Eqs.(1.4.4) and (1.4.5), Eqs.(1.4.6) and (1.4.7), and Eqs.(1.4.10) and (1.4.11)] or non-linear [Eqs.(1.4.8) and (1.4.9) and the set of Eqs.(1.4.16) and (1.4.17)]. It can also have a right hand side which has no explicit time dependence [Eqs.(1.4.8) and (1.4.9)] in which case we have an autonomous system, or a right hand side which has an explicit time dependence [Eqs.(1.4.10) and (1.4.11)] in which case the system is called non-autonomous. The number of first order equations in the system constitutes the dimension of the dynamical system. It should be noted that by re-defining variables a nonautonomous system can made into an autonomous one but the dimensionality changes [Eqs.(1.4.10) and (1.4.11); and Eqs.(1.4.12)-(1.4.14)]. Thus an n-dimensional autonomous dynamical system is characterized by n variables $x_1, x_2, \ldots, x_n$ satisfying

$$\dot{x}_1 = f_1(x_1, x_2, \ldots, x_n)$$
$$\dot{x}_2 = f_2(x_1, x_2, \ldots, x_n)$$
$$\vdots$$
$$\vdots$$
$$\dot{x}_n = f_n(x_1, x_2, \ldots, x_n). \tag{1.4.18}$$

Initial conditions on x_1, x_2, ..., x_n have to be specified and then integrating the above system yields a trajectory which is the set $\{x_i(t)\}$ for all values of t. The space spanned by $\{x_i\}$ is the phase space of the dynamical system and the phase trajectory is a path traced out in this space. Trajectories may be closed, e.g., those for the system represented by Eqs.(1.4.4) and (1.4.5) or open but heading for a definite point in phase space, e.g., those for the system represented by Eqs.(1.4.6) and (1.4.7).

A few representative trajectories are shown in Fig.1.5.

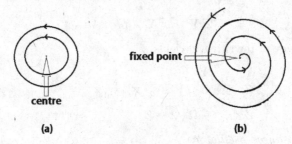

Figure 1.5: (a) Trajectories initially closed, always remain closed. (b) Trajectories end on the fixed point and hence come closer together.

The different trajectories correspond to different initial conditions and it should be noted that a common feature of the above trajectories is that two neighbouring trajectories triggered by two neighbouring initial conditions remain close (Fig.1.5a) or come closer together (Fig.1.5b).

This is what is meant by trajectories being correlated – if initial conditions are close the trajectories remain close as well - "the memory" of initial conditions is always retained. Closed trajectories are also obtained from Eqs.(1.4.10) and (1.4.11) but these are different from those shown in Fig.(1.5a). The closed trajectories (Fig.(1.5a)) for the system of Eqs.(1.4.4) and (1.4.5) are obtained because of a conservation law which leads to $\omega^2 x^2 + y^2 = $ constant, but the closed trajectory of the system of Eqs.(1.4.10) and (1.4.11) is obtained because of the balance between dissipative forces and the periodic external forces and is called a limit cycle. Once again neighbouring trajectories remain close to each other, since independent of the initial conditions, the trajectories fall on the same closed orbit (Fig.1.6).

Qualitatively different solutions are obtained for the system of Eqs.(1.4.16) and (1.4.17) if the various parameters k, f_0 and Ω are properly adjusted. Since an analytical solution is not obtainable in this case, we have to consider even simpler systems to explore these qualitatively different solutions. This cannot be achieved by differential equations but by maps which are obtainable from differential equations through a technique due to Poincare: For an n-dimensional dynamical system, this map (Poincare map) is constructed by considering an $(n-1)$ dimensional hypersurface and noting the successive intersections of an evolving trajectory with this hypersurface. The ith intersection is an $(n-1)$

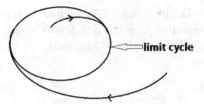

Figure 1.6: Trajectories end up on the fixed closed orbit, the limit cycle.

dimensional vector $\{x_j^{(i)}\}$, where j ranges from 1 to $(n-1)$. The map is constructed as the recurrence relation

$$\{x_{i+1}^{(j)}\} = F_j(\{x_i^{(j)}\}). \tag{1.4.19}$$

For a system with dissipation (e.g., damped oscillator) the simplest such map can be one-dimensional as in

$$x_{i+1} = f\{x_i\}. \tag{1.4.20}$$

For a conservative system, the simplest nontrivial map is generally two-dimensional. The conservation law of the dynamical system must lead to a conserved quantity in the map and the quantity which can be preserved in two dimensions under a mapping is the area. So for Hamiltonian systems, the simplest mapping is a two-dimensional area-preserving one.

At first we discuss the dissipative system, where a simple one-dimensional mapping of the sort shown in Eq.(1.4.20) can result. The simplest such mapping exhibiting complicated behaviour is the Bernoulli shift, defined by

$$x_{n+1} = 2x_n \pmod 1$$

i.e.,

$$x_{n+1} = \begin{cases} 2x_n & \text{if } 0 \le x_n \le \frac{1}{2} \\ 2x_n - 1 & \text{if } \frac{1}{2} \le x_n \le 1 \end{cases}. \tag{1.4.21}$$

To appreciate the action of this map let us write any number between 0 and 1 in the binary representation as the infinite series:

$$a_0 \left(\frac{1}{2}\right)^0 + a_1 \left(\frac{1}{2}\right)^1 + a_2 \left(\frac{1}{2}\right)^2 + \ldots\ldots$$

where the as can be 0 or 1 and $a_0 = 0$. If the number has to be greater than half, then $a_1 = 1$ while for a number less than half $a_1 = 0$. An arbitrary number, thus can be represented by the string $(a_1, a_2, a_3 \ldots)$. The action of the map on this string is clearly to shift the string one step to the left, i.e, $(a_1, a_2, a_3 \ldots)$ goes to $(a_2, a_3, a_4, \ldots)$. The extreme sensitivity to initial conditions of this operation

can now become apparent. Let us assume that we have two starting points very
close to each other, i.e., the strings have the first few as identical. Specifically, let
us suppose that the strings differ at the 100th place, i.e., we have two starting
points x_0 and $\bar{x}_0$ such that

$$
\begin{align}
x_0 &= (a_1, a_2, a_3, \ldots, a_{100}) \tag{1.4.22}\\
\bar{x}_0 &= (a_1, a_2, a_3, \ldots, \bar{a}_{100} \ldots). \tag{1.4.23}
\end{align}
$$

Clearly x_0 and $\bar{x}_0$ differ by $1/2^{100}$. After 100 iterations by the Bernoulli
shift operation

$$
\begin{align}
x_{100} &= (a_{100}, a_{101}, a_{102}, \ldots) \tag{1.4.24}\\
\bar{x}_{100} &= (\bar{a}_{100}, \bar{a}_{101}, \bar{a}_{102}, \ldots) \tag{1.4.25}
\end{align}
$$

and these numbers differ by $O(1)$. This is the extreme sensitivity to initial conditions that for certain parameter values can be an intrinsic feature of a dynamical system (the system has got to be non-linear and at least three-dimensional
if it is autonomous and at least two-dimensional if non-autonomous). The extreme sensitivity to initial conditions can be formalized by an index called the
Lyapunov index λ. For the one dimensional map $x_{n+1} = f(x_n)$ we define it as

$$
\begin{align}
\lambda &= \lim_{\epsilon \to 0} \lim_{N \to \infty} \frac{1}{N} \ln \left| \frac{f^N(x_0 + \epsilon) - f^N(x_0)}{\epsilon} \right| \\
&= \lim_{N \to \infty} \frac{1}{N} \ln \left| \frac{df^N(x_0)}{dx} \right| \\
&= \lim_{N \to \infty} \frac{1}{N} \ln \left| \prod_{i=0}^{N-1} \left(\frac{df}{dx} \right)_{x_i} \right| \\
&= \lim_{N \to \infty} \frac{1}{N} \sum_{i=0}^{N-1} \left| \ln \left(\frac{df}{dx} \right)_{x_i} \right|. \tag{1.4.26}
\end{align}
$$

In the above, $f^N(x)$ is the N times iterated function $[f^2(x) = f\{f(x)\}$
and so on] and the successive iterates of x_0 are $x_1, x_2, \ldots$. Clearly from the first
line of Eq.(1.4.26), $f^N(x_0 + \epsilon)$ will differ by orders of magnitude from $f^N(x_0)$ if
$\lambda > 0$. Thus if the mapping shows sensitivity to initial conditions, the Lyapunov
index $\lambda > 0$. If $\lambda \leq 0$, then we have the iterates maintaining their separation
or coming closer. For the Bernoulli shift discussed above it should be clear that
$\lambda = \ln 2 > 0$, which is consistent with the sensitivity to initial conditions that
we have demonstrated.

We now demonstrate that the Bernoulli shift generates a random sequence.
To do so we consider a well known random sequence - the coin toss. We will
show that given any sequence of heads and tails in a coin toss experiment, the

Bernoulli shift can reproduce it. To do so we make a convention: if the coin toss produces heads, then a number greater than half will be generated and if a coin toss produces tails, then a number less than half will be generated. Now let an arbitrary sequence be

$$H\,T\,T\,H\,H\,H\,T\,T\,H\,T\,H\,T\,H\,T\,H\,H\,T\,T\,T\ldots\,.$$

Accordingly, we start with the initial number for generating a Bernoulli sequence as

$$1\,0\,0\,1\,1\,0\,0\,1\,0\,1\,0\,1\,1\,0\,0\,0\ldots\,.$$

This number is greater than half and corresponds to H. Now a shift moves things to the left by a step and we generate a T since the new number is less than half. As we go on generating the new numbers, the random toss sequence is generated.

The dynamics described above is called chaotic – the main characteristic is the existence of a positive Lyapunov index – this is what leads to neighbouring initial conditions being stretched apart – the points do not go away to infinity since the map folds (in Eq.(1.4.20), the modulo one in the Bernoulli shift is responsible for the folding). They remain confined in a finite interval but an infinitesimal separation becomes a finite separation after sufficiently many time steps. The chaotic dynamics leads to loss of correlation (e.g., the effect of the randomness in the Bernoulli shift) and to the iterates being spread out over an interval often with equal probability.

The systems that one deals with in statistical mechanics are mainly Hamiltonian systems and hence the mapping, if it exists, would exhibit some invariance properties. The simplest nontrivial mapping with area-preserving property is the two-dimensional map:-

$$\left.\begin{array}{rcl} x_{n+1} &=& 2x_n \\ y_{n+1} &=& \frac{y_n}{2} \end{array}\right\} \qquad \text{if } 0 \le x_n \le \frac{1}{2}, 0 \le y_n \le 1$$

$$(1.4.27)$$

$$\left.\begin{array}{rcl} x_{n+1} &=& 2x_n - 1 \\ y_{n+1} &=& \frac{1}{2}(1 + y_n) \end{array}\right\} \qquad \text{if } \frac{1}{2} \le x_n \le 1, 0 \le y_n \le 1.$$

It is obvious that the map is area preserving. It is equally plain that there is stretching in the x-direction identical to the Bernoulli shift and then a folding back. Initial points separated in the x-direction consequently rapidly diverge with a positive Lyapunov index ln 2. However, for separations in the y-direction alone the Lyapunov index is ln 2. The two-dimensional system has two Lyapunov indices, but it suffices that only one of them be positive for the dynamics to be chaotic. In general, for an n-dimensional system, there are n Lyapunov indices. To exhibit chaotic behaviour the system must at least have the largest Lyapunov index positive.

To analyze this map in the manner in which we treated Eq.(1.4.21), let us write the initial point x_0, y_0 in the binary notation. We choose to write

$$x_0 = \frac{a_0}{2} + \frac{a_1}{2^2} + \frac{a_2}{2^3} + \dots \tag{1.4.28}$$

$$y_0 = \frac{a_{-1}}{2} + \frac{a_{-2}}{2^2} + \frac{a_{-3}}{2^3} + \dots \tag{1.4.29}$$

We note that $a_0 = 0$ if $x_0 < \frac{1}{2}$ and $a_0 = 1$ if $x_0 > \frac{1}{2}$. It is clear from our previous discussion that one operation of the map of Eq.(1.4.27) takes us to (x_1, y_1) such that

$$x_1 = \frac{a_1}{2} + \frac{a_2}{2^2} + \frac{a_3}{2^3} + \dots \tag{1.4.30}$$

$$y_1 = \frac{a_0}{2} + \frac{a_{-1}}{2^2} + \frac{a_{-2}}{2^3} + \dots. \tag{1.4.31}$$

If we write the initial point in the slightly unconventional form $(y_0; x_0)$, then the binary representation would yield

$$(\dots\dots a_{-3}, a_{-2}, a_{-1}; a_0, a_1, a_2, a_3 \dots\dots)$$

with the semicolon separating the two infinite series for y_0 and x_0. The advantage of writing the initial point in this form is that (y_1, x_1), the first iterate, becomes

$$(\dots\dots a_{-2}, a_{-1}, a_0; a_1, a_2, a_3 \dots\dots).$$

The entire sequence of as has moved one place to the left. The action of the map can once again be portrayed as a left shift.

To understand the complication associated with the map of Eq.(1.4.27), let us look at the properties of the origin which is a fixed point of the map. The two eigenvalues for motion around the fixed point are 2 and 1/2 which makes the origin an unstable fixed point (a hyperbolic point). The fixed point has the binary representation $(0000\dots)$ i.e., an infinite string of zeros. We can easily find a set of points that will approach the origin arbitrarily close in the future. From the left shift rule of the dynamics, these set of points are seen to be of the form

(something; something else, $000\dots000\dots$).

where the zeros from an infinite string. The "something" are arbitrary finite strings. We can also find a set of points which in the ever receding past were very close to the origin. These set of points can be written as

($\dots000\dots$ something; something else).

Once again the string of zeros is infinite while the "something" is an arbitrary finite string. The former set of points is the stable manifold of the origin

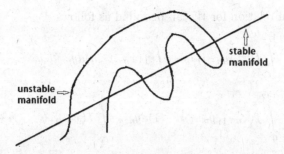

Figure 1.7: Homoclinic tangle.

while the latter is the unstable manifold. Now, it can be established that a stable manifold cannot intersect with itself and nor can an unstable manifold intersect with itself. Should a stable manifold intersect itself, it follows from the continuity of the map that this intersection point will map to another intersection point which in turn will lead to another and as the iteration advances one would be converging on the origin. But the stable manifold at the origin is tangent to an eigenvector of the linear stability matrix at the origin and that rules out the convergence to the origin on the basis of the self intersection of the stable manifold. Similar arguments hold for the unstable manifold, except the time now goes backwards. However, the unstable and stable manifold can intersect. If the two manifolds belong to the same fixed point or periodic orbit, then it is a homoclinic intersection and if they belong to different fixed points or periodic orbits then it is a heteroclinic intersection. It is clear that for the origin above, all points of the form

$$(\ldots 00 \ldots \text{ something; something else;} \ldots 000 \ldots 00 \ldots)$$

are homoclinic points. At all such points the stable and unstable manifolds intersect but there can be no intersection in the two manifolds.

This brings up an amazing tangled geometry called the homoclinic tangle and is the root cause behind the chaos and mixing exhibited by the map of Eq.(1.4.27). An artist's impression of the homoclinic tangle is shown in Fig 1.7.

It would be interesting to find out if under the above chaos-exhibiting mapping, something resembling the postulate of equal a priori probability emerges. Consequently we ask whether an arbitrary probability distribution $f(x_0, y_0, n = 0)$ defined on the unit square with the normalization $\int f(x_0, y_0, n) dx_0 dy_0 = 1$ will redistribute itself. Because of Liouville's theorem, it is clear that $f(x_n, y_n, n = 0)$ cannot change, i.e., $f(x_n, y_n, n) = f(x_{n-1}, y_{n-1}, n - 1)$. So, we choose to work with a derived probability distribution, defined as

$$W(x_n, n) = \int\limits_0^1 f(x_n, y_n, n) dy_n. \qquad (1.4.32)$$

The recursion relation for W can be found as follows:

$$
\begin{aligned}
W(x_n, n) &= \int_0^{1/2} f(x_n, y_n, n)dy + \int_{1/2}^1 f(x_n, y_n, n)dy_n \\
&= \int_0^{1/2} f(x_{n-1}, y_{n-1}, n-1)dy_n + \int_{1/2}^1 f(x_{n-1}, y_{n-1}, n-1)dy_n \\
&= \int_0^{1/2} f\left(\frac{x_n}{2}, 2y_{n-1}, n-1\right) dy_n \\
&\quad + \int_{1/2}^1 f\left(\frac{1+x_n}{2}, y_n, n-1\right) dy_n \\
&= \left[W\left(\frac{x_n}{2}, n-1\right) + W\left(\frac{1+x_n}{2}, n-1\right) \right].
\end{aligned}
\tag{1.4.33}
$$

As $n \to \infty$, one obtains $W(x_n, n) \to 1$. This is an example of equal a priori probability in a restricted sense.

The main postulates of statistical mechanics seem to depend on the microscopic dynamics being chaotic. So it is important to ask, how natural is it for a Hamiltonian system to be chaotic?

1.5   The Prevalence of Chaos

We first note that for an integrable system, the motion in phase space cannot be ergodic. For a $2N$-dimensional system, integrability means setting up a canonical transformation to the action angle variables with the N action variables J as constants of motion. The trajectory thus lies on an N-dimensional torus and certainly cannot cover the $2N-1$ dimensional energy hypersurface. Thus for a Hamiltonian $H(\{q\}, \{p\})$, which is integrable, we seek a canonical transformation to a set $(\{J\}, \{\theta\})$, such that $H = H_0(\{J\})$. The variable $J = \frac{1}{2\pi} \oint pdq$, where the integral is over a closed phase space trajectory. If this can be done,

$$
\begin{aligned}
\dot{J} &= \frac{\partial H_0}{\partial \theta} = 0 \\
\dot{\theta} &= \frac{\partial H_0}{\partial J} = \omega(\{J\}).
\end{aligned}
$$

The resulting integrals are

$$
J = \text{constant} \tag{1.5.1}
$$

and

$$
\theta = \omega(J)t + \text{constant}. \tag{1.5.2}
$$

The canonical transformation is generated by $S(q, J)$ which satisfies

$$p = \frac{\partial S}{\partial q}.$$

As an example consider the harmonic oscillator,

$$H = \frac{1}{2}(p^2 + \omega^2 q^2).$$

We use the generator S to write,

$$\frac{1}{2}\left[\left(\frac{\partial S}{\partial q}\right)^2 + \omega^2 q^2\right] = H_0(J)$$

so that

$$\left(\frac{\partial S}{\partial q}\right)^2 = 2(H_0 - \omega^2 q^2)$$

and J is determined by

$$J = \frac{1}{2\pi}\oint \frac{\partial S}{\partial q}dq = \frac{H_0}{\omega}$$

leading to

$$H_0 = \omega J$$

and the equations of motion

$$J = \text{constant}$$
$$\theta = \omega t + \text{constant}. \tag{1.5.3}$$

For a $2N$-dimensional integrable system, there would be effectively N constants $J_1, J_2, \ldots, J_N$ with N frequencies $\omega_1, \omega_2, \ldots$ which, in general, would depend on the $\{J\}$ and thus the motion would take place on an N-dimensional torus.

For $N = 2$, the corresponding torus is shown in Fig.1.8. It is obvious that trajectories are confined on an N-dimensional torus.

We now ask what happens if this unperturbed system with trajectories on an N-torus is subject to a perturbation i.e., to $H_0(\vec{J})$ we add $\epsilon H_1(\vec{J}, \vec{\theta})$. If the system is still integrable, then we will be able to define a new set of action variables $\{\vec{J}'\}$, and a new generating function $S(\vec{J}', \vec{\theta})$, such that $J_i' = \frac{\partial S'}{\partial \theta_i}$, $\theta_i = \frac{\partial S'}{\partial J_i'}$ and

$$H\left(\frac{\partial S'}{\partial \theta_i}, \theta_i\right) = \bar{H}_0(J_i'). \tag{1.5.4}$$

We anticipate the expansion,

$$S'(\vec{J}', \vec{\theta}) = \vec{\theta}.\vec{J} + \epsilon S_1(\vec{J}', \theta).$$

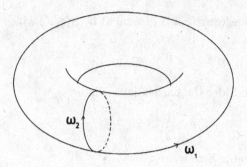

Figure 1.8: A 2-torus.

The canonical transformation equation $J_i' = \frac{\partial S'}{\partial \theta_i}$ leads to $J_i' = J_i + \epsilon \frac{\partial S_1}{\partial \theta_i}$.
Inserting this expansion in Eq.(1.5.4) and working to $O(\epsilon)$, get

$$
\begin{aligned}
H &= H_0(\vec{J}) + H_1(\vec{J}, \vec{\theta}) \\
&= H_0\left(\vec{J}' - \epsilon \frac{\partial S_1}{\partial \vec{\theta}}\right) + \epsilon H_1(\vec{J}, \theta) \\
&= H_0(\vec{J}') - \epsilon \frac{\partial H_0}{\partial \vec{J}'} \cdot \frac{\partial S_1}{\partial \vec{\theta}} + \epsilon H_1(\vec{J}, \vec{\theta}) \\
&= H_0(\vec{J}') - \epsilon \frac{\partial H_0}{\partial \vec{J}} \cdot \frac{\partial S_1}{\partial \vec{\theta}} + \epsilon H_1(\vec{J}, \vec{\theta}).
\end{aligned}
\tag{1.5.5}
$$

For integrability, the right hand side has to be independent of θ, and hence
the coefficient of ϵ must vanish. Thus to $O(\epsilon)$,

$$
\vec{\omega} \cdot \frac{\partial S_1}{\partial \vec{\theta}} = H_1(\vec{J}, \theta)
\tag{1.5.6}
$$

where $\vec{\omega} = \partial H_0 / \partial \vec{J}$ are the frequencies of the unperturbed system. Expanding
S_1 and H_1 in Fourier series as

$$
\begin{aligned}
S_1(\vec{J}, \theta') &= \sum_K S_{1,\bar{K}}.(\vec{J}) e^{2\pi i \vec{K}.\vec{\theta}'} \\
H_1(\vec{J}, \theta') &= \sum_{\vec{K}} H_{1,\bar{K}}.(\vec{J}) e^{2\pi i \vec{K}.\vec{\theta}'}
\end{aligned}
\tag{1.5.7}
$$

where the components of the vector $\bar{K}$ are integers, we obtain

$$
S(\vec{J}, \vec{\theta}) = \vec{\theta}.\vec{J} + i\epsilon \sum_{\vec{K}} \frac{H_{l,\vec{K}}(\vec{J})}{2\pi \vec{K}.\vec{\omega}(\vec{J})} + O(\epsilon^2).
\tag{1.5.8}
$$

Clearly Eq.(1.5.8) does not work if $\vec{K}.\vec{\omega} = 0$ for some values of $\vec{K}$. The
values of $\vec{J}$ so obtained define the resonant tori. These are typically the tori

which are destroyed for any small $\epsilon > 0$. The resonant tori are dense in the phase space of the unperturbed Hamiltonian. There are however, a very large number of strongly non-resonant tori. These satisfy

$$\left|\vec{K}.\vec{\omega}\right| > C|m|^{n+1} \tag{1.5.9}$$

where $|m| = |m_1| + |m_2| + \ldots + |m_N|$ and C is independent of m. For $\vec{\omega}$ satisfying Eq.(1.5.9), the series for S converges if H_1 is analytic in θ. Under these conditions, similar series for $S_2, S_3 \ldots$ etc. also converge. One would still have to argue the convergence of the sum of $S_1, S_2, \ldots$ etc. to the desired action S'. This requires a more sophisticated perturbation theory than we have given. A superconvergent perturbation theory leads to the KAM (Kolmogorov-Arnold-Moser) theorem that for small ϵ most of the tori of the unperturbed integrable Hamiltonian survive. This would violate the ergodic hypothesis, but we note that since the resonant tori are dense, arbitrarily near the surviving tori there will be regions of phase space where the trajectories are not on the surviving tori. It is here that the trajectories are chaotic and the success of statistical mechanics lies on the existence of these dense regions containing chaotic orbits. For larger ϵ, of course, the surviving tori are destroyed and ergodicity etc. are obvious. It should also be abundantly clear that for the success of equilibrium statistical mechanics, it is imperative to consider interacting systems, however weak the interactions may be. For purposes of calculation of thermodynamic quantities, interactions may be neglected whenever appropriate but the foundation of the subject of statistical mechanics depends on the existence of interactions.

1.6 Quantum Dynamics

What we have discussed so far has to do with the passage from classical dynamics to a statistical description. Quite often, classical dynamics is not an adequate description of the system. The system may need to be governed by quantum mechanics. In this section, we will show how the quantum evolution too will lead to a statistical description, very similar to the one already discussed. The state of the quantum system, unlike its classical counterpart, cannot be described by prescribing the coordinates and momenta of the individual particles because of the uncertainty principle which lies at the heart of quantum mechanics. Instead in quantum mechanics, the state of the system is prescribed by the wave function which is a function of all the coordinates involved, i.e. for a N-particle system, the state will be prescribed by a wave function $\psi(q_1, q_2, \ldots, q_{3N}, t)$ in a three-dimensional space.

In reality, we will always have our system placed in an environment and the wave function will be the wave function of the system together with the environment. If the environment coordinates are $\{Q_\alpha\}$, then the wave function of the system and environment will be some $\psi(\{q_i\}, \{Q_\alpha\}, t)$. If the system is described by a Hamiltonian H whose eigenfunctions are the set $\{\phi_n(q_i)\}$, then we have the complete set expansion

$$\psi(\{q_i\}, \{Q_\alpha\}, t) = \sum C_n(Q_\alpha, t)\phi_n(q_i) \tag{1.6.1}$$

with the wave function ψ normalized to unity.

Any physical quantity of the system will have a representation in terms of a Hermitian operator $\hat{O}$. The measured value of the quantity will be a time average of the quantum expectation value over a long time exactly as discussed in Sec.1.2 and we have

$$\langle O \rangle = \lim_{T \to \infty} \frac{1}{T} \int_0^T dt \prod_{i=1}^{3N} dq_i \prod_\alpha dQ_\alpha \psi^*(\{q_i\}, \{Q_i\}, t) \hat{O} \psi(\{q_i\}, \{Q_i\}, t)$$

$$= \sum \overline{C_n{}^* C_m} \int \prod_{i=1}^{3N} dq_i \phi_n{}^*(\{q_i\}) \hat{O} \phi_m(\{q_i\}) \tag{1.6.2}$$

where

$$\overline{C_n{}^* C_m} = \lim_{T \to \infty} \frac{1}{T} \int_0^T dt \int \prod_\alpha dQ_\alpha C_n{}^* C_m. \tag{1.6.3}$$

If we want to make any further progress, we need to make a statement about $\overline{C_n{}^* C_m}$. Suppose our system is almost isolated. This implies its energy lies between E and $E + \Delta$, where $\Delta \ll E$. Our quantum system characterized by the Hamiltonian H can reside in any one of the eigenstates ϕ_n where

$$H\phi_n = E_n \phi_n. \tag{1.6.4}$$

We make the assumption

$$\overline{C_n{}^* C_m} = 0 \quad \text{if} \quad m \neq n \tag{1.6.5}$$

called the random phase approximation and also

$$\overline{C_n{}^* C_n} = 1 \quad \text{if} \quad E < E_n < E + \Delta \tag{1.6.6}$$

which is the analogue of the equal a-priori probability.

The wave function of the system along with the environment can now be thought of as $\sum_n a_n \phi_n$ where $|a_n|^2 = 1$ if $E < E_n < E + \Delta$ and zero otherwise. The phases of the complex numbers a_n are random numbers and the effect of the environment is included in an average way. The crucial step behind the random phase approximation is the existence of the environment. The interactions allow the postulate to make sense. The expectation value of the operator can be written as

$$\langle \hat{O} \rangle = \sum_n |a_n|^2 \prod_i \int \phi_n{}^*(\{q_i\}) \hat{O} \phi_n(\{q_i\}) dq_i \tag{1.6.7}$$

and this triggers the ensemble picture. We consider N identical copies of the system. Each of these identical copies is in the eigenstate ϕ_n. In the contribution to the macrostate, each of these copies provide their own phase – the different microstates of the same macrostate in the classical case. We can now imagine summing over all the available states with energy lying between E and $E + \Delta$. In classical mechanics, this was the volume of the phase space [see Eq.(1.1.11)]. In quantum mechanics, we need to do a sum over the discrete states. However, for the macroscopic system, the energy levels E_n will be very closely spaced and consequently we will be able to write

$$\Gamma(E, \Delta) = \sum_{E < E_n < E + \Delta} |a_n|^2 = \Gamma(E)\frac{\Delta}{E} \qquad (1.6.8)$$

exactly as in the classical case. Hereafter, Boltzmanns hypothesis can be invoked and all the remaining issues remain the same. We simply need to note the correspondence

$$\frac{1}{N!}\frac{1}{h^{3N}} \int d^{3N}q\, d^{3N}p \rightarrow \sum_n \qquad (1.6.9)$$

where 'n' is the quantum state.

We close this introductory chapter with a discussion of a different way of specifying a quantum state in statistical mechanics. This is the so called density operator which is an operator written as

$$\rho = \sum_i p_i |\psi_i\rangle\langle\psi_i| \qquad (1.6.10)$$

where $|\psi_i\rangle$ is a quantum state in which the system is found with a probability p_i. The states $|\psi_i\rangle$ need not be orthogonal. In an orthonormal basis $|u_n\rangle$ we have the matrix (density matrix) with element

$$\rho_{mn} = \sum_i p_i \langle u_m|\psi_i\rangle\langle\psi_i|u_n\rangle. \qquad (1.6.11)$$

The expectation value of an operator $\hat{O}$ in such a mixed state will be given by

$$
\begin{aligned}
\langle\hat{O}\rangle &= \sum_i p_i \langle\psi_i|\hat{O}|\psi_i\rangle \\
&= \sum_i \sum_{m,n} p_i \langle\psi_i|u_m\rangle\langle u_m|\hat{O}|u_n\rangle\langle u_n|\psi_i\rangle \\
&= \sum_{m,n} \rho_{nm}\hat{O}_{mn} \\
&= \sum_n (\rho\hat{O})_{nn} \\
&= \text{Tr}\,(\rho\hat{O}) \qquad (1.6.12)
\end{aligned}
$$

The p_i that we are considering here are precisely the $|a_i|^2$ that are appearing in Eq.(1.6.7). The random phase approximation carried out earlier gives the density matrix description for the mixed state that we have considered here. The density matrix has the potential for making $\langle \hat{O} \rangle$ look like an average in phase space. To see this, we define the Weyl transform of an operator $\hat{A}$ (in coordinate space representation) as

$$\tilde{A}(x,p) = \int e^{-ipy/\hbar} \left\langle x + \frac{y}{2} \middle| \hat{A} \middle| x - \frac{y}{2} \right\rangle dy. \tag{1.6.13}$$

We consider the integral $\int \tilde{A}(x,p)\tilde{B}(x,p)dxdp$ for the product of the two operators and find

$$
\begin{aligned}
\int \tilde{A}(x,p)\tilde{B}(x,p)dxdp &= \int dxdp \int e^{-ipy/\hbar} \left\langle x + \frac{y}{2} \middle| \hat{A} \middle| x - \frac{y}{2} \right\rangle dy \\
&\quad \times \int e^{-ipz/\hbar} \left\langle x + \frac{z}{2} \middle| \hat{B} \middle| x - \frac{z}{2} \right\rangle dz \\
&= \int\int \left\langle x + \frac{y}{2} \middle| \hat{A} \middle| x - \frac{y}{2} \right\rangle \left\langle x - \frac{y}{2} \middle| \hat{B} \middle| x + \frac{y}{2} \right\rangle dydx \\
&= \int \langle u | \hat{A}\hat{B} | u \rangle du \\
&= \mathrm{Tr}\left(\hat{A}\hat{B} \right) \tag{1.6.14}
\end{aligned}
$$

where we have defined $x + \frac{y}{2} = u$ and $x - \frac{y}{2} = v$ and have used the completeness over $|v\rangle$.

Thus $\mathrm{Tr}\,(\rho\hat{O})$ can be written as a phase space integral and we have the formally interesting answer

$$\langle \hat{O} \rangle = \int dxdp\,\tilde{\rho}(x,p)\tilde{O}(x,p). \tag{1.6.15}$$

1.7   Problems for Chapter 1

1. States in phase space: Consider a particle confined in a large cube of side L. The allowed energies of the particle are characterized by three integers n_1, n_2, n_3 and expressed as $E = \frac{\pi^2\hbar^2}{2mL^2}(n_1^2 + n_2^2 + n_3^2)$. A particular set of values of n_1, n_2 and n_3 corresponds to a microstate. You are required to find the number of states accessible to the particle if the energy lies between E and $E + \Delta E$ in the limit of $L \to \infty$. (In this limit the number reduced to the volume of a spherical annulus. Your answer should reduce to $V.d^3p/\hbar^3$.

2. D-dimensional hyper-sphere $\sum_{i=1}^{D} x_i^2 = R^2$: Find the surface area of a unit sphere in D-dimensions. (Consider a D-dimensional space and evaluate the integral $\int\limits_{-\infty}^{\infty} dx_1 \int\limits_{-\infty}^{\infty} dx_2 \ldots \int\limits_{-\infty}^{\infty} dx_D e^{-(x_1^2 + x_2^2 + \ldots + x_D^2)}$ in two different ways.)

2

The Microcanonical Ensemble

2.1 Examples

An isolated system forms the basis of a microcanonical ensemble. In this case, the system cannot exchange energy or particle, with the environment. We characterize it by a fixed volume V, the number of particles N and by the energy E. The primary quantity that is needed is the number of microstates. The energy can actually lie between E and $E + \triangle$ with $\triangle \ll E$, since it is impossible to screen off all interactions between the system and environment in practice (gravity, for example, cannot be screened). One can of course simply consider the area of the constant energy hypersurface. In the thermodynamic limit ($N \to \infty$), the two slightly differing pictures give the same answer. We now need to use the postulates of statistical mechanics laid down in the last chapter to calculate the thermodynamic variables for this system. We will do so for some definite examples.

2.1.1 The Ideal Gas

The system consists of N molecules moving in a volume V without interacting with each other. The total energy is simply the total kinetic energy which can be written as

$$E = \frac{p_1^2}{2m} + \frac{p_2^2}{2m} + \cdots\cdots\cdots + \frac{p_N^2}{2m}. \tag{2.1.1}$$

While E is virtually fixed (only a negligible variation), the microstates are fixed by a particular distribution of each of the components of the vectors $\{\vec{p_i}\}$. Thus all points on the surface of the $3N$ dimensional sphere of Eq.(2.1.1) are possible microstates. They form the members of the microcanonical ensemble. All the microstates are equally probable. To find the probability, we need to calculate the number of microstates. The number is simply the volume of the annulus between two spherical surfaces characterized by the total energy E and $E + \triangle$. Of course for every point on the sphere in the momentum part of the phase space there will be a volume of the coordinate part of the phase space. The total number of microstates is

$$\Gamma(E, N, V) = \frac{1}{h^{3N}} \int \cdots \int \left(\begin{array}{c} \textit{allowed region} \\ \textit{in phase space} \end{array} \right) d^3q_1 \ldots d^3q_N . d^3p_1 \ldots d^3p_N.$$

$$(2.1.2)$$

This is the same expression as written down in Eq.(1.3.4). The allowed region for each q_i can range over the entire linear dimension unaffected by any other q_i or p_i. This is the effect of absence of interaction. Consequently

$$\Gamma(E, N, V) = \frac{V^N}{h^{3N}} \int \cdots \int \left(\begin{array}{c} \textit{allowed region in} \\ \textit{momentum space} \end{array} \right) d^3p_1 \ldots d^3p_N.$$

$$(2.1.3)$$

The allowed region in momentum space is the annulus defined by

$$E \le \sum_{i=1}^{N} \frac{p_i^2}{2m} \le E + \Delta. \tag{2.1.4}$$

We simply need to follow the calculation in Section 1.3 to obtain as in Eq.(1.3.8).

$$\Gamma(E, N, V) = \left[V \left(\frac{2m\pi E}{h^2} \right)^{3/2} \right]^N \frac{1}{\Gamma(3N/2)} \cdot \frac{\Delta}{E}. \tag{2.1.5}$$

For N identical particles, we need an additional factor of $(N!)^{-1}$. The probability of each microstate occurring is $[\Gamma(E, N, V)]^{-1}$. Our goal is to calculate the thermodynamic quantities. For this we need to use the Boltzmann hypothesis in the form $S = k \ln(\Gamma/N!)$

$$S = k \left[\frac{3N}{2} . \ln E + N . \ln V + \text{ constants} \right]. \tag{2.1.6}$$

The temperature T is now introduced as

$$\frac{1}{T} = \left(\frac{\partial S}{\partial E} \right)_{N,V} = \frac{3Nk_B}{2E}. \tag{2.1.7}$$

We immediately find

$$E = \frac{3}{2} N k_B T \tag{2.1.8}$$

leading to the constant volume specific heat

$$C_V = \frac{3}{2} N k_B. \tag{2.1.9}$$

To find the equation of state, we need to return to the laws of thermodynamics in the form

$$TdS = dE + PdV. \tag{2.1.10}$$

This leads to

$$P = T\left(\frac{\partial S}{\partial V}\right)_E \tag{2.1.11}$$

and from Eq.(2.1.6)

$$P = Nk_BT/V. \tag{2.1.12}$$

which is the equation of state that we wanted.

2.1.2 An ideal solid

We imagine the ideal solid as atoms arranged on a three dimensional lattice.

Each atom sitting at the lattice point is capable of small amplitude vibrations around the site. For the atom sitting at the ith site, the energy of the vibrations about the site is $E_i = \frac{1}{2}m\omega^2 x_i^2 + \frac{p_i^2}{2m}$. If we assume that all the vibrators have the same frequency and the same mass, then the total energy is

$$E = \sum_{i=1}^N \left(\frac{1}{2}m\omega^2 x_i^2 + \frac{p_i^2}{2m}\right). \tag{2.1.13}$$

Defining $q_i = m\omega x_i$ we can write

$$E = \sum_{i=1}^N \left(\frac{q_i^2}{2m} + \frac{p_i^2}{2m}\right). \tag{2.1.14}$$

The phase space volume is as before

$$
\begin{aligned}
\Gamma &= \frac{1}{h^{3N}} \int \dots\dots\dots \int d^3x_1 \dots d^3x_N d^3p_1 \dots d^3p_N \\
&= \frac{1}{h^{3N}} \left(\frac{1}{m\omega}\right)^{3N} \int \dots\dots\dots \int d^3q_1 \dots d^3q_N \\
&\times \ d^3p_1 \dots d^3p_N
\end{aligned} \tag{2.1.15}
$$

where, the integrals have been taken over the *allowed region* defined by Eq.(2.1.14). This corresponds to the surface of a sphere in a $6N$ dimensional space and the surface area will be given by $2\pi^{6N/2}(2mE)^{(6N-1)/2}\Gamma\left(\frac{6N}{2}\right)$. We can now write the volume of phase space as

$$\Gamma(E, N) = \left(\frac{2E}{m\omega^2}\right)^{3N/2} \left(\frac{2mE}{h^2}\right)^{3N/2} \frac{\pi^{3N}}{\Gamma(3N)} \frac{\Delta}{E}. \tag{2.1.16}$$

Comparing with Eq.(2.1.5) we note that the volume available to the free particle V has been replaced by the confining volume $\left(\frac{2E}{m\omega^2}\right)^{3/2}$ for the oscillator. Writing $S = k_B \ln \Gamma$ and writing $\frac{1}{T} = \left(\frac{\partial S}{\partial E}\right)_N$, we immediately arrive at

$$E = 3Nk_BT \qquad (2.1.17)$$

leading to the specific heat

$$C = 3Nk_B \qquad (2.1.18)$$

which is the familiar Dulong Petit law for the solids. We note that this counting does not require the correction factor $N!$ because the positions of the oscillators cannot be interchanged.

2.1.3 General oscillators

If we choose to write the constraint "over all allowed region" in formal terms, then Eq.(2.1.2) will become

$$\Gamma(E,V,N) = \frac{1}{h^{3N}} \int d^3q_1 \dots d^3q_N \int d^3p_1 \dots d^3p_N$$

$$\times \quad \delta\left(E - \frac{p_1^2 + p_2^2 + \dots + p_N^2}{2m}\right) \qquad (2.1.19)$$

and Eq.(2.1.15) will become

$$\Gamma(E,V,N) = \frac{1}{h^{3N}} \left(\frac{1}{m\omega}\right)^{3N} \int d^3q_1 \dots d^3q_N \int d^3p_1 \dots d^3p_N$$

$$\times \quad \delta\left(E - \frac{p_1^2 + \dots + p_N^2 + q_1^2 + \dots + q_N^2}{2m}\right). \qquad (2.1.20)$$

Focussing on Eq.(2.1.19), the q-integrations are trivially carried out, and since the 'radial' coordinate p can be used in $3N$-dimensional space for doing the momentum integrations, we can write the volume element $d^3p_1 \dots d^3p_N$ in terms of the radial coordinates in a $3N$-dimensional space as $2\pi^{3N/2}dp/\Gamma(3N/2)$ and we get

$$\Gamma(E,N,V) = \frac{1}{h^{3N}}.V^N.\frac{2\pi^{3N/2}}{\Gamma(3N/2)}.\int_0^\infty p^{3N-1}.\delta\left(E - \frac{p^2}{2m}\right) dp$$

$$= \frac{1}{h^{3N}}.V^N.\frac{2\pi^{3N/2}}{\Gamma(3N/2)}.(2m)^{\frac{3N-1}{2}}.\sqrt{2m}\int_0^\infty X^{3N-1}\delta(E - X^2)dX$$

$$= \frac{V^N}{\Gamma(3N/2)}.\left(\frac{2\pi mE}{h^2}\right)^{3N/2}\frac{1}{E}. \qquad (2.1.21)$$

This answer differs from Eq.(2.1.5) in the energy spread Δ, which is required to make Γ dimensionless. However, these terms are irrelevant in the thermodynamic limit and hence, for all practical purposes, Eqs.(2.1.5) and (2.1.21) are identical.

We could have evaluated the p-integral in Eq.(2.1.19) in a step-by-step process which can be useful for other problems. It suffices to do this in a one-dimensional space where we need to evaluate the integral

$$
\begin{aligned}
I &= \int_{-\infty}^{\infty} dp_N \ldots \int_{-\infty}^{\infty} dp_1 . \delta\left(E - \frac{p_1^2 + \ldots + p_N^2}{2m}\right) \\
&= \int_{-\infty}^{\infty} dp_N \ldots \int_{-\infty}^{\infty} dp_2 \int_{-\infty}^{\infty} dp_1 . \delta\left(E - \frac{p_2^2 + \ldots + p_N^2}{2m} - \frac{p_1^2}{2m}\right) \\
&= \int_{-\infty}^{\infty} dp_N \ldots \int_{-\infty}^{\infty} dp_2 \int_{-\infty}^{\infty} dp_1 . \delta\left(E_1 - \frac{p_1^2}{2m}\right) \\
&= \int_{-\infty}^{\infty} dp_N \ldots \int_{-\infty}^{\infty} dp_2 . \sqrt{\frac{m}{2}} . \frac{1}{\sqrt{E_1}} \\
&= \int_{-\infty}^{\infty} dp_N \ldots \int_{-\infty}^{\infty} dp_3 . \int_{-\sqrt{2mE_2}}^{\sqrt{2mE_2}} dp_2 . \frac{1}{\sqrt{E_2 - \frac{p_2^2}{2m}}} \quad (2.1.22)
\end{aligned}
$$

where

$$
E_2 = E - \frac{p_3^2 + \ldots + p_N^2}{2m} .
$$

The integral over p_2 is over a restricted region of space and will have the structure $C_2 \sqrt{2m}$, where C_2 is a constant. Similarly, the range of p_3 is $2\sqrt{2mE_3}$ where $E_3 = E - (p_4^2 + \ldots + p_N^2)/2m$ (note the sequence: $E_i - \frac{p_i^2}{2m} = E_{i-1}$) and hence, we are reduced to

$$
\begin{aligned}
I &= 2C_2 \sqrt{2m} \int_{-\infty}^{\infty} dp_N \ldots \int_{-\infty}^{\infty} dp_5 . \int_{-\sqrt{2mE_4}}^{\sqrt{2mE_4}} dp_4 \sqrt{E_4 - \frac{p_4^2}{2m}} \\
&= 2C_2 C_3 . \sqrt{2m} \sqrt{2m} . E_4
\end{aligned}
$$

where C_3 is another numerical constant. The integration over p_5 now gives $E_5^{3/2}$ times a number and the subsequent one gives $E_6^{4/2}$. Thus, carrying on down the line, we reach an integration of the expression $(E - p_N^2/2m)^{(N-3)/2}$ over p_N, giving an answer proportional to $E^{(N-2)/2}$, which is the correct answer. The lesson to be learnt from the above discussion is that, the integral with which we started [Eq.(2.1.22) above] will scale as $E^{\frac{N}{2}-1}$ exactly as power count suggests. We can now use the result to handle any oscillator.

If the potential energy is $V(x) = \lambda |x|^n$, then, restricting ourselves to one dimension, the phase space volume is to be obtained from,

$$
\begin{aligned}
\Gamma &= \int_{-\infty}^{\infty} \frac{dp_1 \ldots dp_N . dx_1 \ldots dx_N}{h^N} \\
&\times \delta\left[E - \frac{p_1^2 + \ldots + p_N^2}{2m} \lambda(|x_1|^n + \ldots + |x_N|^n)\right]
\end{aligned}
$$

If we define

$$E' = E - \lambda(|x_1|^n + \ldots + |x_N|^n)$$

then

$$
\begin{aligned}
\Gamma &= \int_{-\infty}^{\infty} dx_1 \ldots dx_N \int \frac{d^N p}{h^N} \delta\left(E' - \frac{p_1^2 + \ldots + p_N^2}{2m}\right) \\
&\propto \int_{-\infty}^{\infty} dx_1 \ldots dx_N (E')^{\frac{N-2}{2}}.
\end{aligned}
\tag{2.1.23}
$$

Every subsequent x-integration adds one $E^{1/n}$ to the result and hence eventually

$$\Gamma \propto E^{\frac{N}{2} + \frac{N}{n} - 1} \simeq E^{N\left(\frac{1}{2} + \frac{1}{n}\right)}. \tag{2.1.24}$$

The temperature is therefore given by

$$\frac{1}{T} = k_B N \frac{\partial}{\partial E}\left(\frac{1}{2} + \frac{1}{n}\right) \ln E = \left(\frac{1}{2} + \frac{1}{n}\right) \frac{k_B N}{E}. \tag{2.1.25}$$

This leads to a specific heat of

$$C = \left(\frac{1}{2} + \frac{1}{n}\right) k_B N. \tag{2.1.26}$$

2.1.4   Magnetic moments in an external field

We consider a set of N magnetic moments each of magnetic moment μ capable of orienting parallel or antiparallel to an externally applied field B. The energy in the parallel configuration is $-\mu B$ and in the antiparallel configuration μB. We need to count the total number of configurations of these N magnets which have a total energy E. If in this configuration N_1 particles are aligned parallel and N_2 aligned antiparallel, then the total energy is

$$E = -\mu B(N_1 - N_2) \tag{2.1.27}$$

while

$$N = N_1 + N_2 \tag{2.1.28}$$

leading to

$$
\begin{aligned}
N_1 &= \frac{1}{2}\left(N - \frac{E}{\mu B}\right) \\
N_2 &= \frac{1}{2}\left(N - \frac{E}{\mu B}\right).
\end{aligned}
\tag{2.1.29}
$$

The number of configurations is the number of ways in which N_1 can be chosen from out of N and is

$$\Gamma(N_1, N_2) = \frac{N!}{N_1!(N - N_1)!} = \frac{N!}{N_1!N_2!}$$

$$= \frac{N!}{\frac{1}{2}\left(N - \frac{E}{\mu B}\right)! \frac{1}{2}\left(N + \frac{E}{\mu B}\right)!} \tag{2.1.30}$$

The entropy follows using Stirling's formula as

$$S(E, N, B) = k_B \ln \Gamma(E, N, B)$$

$$= k_B \left\{ N \ln N - N - \frac{1}{2}\left(N - \frac{E}{\mu B}\right) \ln\left[\frac{1}{2}\left(N - \frac{E}{\mu B}\right)\right]\right.$$

$$+ \frac{1}{2}\left(N - \frac{E}{\mu B}\right) - \frac{1}{2}\left(N + \frac{E}{\mu B}\right) \ln\left[\frac{1}{2}\left(N + \frac{E}{\mu B}\right)\right]$$

$$\left. + \frac{1}{2}\left(N + \frac{E}{\mu B}\right)\right\}$$

$$= k_B \left\{ N \ln 2 - \frac{1}{2}N\left(1 - \frac{E}{\mu NB}\right) \ln\left(1 - \frac{E}{\mu NB}\right)\right.$$

$$\left. - \frac{1}{2}N\left(1 + \frac{E}{\mu NB}\right) \ln\left(1 + \frac{E}{\mu NB}\right)\right\} \tag{2.1.31}$$

The temperature follows from the usual relation

$$\frac{1}{T} = \left(\frac{\partial S}{\partial E}\right)_{N,B} = \frac{k_B}{2\mu B} \ln\left(1 - \frac{E}{\mu NB}\right) - \frac{k_B}{2\mu B} \ln\left(1 + \frac{E}{\mu NB}\right),$$

or

$$\frac{2\mu B}{k_B T} = \ln\left(\frac{1 - E/\mu NB}{1 + E/\mu NB}\right). \tag{2.1.32}$$

To find the magnetization M, we need to add the magnetic term to the entropy equation of Eq.(2.1.10), which gets modified to

$$TdS = d\tilde{E} + PdV - BdM$$

$$= d(\tilde{E} - BM) + PdV + MdB$$

$$= dE + PdV + MdB \tag{2.1.33}$$

so that

$$\frac{M}{T} = \left(\frac{\partial S}{\partial B}\right)_E. \tag{2.1.34}$$

In the above $\tilde{E}$ is the normal internal energy and E is the internal energy in the presence of the magnetic field. Straightforward algebra yields

$$\frac{M}{T} = k_B \frac{E}{2\mu B^2}\left[\ln\left(1 + \frac{E}{2\mu B^2}\right) - \ln\left(1 - \frac{E}{2\mu B^2}\right)\right]$$

$$= -\frac{E}{BT}, \tag{2.1.35}$$

where in the last line Eq.(2.1.32) has been used. Thus

$$M = -\frac{E}{B} = \mu N \tanh \frac{\mu B}{k_B T} \tag{2.1.36}$$

where we have used Eq.(2.1.32) once again.

It is important to check two limits

1. $\mu B \gg k_B T$, in this case all spins will be aligned and hence $M \to \mu N$ as expected.

2. $\mu B \ll k_B T$, in this case up spins and down spins appear with almost equal probability and that should make M vanish. Thus, in this limit,

$$M \approx \frac{\mu^2 N B}{k_B T}. \tag{2.1.37}$$

The susceptibility at this high temperature limit is

$$\chi = \frac{\mu^2 N}{k_B T} \tag{2.1.38}$$

which is Curie's law.

2.1.5   The quantum solid

Our next example is the ideal solid of Fig.2.1 once again with the difference that the dynamics of the oscillators are assumed to be governed by quantum mechanics (as would be the case at low temperatures, when the de Broglie wavelength $h/\sqrt{2\pi m k_B T}$ becomes large). The allowed energies of such an oscillator is

$$E_{n_1, n_2, n_3} = \left(n_1 + n_2 + n_3 + \frac{3}{2} \right) \hbar\omega \tag{2.1.39}$$

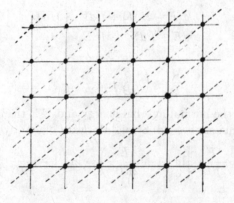

Figure 2.1: An ideal solid

where n_1, n_2 and n_3 are integers which run from zero to infinity independent of each other. Consequently, we can treat the three dimensions independent of each others. We will carry out the explicit calculations in one dimension, where the energy is determined by only one integer and is written as

$$E_n = \left(n + \frac{1}{2}\right) \hbar\omega. \tag{2.1.40}$$

If there are N such oscillators and the total energy is E, then specification of a configuration would require knowing the integers n_i associated with the ith oscillator and thus we have the condition

$$\sum_{i=1}^{N} \left(n_i + \frac{1}{2}\right) \hbar\omega = E. \tag{2.1.41}$$

Since there are N oscillators, this leads to the constraint

$$\sum_{i=1}^{N} n_i = \frac{E}{\hbar\omega} - \frac{N}{2} = M \tag{2.1.42}$$

where M is clearly an integer. The number of configuration comes form the set of n_i satisfying the relation given above. The interpretation of Eq.(2.1.42) is that there are N boxes and M particles that have to be put in these boxes. The number of configurations is the number of ways in which this partitioning can be done. The boxes are identical and so are the particles. We can create N boxes by drawing $(N-1)$ lines as shown in Fig.2.2.

Figure 2.2: Particles in boxes.

The particles are shown by dots. The number of dots between two successive lines is the number of particles in that box. All possible occupation are found by considering all permutations of the $(M + N - 1)$ objects.

This is clearly $(M + N - 1)!$. But the M particles and $(N - 1)$ lines are identical. Hence $M!$ and $(N - 1)!$ configurations of these identical objects have to be removed from the total number. Thus

$$\Gamma(M, N) = \frac{(M + N - 1)!}{M! \, (N - 1)!} \tag{2.1.43}$$

The entropy is found using Boltzmann's hypothesis and Stirling's formula

$$\begin{aligned}
S &= k_B \ln \Gamma \\
&= k_B \left[(M + N - 1) \ln(M + N - 1) - (M + N - 1) \right. \\
&\quad - M \ln M + M - (N - 1) \ln(N - 1) + N - 1] \\
&= k_B [(M + N - 1) \ln(M + N - 1) - M \ln M - (N - 1) \ln(N - 1)].
\end{aligned}$$

The temperature is now found from

$$\frac{1}{T} = \left(\frac{\partial S}{\partial E}\right)_N$$

$$= k_B \left[\ln(M + N - 1) - \ln M\right] \cdot \frac{\partial M}{\partial E}$$

$$= \frac{k_B}{\hbar\omega} \left[\ln\left(1 + \frac{N - 1}{M}\right)\right]. \qquad (2.1.44)$$

Since $N \gg 1$, the above leads to

$$1 + \frac{N}{M} = e^{\hbar\omega/k_B T}$$

and thus

$$E = N \left[\frac{\hbar\omega}{e^{\hbar\omega} - 1} + \frac{\hbar\omega}{2}\right]. \qquad (2.1.45)$$

This is the one dimensional result. For the three dimensional situation of Eq.(2.1.39), there are three independent integers n_1, n_2 and n_3 and thus the number of boxes in Eq.(2.1.42) simply goes up to $3N$ and Eq.(2.1.45) becomes

$$E = 3N \left[\frac{\hbar\omega}{e^{\hbar\omega} - 1} + \frac{\hbar\omega}{2}\right]. \qquad (2.1.46)$$

Quantum effects tend to disappear at high temperature i.e $k_B T \gg \hbar\omega$ and in this limit, Eq.(2.1.46) becomes

$$E \approx 3Nk_B T \qquad (2.1.47)$$

in exact agreement with the classical result of Eq.(2.1.17). At low temperature $k_B T \ll \hbar\omega, e^{\hbar\omega/k_B T} \gg 1$ and

$$E \approx 3\hbar\omega e^{-\hbar\omega/k_B T}$$

leading to the specific heat as

$$C \approx 3Nk_B \left(\frac{\hbar\omega}{k_B T}\right)^2 e^{-\hbar\omega/k_B T}. \qquad (2.1.48)$$

This specific heat vanishes at low temperatures in qualitative agreement with the experimental results. However, the low temperature behavior of Eq.(2.1.48) implies an exponentially small specific heat, while the experiments show a power law behavior for $k_B T \ll \hbar\omega$. The above result is known as the Einstein model of solids, where all the atoms are taken to vibrate with the same frequency. A more realistic model recognizes the existence of different modes of vibration of the solid and hence the existence of different frequencies.

For every frequency ω the energy content is as shown in Eq.(2.1.46). If $g(\omega)d\omega$ is the number of modes between ω and $\omega + d\omega$, then the total energy is

$$E = 3N \int g(\omega) \left[\frac{\hbar\omega}{e^{\hbar\omega} - 1} + \frac{\hbar\omega}{2} \right] \qquad (2.1.49)$$

and an enumeration of modes (to be done in a later chapter) leads to the correct T^3 law for the specific heat. This is the Debye model of solids.

2.1.6 The quantum free particle

In subsection (A) above we talked about the ideal gas which is a collection of classical free particles in a volume V. In this section we would like to tackle the question of quantum free particles in a volume V, which we take to be a cubical box of length L. The allowed energy levels of the ith particle in such a box is

$$E_i = (n_{i1}{}^2 + n_{i2}{}^2 + n_{i3}{}^2).\frac{\pi^2\hbar^2}{2mL^2}$$

where n_{i1}, n_{i2} and n_{i3} are integers. For a total of N particles the total energy is given by

$$E = \sum_{i=1}^{N} E_i = \sum_{i=1}^{N}(n_{i1}{}^2 + n_{i2}{}^2 + n_{i3}{}^2).$$

This imposes a condition on the integers n_{i1}, n_{i2} and n_{i3} as,

$$\sum_{i=1}^{N}(n_{i1}{}^2 + n_{i2}{}^2 + n_{i3}{}^2) = \frac{2mEL^2}{\pi^2\hbar^2}$$

The right hand side is fixed and obviously an integer and the combinatorics is the solution of the problem: in how many ways can a given integer be expressed as a sum of squares of a set of given integers? The set of number of integers that we are considering here is $3N$ (the three refers to the Euclidean dimension of the box). Using some standard mathematical notation, this number is Y_{3N} where $3N$ is the number of integers $m_1, m_2, \ldots, m_{3N}$ whose squares are being taken to add up to an integer G, i.e.,

$$\sum_{i=1}^{3N} m_i{}^2 = G = \frac{2mEL^2}{\pi^2\hbar^2}.$$

The answer is given by a theorem due to Hardy and Littlewood which asserts that

$$Y_G(3N) = I(G).J(G) + O(G^{3N/4})$$

with

$$J(G) = \frac{\pi^{3N/2}}{2^{3N}\Gamma(3N/2)} \cdot G^{(3N-2)/2}$$

and for $3N > 5$, there exists positive constants $C_1 = C_1(3N)$ and $C_2 = C_2(3N)$, such that $C_1 < I(G) < C_2$.

Thus, $I(G)$ is a constant whose precise value is not known *a priori*, but that is not vital for us. We thus find,

$$\Gamma(E) \propto L^{(3N-2)} \left(\frac{2mE}{\hbar^2} \right)^{(3N-2)/2} C(N).$$

For $N \gg 1$, this gives $(V = L^3)$

$$S(E, N, V) = k_B \left[N \ln V + \frac{3}{2} NE + N(\text{constants}) \right]$$

This gives, as expected, the molar specific heat $C = \frac{3}{2}R$. This should happen since, in this very classical setting where, for a large system size the energies are almost continuously distributed, the classical result should be reproduced.

The above examples make it clear that in principle all systems can be handled within the microcanonical ensemble. As is obvious we have restricted ourselves to non-interacting systems. Interacting systems are tough to deal with within this ensemble because it automatically puts constraints on the positions of the particles and the unconstrained sums that we have been carrying out on the coordinate variable will no longer be possible. This brings in technical problems and it is much easier to handle these systems in the other ensembles, where the system can interchange energy or energy and particle with the surroundings. We emphasize that these new ensembles will not require any new postulates.

2.2   Additivity of Entropy and Microcanonical Ensemble

We now consider an isolated system where the energy lies between E and $E + \Delta$, (where $\Delta \ll E$) and enumerate the number of microstates satisfying the total energy constraint. The equal a priori probability hypothesis asserts: All microstates compatible with the given macrostate (energy between E and $E + \Delta$) are equally probable.

An ensemble satisfying this energy constraint is a microcanonical ensemble.

We first verify that if two microcanonical ensembles are put in contact, the entropies add. System '1' has energy lying between E_1 and $E_1 + \Delta$ system '2' has energy between E_2 and $E_2 + \Delta$.

When put in contact let the final ensemble have an energy lying between E and $E+2\Delta$. If $\Gamma(E_1,\Delta)$ is the number of available states for '1' and $\Gamma(E_2,\Delta)$ the number of states for '2', then the number of states where the energy lies between E_1+E_2 and $E_1+E_2+2\Delta$ is $\Gamma(E_1,\Delta).\Gamma(E_2,\Delta)$. The total number of states $\Gamma(E,\Delta)$ available to the final ensemble is then given by

$$\Gamma(E,\Delta) = \sum_{E_1+E_2<E<E_1+E_2+2\Delta} \Gamma(E_1,\Delta)\Gamma(E_2,\Delta). \qquad (2.2.1)$$

Let $E_1 = \bar{E}_1$ and $E_2 = \bar{E}_2$ be the values of E_1 and E_2 for which the product $\Gamma(E_1,\Delta).\Gamma(E_2,\Delta)$ has its maximum value. Then, if we replace the sum by one term,

$$\Gamma(E,\Delta) \geq \Gamma(\bar{E}_1,\Delta)\Gamma(\bar{E}_2,\Delta). \qquad (2.2.2)$$

The number of terms in the sum of the right hand side of Eq.(2.2.1) is of the order of E/Δ. This follows from the fact that E_1 (or E_2) must be bounded from below for stability and if we call this lower bound zero, then the range of E_1 is from 0 to E, approximately in steps of Δ and hence the foregoing estimate of the number of terms. Clearly,

$$\Gamma(E,\Delta) \leq \frac{E}{\Delta}\Gamma(\bar{E}_1,\Delta)\Gamma(\bar{E}_2,\Delta) \qquad (2.2.3)$$

and thus

$$\Gamma(E_1,\Delta)\Gamma(\bar{E}_2,\Delta) \leq \Gamma(E,\Delta) \leq \frac{E}{\Delta}\Gamma(\bar{E}_1,\Delta)\Gamma(\bar{E}_2,\Delta)$$

or

$$\ln \Gamma(\bar{E}_1,\Delta)\Gamma(\bar{E}_2,\Delta) \leq \ln \Gamma(E,\Delta) \leq \ln \Gamma(\bar{E}_1,\Delta)\Gamma(\bar{E}_2,\Delta) + \ln \frac{E}{\Delta}. \qquad (2.2.4)$$

Now if system '1' has N_1 particles and system '2' has N_2, then $\ln \Gamma(\bar{E}_1,\Delta)$ is $O(N_1)$ and $\ln\Gamma(\bar{E}_2,\Delta)$ is $O(N_2)$ while $\ln\Gamma(E,\Delta)$ is $O(N_1+N_2)$. In the thermodynamic limit, $\ln(E/\Delta)$ is consequently negligible as it can at best be of $O(\ln N)$. Thus Eq.(2.2.4) becomes

$$k_B \ln \Gamma(E,\Delta) = k_B \ln \Gamma(\bar{E}_1,\Delta) + k_B \ln \Gamma(\bar{E}_2,\Delta)$$

or

$$S_{\text{combined}} = S_1 + S_2 \qquad (2.2.5)$$

where S_1 and S_2 are the entropies of the individual systems '1' and '2'.

The above exercise gives more information than exhibiting the additive property of entropy. Let us explore the consequences of maximizing the product $\Gamma(E_1).\Gamma(E_2)$ for $E_1 = \bar{E}_1$ and $E_2 = \bar{E}_2$, i.e., setting,

$$\delta[\Gamma(E_1)\Gamma(E_2)] = 0 \qquad (2.2.6)$$

subject to the constraint that the total energy E_1+E_2 remains constant (energy conservation for an isolated system on a macroscopic scale), i.e.,

$$\delta E_1 + \delta E_2 = 0 \qquad (2.2.7)$$

Considering small deviations about $\bar{E}_1$ and $\bar{E}_2$,

$$\frac{1}{\Gamma(E_1)}\left[\frac{\partial \Gamma(E_1)}{\partial E_1}\right]_{\bar{E}_1}\delta E_1 + \frac{1}{\Gamma(E_2)}\left[\frac{\partial \Gamma(E_2)}{\partial E_2}\right]_{\bar{E}_1}\delta E_2 = 0$$

or

$$\left[\frac{\partial \ln \Gamma(E_1)}{\partial E_1}\right]_{\bar{E}_1}\delta E_1 + \left[\frac{\partial \ln \Gamma(E_2)}{\partial E_2}\right]_{\bar{E}_2}\partial E_2 = 0. \qquad (2.2.8)$$

Now using Eq.(2.2.7) with a Lagrange multiplier λ, we have

$$\left[\left(\frac{\partial \ln \Gamma(E_1)}{\partial E_1}\right)_{\bar{E}_1} + \lambda\right]\delta E_1 + \left[\left(\frac{\partial \ln \Gamma(E_2)}{\partial E_2}\right)_{\bar{E}_2} + \lambda\right]\delta E_2 = 0 \qquad (2.2.9)$$

leading to

$$\left[\frac{\partial}{\partial E_1}\ln \Gamma(E_1)\right]_{\bar{E}_1} = -\lambda = \left[\frac{\partial \ln \Gamma(E_2)}{\partial E_2}\right]_{\bar{E}_2}. \qquad (2.2.10)$$

Using Boltzmann's hypothesis

$$\left(\frac{\partial S_1}{\partial E_1}\right)_{\bar{E}_1} = \left(\frac{\partial S_2}{\partial E_2}\right)_{\bar{E}_2} \qquad (2.2.11)$$

which, on using the thermodynamic relation $T^{-1} = (\partial S/\partial E)_{V,N}$ leads to $T_1 = T_2$ and thus the important result that if two systems are put in contact and they come to equilibrium, then the entropy is maximized. This leads to the expectation that the equilibrium state will be a maximum entropy state. In the next section we shall see this from a different point of view.

2.3   Shannon Entropy

There is a different way of defining entropy which was introduced by Shannon. If in a probabilistic description of outcomes of an event, the ith outcome occurred with a probability p_i, then the Shannon entropy is defined as

$$S = -k_B \sum_i p_i \ln p_i \qquad (2.3.1)$$

with $\sum p_i = 1$, when summed over all outcomes. If one maximizes S with this constraint using a Lagrange multiplier λ, then

$$0 = \delta S = -\sum_i (\ln p_i + \lambda).\delta p_i. \qquad (2.3.2)$$

Since the δp_i's are now independent, we have $\ln p_i = -\lambda$, or $p_i = $ constant. If there are M possible outcomes, then $p_i = 1/M$. Thus, maximization of entropy has led to equal probability of all microstates! We also see from Eq.(2.3.1) that $S = k_B \ln M$. Since the number outcomes is, in our context, the possible volume Γ if the available states, the relation is simply $S = k \ln \Gamma$, when the dimensional constant is introduced.

Thus, we see that in every way, the microcanonical ensemble provides an adequate description of the macroscopic world. However, for reasons of calculational convenience, we need to introduce non-isolated systems.

2.4 Problems for Chapter 2

1. Consider a long chain consisting of N links each of length a. The links favour an orientation of lining up but a link can orient itself perpendicular to the length of the chain if it has an energy ϵ. We thus have two possible orientations of a link, one in which is longitudinal with zero energy and another where it is transverse with energy ϵ. At any finite temperature find the expected length of the hanging chain.

2. Consider a lattice consisting of N atoms. At a finite temperature T, it is possible for atoms to escape. An atom at a lattice site requires an energy ϵ to move out. In a lattice with N_v vacancies, the total energy of the system is $E = \epsilon N_v$. At a temperature T, find the average number of vacancies in the lattice.

3. Consider a spin-$1/2$ object in a magnetic field. Make a plot of the entropy against the energy over all the allowed values of energy. At what points of this graph is the absolute value of the temperature zero or infinite? In some range of energy the temperature is negative. Is a negative temperature 'colder' or 'hotter' than a positive temperature? (This can be argued by finding the direction of heat flow when two objects at positive and negative temperatures are put in thermal contact.)

4. In the text we showed that all oscillators with the energy $E = \frac{p^2}{2m} + \lambda x^{2n}$ can be handled in the framework of a microcanonical ensemble. If we have a mixture of two potentials, one quadratic and the other quartic, i.e., $E = \frac{p^2}{2m} + \frac{1}{2}m\omega^2 x^2 + \frac{\lambda}{4}x^4$, life is far more difficult. For λ very small, use the microcanonical ensemble to find the first correction to the specific heat $N k_B$, which we get for $\lambda = 0$. (The difficulties of a calculation even for a non-interacting system, as seen in this case, is probably one of the reasons that the microcanonical prescription is not used very much.)

5. Consider Eq.(2.1.31) at fixed values of N and B. Plot S vs. E. You should find the slope of the curve to be negative for $E > 0$. What does this imply for the temperature T? Which body will be hotter – one at positive temperature or the one at negative temperature, i.e., if connected thermally which way will the heat flow?

3

Other Ensembles

3.1 Canonical Ensemble

In the last chapter, we asserted that the postulate of equal a priori probability and the concept of microcanonical ensembles are all that we require to find the probability distribution in situations where the system under consideration can exchange energy and particles with its environment. Expounding this theme and constructing the connection between this statistical description and the thermodynamic one will be the purpose of this chapter. We begin with the situation where our system has a fixed number of particles, but can exchange energy with its environment. The question to be asked is, what is the probability of the system having an energy E, when in equilibrium with its surrounding.

The technique to be used must employ a microcanonical ensemble according to our contention. So we consider our system (say S) together with the environment as an isolated system having a total energy lying between E_{tot} and $E_{\text{tot}} + \Delta$ and with all microstates corresponding to this macrostate equally probable. The probability $\rho(E)$ of our system S being in a microstate with energy E is consequently proportional to the number of states $\Gamma_e(E_{\text{tot}} - E)$ available to the environment, i.e.

$$\rho(E) \propto \Gamma_e(E_{\text{tot}} - E). \tag{3.1.1}$$

The environment is much bigger than the system and hence the environment energy $E_e \cong E_{\text{tot}}$ and $E \ll E_{\text{tot}}$ or E_e. Consequently the E-dependence can be expanded in a Taylor series with only the lowest term retained. With this in mind, we now recast Eq.(3.1.1) as

$$
\begin{aligned}
\rho(E) \quad &\propto \quad \Gamma_e(E_{\text{tot}} - E) \cong \Gamma_e(E_e - E) \\
&= \quad \Gamma_e(E_e) - E\frac{\partial \Gamma_e}{\partial E_e} + O(E^2) \\
&= \quad \Gamma_e(E_e)\left\{1 - E\frac{\partial \ln \Gamma_e}{\partial E_e} + O(E^2)\right\}.
\end{aligned} \tag{3.1.2}
$$

Since $\Gamma_e(E_e - E)$ is an exponential of a large quantity, an expansion of Γ in powers of E is not convenient. Hence one considers $\ln \Gamma$ for expansion purposes. Thus,

$$
\begin{aligned}
\ln \rho(E) &= \ln \Gamma_e(E_e) + \ln \left\{ 1 - E \frac{\partial \ln \Gamma_e}{\partial E_e} + O(E^2) \right\} \\
&= \ln \Gamma_e(E_e) - E \frac{\partial \ln \Gamma_e}{\partial E_e} + O(E^2) \\
&\cong \ln \Gamma_e(E_e) - E \frac{\partial \ln \Gamma_e}{\partial E_e}.
\end{aligned}
\tag{3.1.3}
$$

It is worth noting that the above form is useful only for $\ln \rho$ and not for ρ. We now recognize that the entropy S_e of the environment is $k_B \ln \Gamma_e$ and hence Eq.(3.1.3) can be written as

$$
\ln \rho(E) = \ln \Gamma_e(E_e) - \frac{E}{k_B} \frac{\partial S_e}{\partial E_e}.
\tag{3.1.4}
$$

Making use of the fact that $\partial S_e/\partial E_e = 1/T_e$ specifies the temperature of the environment and that in equilibrium the system temperature $T = T_e$, we find

$$
\ln \rho(E) = \ln \Gamma_e(E_e) - \frac{E}{k_B T}
$$

or

$$
\rho(E) = e^{\Gamma_e(E_e)} . e^{-E/k_B T} = A e^{-E/k_B T}
\tag{3.1.5}
$$

where A is a constant and the last line follows since $\Gamma_e(E_e)$ is a constant so far as S is concerned. We thus find that if the system S is in thermal equilibrium with its environment and is characterized by a temperature T, then the probability of its having an energy E is given by Eq.(3.1.5). The distribution should be normalized to unity, which implies that

$$
A = \left(\sum_{\text{all microstates}} e^{-E/k_B T} \right)^{-1} = Z^{-1}
\tag{3.1.6}
$$

where

$$
Z = \left(\sum_{\text{microstates}} e^{-E/k_B T} \right)
\tag{3.1.7}
$$

is called the canonical partition function of the system. The sum over all microstates is $\sum_E g(E) e^{-E/k_B T}$ where $g(E)$ is the number of microstates of energy E. A sum over states thus becomes a sum over allowed energies.

We now have to rationalize the fact that the thermodynamic state is characterized by a definite energy, while the statistical description above envisages a distribution of energy. If one has a distribution, then a definite number can be obtained by considering the mean of the distribution. The mean will be obtained repeatedly if the fluctuation about the mean is small. That is where the contact with the macroscopic description must lie. The above distribution should give a very small fluctuation about the mean. This is what we now establish.

The mean energy is obviously

$$\langle E \rangle = \frac{1}{Z} \sum_{\text{states}} E e^{-E/k_B T} = k_B T^2 \frac{\partial \ln Z}{\partial T} \tag{3.1.8}$$

A derivative gives the specific heat as

$$
\begin{aligned}
C &= \frac{\partial \langle E \rangle}{\partial T} \\
&= k_B \left[2T \frac{\partial \ln Z}{\partial T} + T^2 \frac{\partial^2 \ln Z}{\partial T^2} \right] \\
&= k_B T \frac{\partial^2}{\partial T^2} (T \ln Z). \tag{3.1.9}
\end{aligned}
$$

Since thermodynamics gives $C = -T(\frac{\partial^2 F}{\partial T^2})_V$, where F is the thermodynamic free energy, this prompts the identification $F = -k_B T \ln Z$.

A different way of expressing the derivative is

$$
\begin{aligned}
\frac{\partial \langle E \rangle}{\partial T} &= \frac{1}{Z} \sum \frac{E^2}{k_B T^2} e^{-E/k_B T} - \frac{1}{Z^2} \frac{\partial Z}{\partial T} \sum E e^{-E/k_B T} \\
&= \frac{1}{k_B T^2} \left[\frac{1}{Z} \sum E^2 e^{-E/k_B T} \right] \\
&\quad - \frac{1}{Z^2} \left(\frac{1}{k_B T^2} \sum E^2 e^{-E/k_B T} \right) \sum E e^{-E/k_B T} \\
&= \frac{1}{k_B T^2} [\langle E^2 \rangle - \langle E \rangle^2] \\
&= \langle (\Delta E)^2 \rangle / k_B T^2 \tag{3.1.10}
\end{aligned}
$$

where $\langle (\Delta E)^2 \rangle$ is the fluctuation about the mean, defined as

$$\langle (\Delta E)^2 \rangle = \langle (E - \langle E \rangle)^2 \rangle = \langle E^2 \rangle - \langle E \rangle^2. \tag{3.1.11}$$

From Eq.(3.1.10),

$$\langle (\Delta E)^2 \rangle = k_B T^2 \frac{\partial}{\partial T} \langle E \rangle = k_B T^2 C \tag{3.1.12}$$

where C is the specific heat, which is O(N), and N is the number of particles in the system.

The relative fluctuation is

$$\frac{\langle (\Delta E)^2 \rangle^{1/2}}{\langle E \rangle} = \frac{k_B^{1/2} C T}{\langle E \rangle} = O\left(\frac{1}{N^{1/2}} \right) \tag{3.1.13}$$

since $C^{1/2}$ is $O(N^{1/2})$ and $\langle E \rangle$ is $O(N)$. In the thermodynamic limit, $N \to \infty$ and hence the relative fluctuation decreases enormously. This is the reason why the statistical description is also an adequate description of the macroscopic thermodynamics. The entropy can be found from the Shannon form given in Eq.(2.3.1). This is an average entropy S and we get

$$
\begin{aligned}
S &= -k_B \sum_i p_i \ln\, p_i \\
&= k_B \sum_E \frac{1}{Z} e^{-\frac{E}{k_B T}} \left(\frac{E}{k_B T} + \ln\, Z \right) \\
&= \frac{1}{TZ} \sum_E E e^{-\frac{E}{k_B T}} + k_B \ln\, Z \\
&= \frac{\langle E \rangle}{T} + k_B \ln\, Z
\end{aligned}
\tag{3.1.14}
$$

leading to

$$
F = -k_B T \ln\, Z.
\tag{3.1.15}
$$

We make the above point and the connection between statistical mechanics and thermodynamics even more transparent by rewriting the canonical distribution through the following considerations:

$$
\begin{aligned}
Z &= \sum_{states} e^{-E/k_B T} \\
&= \int \Gamma(E) e^{-E/k_B T} dE \\
&= \int e^{S(E)/k_B} \cdot e^{-E/k_B T} dE \\
&= \int e^{-(E-TS)/k_B T} dE.
\end{aligned}
\tag{3.1.16}
$$

In the above $\Gamma(E)dE$ is the number of microstates with energy between E and $E + dE$ and we have made use of Boltzmann's hypothesis in the form $S = k_B \ln\, \Gamma(E)$.

We now examine the integrand of Eq.(3.1.16) and note that $E - TS$ is of $O(N)$ and hence for very large N, the integrand will be dominated by the minimum of $E - TS$ (provided of course $E - TS$ has a minimum). The minimum of $E - TS$ is obtained at $E = \bar{E}$, where

$$
1 = T \left(\frac{\partial S}{\partial E} \right)_{\bar{E}}.
\tag{3.1.17}
$$

But this is the thermodynamic relation between entropy and temperature and hence $\bar{E}$ is the thermodynamic energy that one is familiar with. We expand

$(E - TS)$ about $\bar{E}$ to write,

$$
\begin{aligned}
E - TS &= \bar{E} + E - \bar{E} - TS(\bar{E}) - T\left(\frac{\partial S}{\partial E}\right)_{\bar{E}} \Delta E \\
&\quad - \frac{T}{2}\left(\frac{\partial^2 S}{\partial E^2}\right)_{\bar{E}} (\Delta E)^2 + O(\Delta E)^3 \\
&= \bar{E} - TS(\bar{E}) - \frac{1}{2}T\left(\frac{\partial^2 S}{\partial E^2}\right)_{\bar{E}} (\Delta E)^2 \\
&= \bar{F} - \frac{T}{2}\frac{\partial}{\partial E}\left(\frac{1}{T}\right)(\Delta E)^2 \\
&= \bar{F} + \frac{T}{2}\frac{1}{T^2}\frac{1}{\partial E/\partial T}(\Delta E)^2 \\
&= \bar{F} + \frac{(\Delta E)^2}{2TC}
\end{aligned}
\tag{3.1.18}
$$

where $\bar{F} = \bar{E} - TS(\bar{E})$ is the thermodynamic free energy (Helmholtz free energy) and is the minimum value of $E - TS$ provided the correction term in Eq.(3.1.18) is positive which will happen for all $C > 0$, i.e. for positive specific heat. With the above relation, Eq.(3.1.17) can be written as

$$
Z = e^{-\bar{F}/k_B T} \int e^{-(\Delta E)^2/2k_B T^2 C} dE.
\tag{3.1.19}
$$

Thus, the canonical distribution corresponds to a Gaussian fluctuation about the mean (thermodynamic) energy. The integral in Eq.(3.1.19) is clearly $O(C^{1/2})$ which is $O(N^{1/2})$ and hence

$$
\ln Z = -\bar{F}/k_B T + \ln\{O(N^{1/2})\}.
\tag{3.1.20}
$$

The first term on the right hand side is $O(N)$ and hence dominates in the thermodynamic limit. In this limit the minimum free energy is the same as the ensemble average found in Eq.(3.1.14) and the thermodynamic free energy obtained just after Eq.(3.1.9), viz.,

$$
F(\text{thermodynamic free energy}) = -k_B T \ln Z
\tag{3.1.21}
$$

the desired connection between a statistical mechanical quantity Z and the thermodynamic quantity F. It also follows that E is the same as $\langle E \rangle$, which can be identified as the thermodynamical energy. We also note that thermodynamic stability requires $C > 0$. The connection between statistical mechanics and thermodynamics also ensues from using Boltzmann's relation in the form $\langle S \rangle = k_B \langle \ln \rho(E) \rangle$ by noting that $\rho(E)$ is inversely proportional to $\Gamma(E)$ and hence the thermodynamic entropy

$$
\begin{aligned}
S &= \langle S \rangle = -k_B \langle \ln(e^{-E/k_B T}/Z) \rangle \\
&= k_B \ln Z + \langle E \rangle/T \\
&= k_B \ln Z + E/T
\end{aligned}
$$

or

$$
F = -k_B T \ln Z.
$$

3.2   Alternative Derivation

In this section, we take up the study of the canonical distribution from a different point of view. The emphasis now is on the canonical ensemble aspect and the fact that the distribution that we are seeking is an equilibrium distribution is established by appealing to the volumes occupied by different distributions in phase space. We assume our given system S can reside in various energy states characterized by the discrete energies $\epsilon_1, \epsilon_2, \ldots \epsilon_n, \ldots$. This occurrence of discrete energies is usual in quantum mechanics - it should be borne in mind, however, that one is not dealing with quantum statistics as yet - and no quantum mechanical question about the state of the system is being asked. We simply assume the existence of discrete energy levels which is, however, a consequence of quantum mechanics.

Consider N mental copies of our system S and assume that n_1 of them have energy ϵ_1, n_2 have energy ϵ_2 and so on.

Clearly,

$$\sum n_i = N$$
$$\sum \epsilon_i n_i = E \tag{3.2.1}$$

A simple example would be a collection of non-interacting gas molecules in thermal equilibrium. In this case, a single molecule would constitute our system S and the whole collection (all other molecules are mental copies of our chosen one) would be the entire ensemble.

A given configuration (n_i, systems having energy ϵ_i) can be achieved in $P(\{n_i\})$ ways where $P(\{n_i\})$ is clearly given by

$$P(\{n_i\}) = \frac{N!}{\prod_i n_i!}. \tag{3.2.2}$$

How does the set $\{n_i\}$ distribute itself in equilibrium? This is where the criterion for an equilibrium distribution comes in. We will see in our discussions of the non-equilibrium situations that if one waits long enough, an equilibrium distribution is set up - a system prepared in a non-equilibrium state has an extremely large probability of ending up in the equilibrium state. Now, every distribution occupies a volume in phase space. If the natural tendency for an arbitrary distribution is to approach the equilibrium distribution, then it is reasonable to assume that the equilibrium distribution occupies the maximum volume in phase space. Thus the desired distribution is obtained by maximizing P of Eq.(3.2.2) subject to the constraints that the total number of particles and the total energy are fixed at N and E respectively. The maximization is clearly achieved by the set $\{\bar{n}_i\}$ where

$$\bar{n}_i = \bar{A} e^{-\beta \epsilon_i} \tag{3.2.3}$$

[above, we maximize $\ln P = \ln N! - \sum_i \ln n_i! = N \ln N - N - \sum n_i \ln n_i + N = N \ln N - \sum_i n_i \ln n_i$. The differential $\delta \ln P = 0$ leads, after some elementary algebraical manipulation, to $\sum_i (\ln n_i) \delta n_i = 0$. The constraints $\sum_i \delta n_i = \sum_i \epsilon_i \delta n_i = 0$ are taken care of by the Lagrange multipliers α and β and we have written $\bar{A} = e^{-\alpha}$ in Eq.(3.2.3).]

Consequently, the probability p_i, of finding the system in an energy state ϵ_i is given by

$$p_i = A e^{-\beta \epsilon_i} = \left(\sum_i e^{-\beta \epsilon_i} \right)^{-1} e^{\beta \epsilon_i} = \frac{1}{Z} e^{-\beta \epsilon_i} \qquad (3.2.4)$$

where Z, the canonical partition function, is $\sum_i e^{-\beta \epsilon_i}$. To fix the Lagrange multiplier β, we may apply the result to the ideal gas in a volume V and calculate the pressure. This fixes the constant β as $(k_B T)^{-1}$. The distribution of Eq.(3.2.4) now agrees completely with that of Eq.(3.1.5).

We now want to demonstrate that the set $\{\bar{n}_i\}$ of Eq.(3.2.3) indeed maximizes $P(\{n_i\})$. The probability of seeing a given set $(\{n_i\})$ among all possible sets is

$$\Omega(\{n_i\}) = \frac{P(\{n_i\})}{\sum_{\{n_i\}} P(\{n_i\})}. \qquad (3.2.5)$$

In principle, we would like to demonstrate that for $\{n_i\} \to \{\bar{n}_i\}$, $\Omega\{n_i\} \to 1$, but that being difficult, we intend to show that for $\{n_i\} \to \{\bar{n}_i\}$ the fluctuations in $\{n_i\}$ tend to zero. One begins by generalizing Eq.(3.2.2) to

$$P(\{n_i\}) = \frac{N!}{\prod_i n_i!} \prod_i g_i^{n_i} \qquad (3.2.6)$$

where all the g_i are to be set equal to unity at the end of the calculation. We first calculate the average number $\langle n \rangle$ of ensemble systems with energy ϵ_i, as

$$\begin{aligned}
\langle n_i \rangle &= \sum_{(n_i)} n_i \Omega(\{n_j\}) \\
&= g_i \frac{\partial}{\partial g_i} \left[\ln \sum_{\{n_i\}} P(\{n_j\}) \right].
\end{aligned} \qquad (3.2.7)$$

We now note the following identity

$$\begin{aligned}
g_i \frac{\partial}{\partial g_i} \langle n_i \rangle &= g_i \frac{\partial}{\partial g_i} \sum_{\{n_i\}} \frac{n_i P(\{n_j\})}{\sum_{\{n_i\}} P(\{n_k\})} \\
&= \sum_{\{n_i\}} \frac{n_i^2 P(\{n_j\})}{\sum_{\{n_i\}} P(\{n_k\})} - \left[\sum_{\{n_i\}} \frac{n_i P(\{n_k\})}{\sum_{\{n_i\}} P(\{n_j\})} \right]^2 \\
&= \langle n_i^2 \rangle - \langle n_i \rangle^2 \\
&= \langle (\Delta n_i)^2 \rangle. \qquad (3.2.8)
\end{aligned}$$

Hence the ensemble fluctuation in n_i, is given by

$$\langle(\Delta n_i)^2\rangle^{1/2} = \lim_{\{g_i\}\to 1}\left[g_i\frac{\partial}{\partial g_i}\langle\Delta n_i\rangle\right]^{1/2}. \qquad (3.2.9)$$

It should be emphasized that these averages and fluctuations refer to a space of ensembles – averages are taken over various kinds of ensembles that can exist and the averages have nothing to do with the ensemble averages discussed in the previous section which were obtained by working within one ensemble, namely the canonical ensemble.

With the P changed to the form given by Eq.(3.2.6), the method of max-imizing P leads to

$$\bar{n}_i = \bar{A}_i g_i e^{\beta\epsilon_i} = N\frac{g_i e^{-\beta\epsilon_i}}{\sum g_i e^{-\beta\epsilon_i}}. \qquad (3.2.10)$$

If we consider all the g's equal to 1, except g_i, which we take close to unity, then $\bar{n}_i \propto g_i$, and we can make the identification $\bar{n}_i \cong \langle n_i\rangle$ (see Eq.(3.2.7)). This leads to

$$g_i\frac{\partial}{\partial g_i}\langle n_i\rangle = g_i\frac{\partial}{\partial g_i}\bar{n}_i \cong \langle\bar{n}_i\rangle \cong \langle n_i\rangle. \qquad (3.2.11)$$

Thus,

$$\frac{\{(\Delta n_i)^2\}^{1/2}}{\langle n_i\rangle} = \left\{\frac{1}{\langle n_i\rangle}\right\}^{1/2} \qquad (3.2.12)$$

leading to the result that if $\langle n_i\rangle = \bar{n}_i$ then the fluctuation in $\langle n_i\rangle$ can be made arbitrarily small by making $\langle n_i\rangle$ large. Hence in the space of all distributions, almost all distributions are of the canonical form and if one starts from an arbi-trary distribution, then there is an overwhelming probability that the canonical distribution will be obtained.

It is worthwhile pointing out exactly what is ignored in Eq.(3.2.11). To do so we differentiate Eq.(3.2.7) directly and obtain,

$$g_i\frac{\partial}{\partial g_i}\langle n_i\rangle = g_i\frac{\partial}{\partial g_i}\ln\sum_{\{n_j\}}P(\{n_j\}) + g_i\frac{\partial^2}{\partial g_i^2}\ln\sum_{\{n_j\}}P(\{n_j\})$$

$$= \sum_{\{n_j\}}n_j\frac{P(\{n_j\})}{\sum_{\{n_k\}}P(\{n_k\}))} + g_i\frac{\partial^2}{\partial g_i^2}\ln\sum_{\{n_j\}}P(\{n_j\})$$

$$= \sum_{\{n_j\}}\langle n_j\rangle + g_i\frac{\partial}{\partial g_i}\ln\sum_{\{n_j\}}P(\{n_j\}). \qquad (3.2.13)$$

It is the second term on the right hand side of Eq.(3.2.13) that is dropped when we obtain $\langle(\Delta n_i)^2\rangle \cong \langle n_i\rangle$.

To conclude this section, we note that the canonical partition function $Z = \sum_i e^{-\epsilon_i/kT}$ need not always converge. A collection of non-ineracting hydrogen atoms is the classic example, where one is left with a clearly divergent sum of the form $\sum_{n=1}^{\infty} n^2 e^{c/n^2}$ where n is an integer and c is a constant.

3.3 Varying Number of Particles

As a prelude to discussing distributions for systems where the particle number may vary, we first study the macroscopic properties of such systems. Whenever there is an extra variable that can change (i.e., apart from temperature and volume), one has to think of the forces which can cause such changes and the resultant work done. When the number of particles changes, the relevant force is the chemical potential μ and the second law of thermodynamics takes the form

$$TdS = dE + PdV - \mu dN. \tag{3.3.1}$$

The sign of the last term can be appreciated from the fact that if a certain amount of heat is supplied to the system (positive TdS), there will be a propensity for the particles to leave, i.e. a negative dN, provided of course $\mu > 0$. Clearly from the above form of the second law

$$\left(\frac{\partial S}{\partial N}\right)_{E,V} = -\frac{\mu}{T}. \tag{3.3.2}$$

This is a useful relation for obtaining the chemical potential. Using Boltzmann's hypothesis we have obtained the entropy of an ideal gas in Sec.2.3 as

$$S = k_B N \left[\ln\left(\frac{V}{N}\right) + \ln\left(\frac{E}{N}\right)\right]^{3/2} + \frac{3}{2}\ln\left(\frac{2\pi m}{h^2}\right) + \frac{5}{2} \tag{3.3.3}$$

leading to

$$
\begin{aligned}
-\frac{\mu}{T} = \left(\frac{\partial S}{\partial N}\right)_{E,V} &= k_B \left[\ln\left(\frac{V}{N}\right) + \ln\left(\frac{E}{N}\right)^{3/2} + \frac{3}{2}\ln\left(\frac{2\pi m}{h^2}\right)\right] \\
&= k_B \ln\left[\left(\frac{2\pi m k_B T}{h^2}\right)^{3/2} \frac{V}{N}\right]
\end{aligned}
$$

or

$$e^{-\mu/k_B T} = \frac{V}{N}\cdot\left(\frac{2\pi m k_B T}{h^2}\right)^{3/2}. \tag{3.3.4}$$

We can now generate the different thermodynamic potentials. Corresponding to the Helmholtz free energy F, we define a variation $\bar{F}$ as

$$\bar{F} = E - TS - \mu N \tag{3.3.5}$$

with

$$
\begin{aligned}
d\bar{F} &= dE - TdS - SdT - \mu dN - Nd\mu \\
&= -SdT - pdV - Nd\mu.
\end{aligned} \tag{3.3.6}
$$

This leads to

$$N = -\left(\frac{\partial \bar{F}}{\partial \mu}\right)_{T,V} \tag{3.3.7}$$

and also shows that for a thermodynamic system at constant T, V and $\mu, \bar{F}$, is the quantity which has to be minimized at equilibrium. The Gibbs free energy G is defined in the usual manner, but its differential form changes to account for Eq.(3.3.1)

$$
\begin{aligned}
G &= E - TS + PV \\
dG &= dE - TdS + PdV + VdP - SdT \\
&= -SdT + VdP + \mu dN
\end{aligned}
$$
(3.3.8)

leading to

$$
\left(\frac{\partial G}{\partial T}\right)_{P,N} = -S, \quad \left(\frac{\partial G}{\partial T}\right)_{P,N} = V, \quad \left(\frac{\partial G}{\partial T}\right)_{T,P} = \mu.
$$
(3.3.9)

We readily see that a thermodynamic system at constant T, P and N is characterized by the minimum of the Gibbs free energy.

We now set up an interesting connection between G and μ. The free energy can be expressed as function of any three independent variables and we choose P, T and N for our purpose. Clearly, the energy is proportional to N and so must have the form (Note that changing N does not change P or T)

$$
G = N\Phi(P,T)
$$
(3.3.10)

The differential reads

$$
\begin{aligned}
dG &= \left(\frac{\partial G}{\partial P}\right)_{N,T} dP + \left(\frac{\partial G}{\partial T}\right)_{N,P} dT + \Phi dN \\
&= VdP - SdT + \frac{G}{N} dN.
\end{aligned}
$$
(3.3.11)

Comparing with Eq.(3.3.8), we see

$$
G = \mu N
$$
(3.3.12)

the chemical potential is simply the Gibbs free energy per particle. The free energy $\bar{F}$, defined above, now becomes

$$
\begin{aligned}
\bar{F} &= E - TS - \mu N \\
&= E - TS - G \\
&= E - TS - (E - TS + PV) \\
&= -PV.
\end{aligned}
$$
(3.3.13)

From Eq.(3.3.7), we now obtain

$$
N = \left[\frac{\partial(PV)}{\partial \mu}\right]_{T,V}.
$$
(3.3.14)

This relation is extremely useful for calculating the chemical potential of a gas as will become evident later.

3.4   Grand Canonical Ensemble

We will now discuss the probability distribution of a system which has variable energy and a variable number of particles. The thermodynamic state of such a system is characterized by a definite temperature and a definite chemical potential - this is made possible by the system being in equilibrium with a surrounding with which it can exchange energy and particles. Our derivation will be almost identical to the derivation of ρ for the canonical ensemble in Sec. 3.1. We will focus on our system S which is in contact with the environment and the combination of S and its environment constitutes an isolated system characterized by $E_{\text{tot}}, N_{\text{tot}}$ and V_{tot}, for energy, particle number and volume. At any instant, our system S has energy E, particle number N and volume V (fixed) where $E, N, V \ll E_{\text{tot}}, N_{\text{tot}}, V_{\text{tot}}$ respectively. The probability that S has energy E and particle number N, is proportional to the number of microstates for which this occurs. The probability of occurrence of a microstate corresponding to a given E and N is proportional to the number of microstates for which the environment has energy $E_{\text{tot}} - E$ and particle number $N_{\text{tot}} - N$. Thus

$$\rho(E, N) \propto \Gamma_e(E_{\text{tot}} - E, N_{\text{tot}} - N) \cong \Gamma_e(E_e - E, N_e - N). \qquad (3.4.1)$$

In the last step we have assumed that the environment energy and particle number are very nearly equal to the total. Working to $O(E/E_e)$ and $O(N/N_e)$, we get

$$
\begin{aligned}
\rho(E, N) \quad \propto \quad & \Gamma_e(E_e, N_e) - E\left(\frac{\partial \Gamma_e}{\partial E_e}\right)_N - N\left(\frac{\partial \Gamma_e}{\partial E_e}\right)_E \\
= \quad & \Gamma_e(E_e, N_E)\left[1 - E\left(\frac{\partial \ln \Gamma_e}{\partial E_e}\right)_N - N\left(\frac{\partial \ln \Gamma}{\partial E_e}\right)_E\right] \\
= \quad & \Gamma_e(E_e, N_e)\left[1 - \frac{E}{k_B}\left(\frac{\partial S}{\partial E_e}\right)_N - \frac{N}{k_B}\left(\frac{\partial S}{\partial N}\right)_E\right] \\
= \quad & \Gamma_e(E_e, N_e)\left[1 - \frac{E}{k_B T} + \frac{\mu N}{k_B T}\right] \qquad (3.4.2)
\end{aligned}
$$

where the Boltzmann relation and the definition of T and μ have been made use of in lines three and four of Eq.(3.4.2).

Once again working to linear order in E and N

$$\ln \rho(E, N) = \ln \Gamma_e(E_e, N_e) - \frac{(E - \mu N)}{k_B T}$$

leading to

$$\rho(E, N) = A \exp\left(-\frac{(E - \mu N)}{k_B T}\right) \qquad (3.4.3)$$

where A is a constant as far as S is concerned, the system having nothing to do with E_e and N_e. Thus, the grand canonical distribution function is given by

Eq.(3.4.3) and the grand partition function G is obtained by summing over all the microstates, so that

$$Q = \sum_{states} e^{-(E-\mu N)/k_B T} \tag{3.4.4}$$

where, as before, $\sum_{states}$ stands for sum over all microstates. We can now write the distribution function as

$$\rho(E, N) = \frac{1}{Q} e^{-(E-\mu N)/k_B T}. \tag{3.4.5}$$

Once again, the contact with thermodynamics depends crucially on the fact that the distribution is very sharply centred around $\bar{E}$ and $\bar{N}$ in the limit of $\bar{N} \to \infty$ and hence we must demonstrate that the above distribution does lead to very small fluctuations in $\langle E \rangle$ and $\langle N \rangle$.

We begin with the fluctuation in the average particle number, by noting that

$$\langle N \rangle = \frac{1}{Q} \sum_{states} N e^{-\beta(E-\mu N)} \tag{3.4.6}$$

$$\left(\frac{\partial \langle N \rangle}{\partial \mu} \right)_{V,T} = \frac{\beta}{Q} \sum N^2 e^{-\beta(E-\mu N)} - \frac{\beta}{Q^2} \left(\sum N e^{-\beta(E-\mu N)} \right)^2$$

$$= \beta[\langle N^2 \rangle - \langle N \rangle^2]$$

$$= \beta\langle(\Delta N^2)\rangle \tag{3.4.7}$$

where $\Delta N = N - \langle N \rangle$. To proceed further, we relate the derivative $\frac{\partial \langle N \rangle}{\partial \mu}$ to the susceptibility. Our aim of showing vanishingly small fluctuation has of course already been satisfied by Eq.(3.4.7) since $\partial \langle N \rangle / \partial \mu$ is $O(\langle N \rangle)$ and hence $\langle(\Delta N)^2\rangle^{1/2}/\langle N \rangle$ is $O(\langle N \rangle^{1/2})$ and becomes vanishingly small as $\langle N \rangle \to \infty$. In an identical manner the smallness of $\langle(\Delta E)\rangle^{1/2}/\langle E \rangle$ can be established as we will explicitly demonstrate later. Now,

$$\left(\frac{\partial \langle N \rangle}{\partial \mu} \right)_{V,T} = \frac{\left(\frac{\partial \langle N \rangle}{\partial P} \right)_{V,T}}{\left(\frac{\partial \mu}{\partial P} \right)_{V,T}}$$

$$= \frac{N}{V} \left(\frac{\partial \langle N \rangle}{\partial P} \right)_{V,T} \tag{3.4.8}$$

where we have made use of Eq.(3.3.14). Now P is a function of N/V and T and hence differentiating with respect to N and differentiating with respect to V are going to be simply related as

$$\left(\frac{\partial P}{\partial V} \right)_{N,T} = -\frac{N}{V} \left(\frac{\partial P}{\partial N} \right)_{V,T}. \tag{3.4.9}$$

Consequently Eq.(3.4.8) becomes

$$\left(\frac{\partial N}{\partial \mu}\right)_{V,T} = \frac{N}{V}\left(\frac{\partial N}{\partial P}\right)_{V,T}$$

$$= -\frac{N}{V}\cdot\frac{N}{V}\left(\frac{\partial V}{\partial P}\right)_{N,T}$$

$$= -n^2 N\left[\frac{\partial}{\partial P}\left(\frac{V}{N}\right)\right]_{N,T}$$

$$= -n^2 N\chi_T \tag{3.4.10}$$

where $n = N/V$, the number density and $\chi_T = -\left[\frac{\partial}{\partial P}\left(\frac{1}{n}\right)\right]_{N,T}$ is the isothermal susceptibility. Thus,

$$\langle(\Delta N)^2\rangle = k_B T n^2 N\chi_T \tag{3.4.11}$$

and as expected $\langle(\Delta N)^2\rangle^{1/2}/N$ does go to zero for $\langle N\rangle \to \infty$, unless χ_T blows up, which it can, near a second-order phase transition.

3.5 Energy Fluctuation in Grand Canonical Ensemble

To find the energy fluctuation, we proceed by finding (volume is taken to be held fixed)

$$\left(\frac{\partial\langle E\rangle}{\partial \beta}\right)_\mu = \frac{\partial}{\partial \beta}\frac{1}{Q}\sum Ee^{-\beta(E-\mu N)}$$

$$= -\frac{1}{Q}\sum E(E-\mu N)e^{-\beta(E-\mu N)} + \frac{1}{Q^2}\sum Ee^{-\beta(E-\mu N)}$$

$$\times \sum(E-\mu N)e^{-\beta(E-\mu N)}$$

$$= -\langle(\Delta E)^2\rangle + \mu[\langle EN\rangle - \langle E\rangle\langle N\rangle] \tag{3.5.1}$$

and

$$\left(\frac{\partial\langle E\rangle}{\partial \mu}\right)_T = \frac{\beta}{Q}\sum ENe^{-\beta(E-\mu N)} - \frac{\beta}{Q^2}\sum Ee^{-\beta(E-\mu N)}$$

$$\times \sum Ne^{-\beta(E-\mu N)}$$

$$= \beta[\langle EN\rangle - \langle E\rangle\langle N\rangle]$$

$$= \beta\langle(\Delta E)(\Delta N)\rangle. \tag{3.5.2}$$

Combining Eqs.(3.5.1) and (3.5.2), we find

$$\langle(\Delta E)^2\rangle = k_B T^2\left(\frac{\partial\langle E\rangle}{\partial T}\right)_{\mu,V} + \mu k_B T\left(\frac{\partial\langle E\rangle}{\partial \mu}\right)_{T,V}. \tag{3.5.3}$$

Once again $\langle(\Delta E)^2\rangle$ is O(N) and hence $\langle(\Delta E)^2\rangle^{1/2}/\langle E\rangle$ is O($N^{1/2}$) which is vanishingly small as $\langle N\rangle \to \infty$.

It is desirable to put Eq.(3.5.3) in a more usable form. For this, we note that

$$\left(\frac{\partial \langle E\rangle}{\partial T}\right)_{\mu,V} = \left(\frac{\partial \langle E\rangle}{\partial T}\right)_{N,V} + \left(\frac{\partial \langle E\rangle}{\partial N}\right)_{T,V} \cdot \left(\frac{\partial N}{\partial T}\right)_{\mu,V} \tag{3.5.4}$$

and have (noting that $\left(\frac{\partial \langle E\rangle}{\partial t}\right)_{N,V} = C_{N,V}$)

$$
\begin{aligned}
\langle(\Delta E)^2\rangle &= k_B T^2 C_{N,V} + \mu k_B T \left(\frac{\partial \langle E\rangle}{\partial \mu}\right)_{T,V} + k_B T^2 \left(\frac{\partial \langle E\rangle}{\partial N}\right)_{T,V} \cdot \left(\frac{\partial N}{\partial T}\right)_{\mu,V} \\
&= k_B T^2 C_{N,V} + k_B T \left[\left(\frac{\partial \langle E\rangle}{\partial N}\right)_{T,V} \cdot \left(\frac{\partial N}{\partial T}\right)_{\mu,V} \right. \\
&\quad + \left. \mu \left(\frac{\partial \langle E\rangle}{\partial N}\right)_{T,V} \cdot \left(\frac{\partial N}{\partial \mu}\right)_{T,V} \right] \\
&= k_B T^2 C_{N,V} + k_B T^2 \left(\frac{\partial \langle E\rangle}{\partial N}\right)_{T,V} \left[\left(\frac{\partial N}{\partial T}\right)_{\mu,V} + \frac{\mu}{T}\left(\frac{\partial N}{\partial \mu}\right)_{T,V} \right] \\
&= k_B T^2 C_{N,V} + k_B T^2 \left(\frac{\partial \langle E\rangle}{\partial N}\right)_{T,V} \\
&\quad \times \left(\frac{\partial N}{\partial \mu}\right)_{T,V} \left[\frac{\mu}{T} - \left(\frac{\partial \mu}{\partial T}\right)_{N,V} \right].
\end{aligned}
\tag{3.5.5}
$$

We now note that $dE = TdS - PdV + \mu dN$ leads to

$$\left(\frac{\partial E}{\partial N}\right)_{T,V} = T\left(\frac{\partial S}{\partial N}\right)_{T,V} + \mu$$

and a Maxwell relation $\left(\frac{\partial S}{\partial N}\right)_{T,V} = -\left(\frac{\partial \mu}{\partial T}\right)_{N,V}$ can be constructed from $dF = -SdT - PdV + \mu dN$. This allows us to write Eq.(3.5.5) as

$$
\begin{aligned}
\langle(\Delta E)^2\rangle &= k_B T^2 C_{N,V} + k_B T \left(\frac{\partial N}{\partial \mu}\right)_{T,V} \cdot \left(\frac{\partial E}{\partial N}\right)^2_{T,V} \\
&= k_B T^2 C_{N,V} + \langle(\Delta N)^2\rangle \left(\frac{\partial E}{\partial N}\right)^2_{T,V}.
\end{aligned}
\tag{3.5.6}
$$

The first term on the right hand side is a generalization of the canonical ensemble and the second can be looked upon as each exchanged particle carrying an energy of $\left(\frac{\partial E}{\partial N}\right)_{T,V}$.

3.6 Physical Interpretation of Q

The validity of the thermodynamic distribution having been established for the macroscopic system, we now establish the passage from the grand partition function to a thermodynamic variable. Before doing so it is worth pointing out that Eq.(3.5.2) shows that the fluctuations in E and N are not independent of each other and gives the strength of the correlation of the fluctuations.

Boltzmann's formula once more provides the quantitative connection with thermodynamics:

$$
\begin{aligned}
S &= -k_B \left\langle \ln \left(\frac{1}{Q} e^{-\beta(E-\mu N)} \right) \right\rangle \\
&= \frac{1}{T} \langle E - \mu N \rangle + k_B \ \ln \ Q
\end{aligned}
$$

or

$$
TS = E - \mu N + k_B T \ \ln \ Q
$$

or

$$
-k_B T \ \ln \ Q = E - TS - \mu N = \bar{F} = -PV
$$

leading to

$$
PV = k_B T \ \ln \ Q. \tag{3.6.1}
$$

We have made use of Eq.(3.3.13) in the final step. Combining Eqs.(3.6.1) and (3.3.14), we can obtain the equation of state of a gas.

We finally show that the grand canonical distribution is equivalent to a Gaussian distribution in fluctuation in energy and particle number. To do so, we write the grand canonical partition function as

$$
\begin{aligned}
Q &= \int e^{-\beta(E-\mu N)} \Gamma(E, N) dE dN \\
&= \int e^{-(E-TS-\mu N)/k_B T} dE dN. \tag{3.6.2}
\end{aligned}
$$

Now the argument of the exponential in the above integrand is $O(N)$ and in the limit of $\langle N \rangle \to \infty$ will be dominated by the minimum of $E - TS - \mu N$, provided the expression does have a minimum. The conditions for extremization are

$$
1 = T \left(\frac{\partial S}{\partial E} \right)_{V,N} \tag{3.6.3}
$$

and

$$
-\mu = T \left(\frac{\partial S}{\partial N} \right)_{V,E} \tag{3.6.4}
$$

The solutions of the above equations give $E = \bar{E}$ and $N = \bar{N}$. Now, we write

$$
\begin{aligned}
E - TS - \mu N &= \bar{E} - \mu\bar{N} - TS(\bar{E}, \bar{N}) \\
&\quad - \frac{T}{2}\left[\left(\frac{\partial^2 S}{\partial E^2}\right)_{N,V}(\Delta E)^2 + 2\frac{\partial^2 S}{\partial E \partial N}\Delta E \partial N\right. \\
&\qquad \left. + \left(\frac{\partial^2 S}{\partial N^2}\right)_{E,V}(\Delta N)^2\right] \\
&= -PV + \frac{T}{2}\left[\alpha(\Delta E)^2\right. \\
&\qquad \left. + 2\beta(\Delta E)(\Delta N) + \gamma(\Delta N)^2\right]
\end{aligned}
\tag{3.6.5}
$$

where we have recognized that $\bar{E}$ and $\bar{N}$ are the thermodynamic values of E and N [see Eqs.(3.6.3) and (3.6.4)] and set

$$
\alpha = -\frac{\partial^2 S}{\partial E^2}, \quad \beta = -\frac{\partial^2 S}{\partial E \partial N}, \quad \gamma = -\frac{\partial^2 S}{\partial N^2}.
\tag{3.6.6}
$$

For $E - TS - \mu N$ to have a minimum, the quadratic form in the square brackets in Eq.(3.6.5) must be positive definite, which can be achieved if $\alpha > 0$, and $\alpha\gamma > \beta^2$. We can now write

$$
\begin{aligned}
Q &= e^{PV/k_B T}\int d(\Delta E)d(\Delta N) \\
&\quad \times \exp\left[-\frac{1}{2k_B}\left\{\alpha(\Delta E)^2 + 2\beta(\Delta E)(\Delta N) + \mu(\Delta N)^2\right\}\right].
\end{aligned}
\tag{3.6.7}
$$

The integral is $O(N)$ and hence taking the logarithm in Eq.(3.6.5),

$$
k_B T \ln Q = PV + O(\ln N).
\tag{3.6.8}
$$

As $\langle N \rangle \to \infty$, Eq.(3.6.1) is regained as PV is $O(N)$ and completely dominates. We can use Eq.(3.6.7) to obtain the mean value of the fluctuation $(\Delta E)^2, (\Delta N)^2$ and $(\Delta E)(\Delta N)$ and, as promised, Eq.(3.6.7) shows that the grand canonical distributions can be viewed as quadratic distribution of fluctuations in E and N.

3.7   Problems for Chapter 3

1. Consider a function $g(x) = e^{Nf(x)}$, where N is a number which is of the order of Avogadro Number, i.e., N is very large. Imagine doing a Taylor expansion about $x = x_0$. Write $f(x)$ correct to first order in δx about x_0. Now, is it possible to write $g(x)$ to first order with confidence? This exercise allows you to appreciate the difficulty in expanding Eq.(3.1.2). In that equation, do a formal expansion keeping all terms and then show that if $\frac{\partial S}{\partial E} = \frac{1}{T}$ is fixed (canonical ensemble is a constant temperature ensemble), the distribution of Eq.(3.1.5) follows.

2. Consider a body placed in an external field, such that its energy $E(p, q)$ can be written as

$$E(p, q) = E_0(p, q) + V(p, q)$$

where $V(p, q)$ is a 'potential' representing the contribution of the external field to the energy. If F is the total free energy and F_0 the free energy in the absence of V, then show that

$$e^{-(F - F_0)/k_B T} = \langle e^{-V/k_B T} \rangle_0,$$

the expectation value being taken with respect to the distribution corresponding to E_0.

3. Find the canonical distribution function for a body rotating with angular velocity Ω.

4. Consider an ensemble where energy E and the volume V fluctuates (isothermal isobaric ensemble). Find the connection between partition function and a thermodynamic free energy for the ensemble. Calculate the partition function for an ideal gas by first integrating out the kinetic energy part and then performing the volume integration. Obtain the equation of state.

5. For the isothermal-isobaric ensemble of the previous example, express the mean square volume fluctuation in terms of the isothermal compressibility.

6. Consider a system distributed over its accessible states r in accordance with an arbitrary probability distribution P_r and let its entropy be defined as $S = -k_B \sum P_r \ln P_r$. We have the normalization $\sum P_r = 1$. For a canonical distribution $P_r^{(0)} = e^{-\beta E_r} / \sum e^{-\beta E_r}$. We constrain the mean energies to be the same, i.e.

$$\sum P_r^{(0)} E_r = \sum P_r E_r = E.$$

The entropy for the canonical distribution is $S_0 = -k_B \sum P_r^{(0)} \ln P_r^{(0)}$. Show that $S - S_0 = k \sum p_r \ln P_r^{(0)}$ and that $S_0 \leq S$, with the equality holding for $P_r = P_r^{(0)}$. Hence for a given mean energy, the canonical distribution maximizes the entropy.

4

Non-interacting Systems

4.1   Calculational Procedure for Canonical Ensemble

This chapter deals with a system which is a collection of N non-interacting particles. The energy of such a system can be written as

$$E(\{q_j\}, \{p_j\}) = \sum_{i=1}^{N} E_i(\{q_j\}, \{p_j\}) \tag{4.1.1}$$

where E_i is the energy of the ith particle of the collection. This leads to an immense simplification in the calculation of the partition function Z in the canonical ensemble (it should be remembered that Z is all that one needs to calculate, all thermodynamic properties follow from it). For a system characterized by continuous energy variation as functions of $(\{q_j\}, \{p_j\})$, the N-particle partition function is

$$Z_N = \frac{1}{h^N} \int dq_1 \ldots dq_N \int dp_1 \ldots dp_N e^{-E(\{q_j\}, \{p_j\})/k_B T} \tag{4.1.2}$$

and for the discrete set of allowed energy levels,

$$Z_N = \sum_n e^{E_n/k_B T} \tag{4.1.3}$$

where E_n are the allowed energy levels of the system. For the continuous variation, introducing Eq.(4.1.2), we can write

$$Z_N = \prod_{i=1}^{N} \left[\frac{1}{h} \int dq_i dp_i e^{-E_i(q_i, p_i)/k_B T} \right] \tag{4.1.4}$$

and since all the E's are the same functions of q and p, each separable integral gives the same contribution and hence

$$Z_N = (Z_S)^N \tag{4.1.5}$$

where Z_S is the single particle partition function given by

$$Z_S = \int \frac{dqdp}{h} e^{-E(q,p)/k_B T}. \tag{4.1.6}$$

This tacitly assumes one dimensional space. In D-dimensional space, we will have

$$Z_S = \int \frac{d^D q d^D p}{h^D} e^{-E(\vec{q},\vec{p})/k_B T}. \tag{4.1.7}$$

and Eq.(4.1.5) will still hold. For identical particles the system will be such that the motion of the particles allows them to interchange positions in the course of time and then it will be impossible to differentiate between configurations as to which particle positions and momenta are interchanged. This would cut down the available phase space by a factor of $N!$. In such cases,

$$Z_N = \frac{Z_S^N}{N!} \tag{4.1.8}$$

as identical particles are capable of interchanging coordinates. In the case of discrete energy levels, a given energy $E = \sum_{i=1}^{N} E_i$, where E_i represents the allowed energies of the ith particle. Once again the sum over all configurations will be allowing particle 1 to explore all energy states independent of the other (non-interacting) particles. A similar procedure holds for all the other particles and hence again $Z_N = (Z_S)^N$, where $Z_S = \sum_n e^{-\epsilon_n/k_B T}$, ϵ_n being the allowed single particle energies. The division by $N!$ is carried out for identical particles whose positions and momenta can be interchanged.

Summarising, all calculations of non-interacting particles are carried out following these guidelines:

1. Calculate the single particle partition function Z_S, which is given by Eq.(4.1.7), or $Z_S = \sum_n e^{-\epsilon_n/k_B T}$, as the case may be, i.e., continuous or discrete energy levels.

2. Now the N-particle answer comes from Eq.(4.1.5).

3. Divide this by $N!$, if the particle positions and momenta can be interchanged.

4.2 Ideal Gas

In this chapter we will illustrate the use of the ensembles, general statistical mechanical concepts and their connection with thermodynamics by taking up examples of simple non-interacting classical systems. The best known of such systems is the ideal gas – a collection of N identical molecules confined in a volume V at a temperature T. The general problem is to calculate the pressure P as a function of N, V and T. This relation is called the equation of state.

The energy of the N-particle system is

$$E = \sum_{i=1}^{N} \frac{p_i^2}{2m} \qquad (4.2.1)$$

where it is to be understood that each p_i^2 is $p_i^2 = p_{ix}^2 + p_{iy}^2 + p_{iz}^2$, when we are dealing with three- dimensional space. Fixing N, we use the canonical distribution to write the partition function as

$$
\begin{aligned}
Z &= \int \exp\left[-\sum \frac{p_i^2}{2mk_BT}\right] d\Gamma \\
&= \int \exp\left[-\sum \frac{p_i^2}{2mk_BT}\right] \frac{d^{3N}p\, d^{3N}q}{h^{3N}} \cdot \frac{1}{N!} \\
&= \frac{V^N}{N!} \frac{1}{h^{3N}} \left[\int e^{-p_i^2/2mk_BT} d^3p\right]^N \\
&= \frac{V^N}{N!} \left(\frac{2m\pi k_BT}{h^2}\right)^{3N/2}. \qquad (4.2.2)
\end{aligned}
$$

Note that in the second step the $N!$ has entered the denominator for correct counting when the particles are indistinguishable. It should be pointed out that a single gas molecule could be taken to constitute one of the systems in a canonical ensemble with the rest of the molecules providing the environment and in that case the single molecule partition Z_S would be

$$
\begin{aligned}
Z_S &= \int \frac{d^3q\, d^3p}{h^3} e^{-p^2/2mk_BT} \\
&= \left(\frac{2\pi m k_BT}{h^2}\right)^{\frac{3}{2}} V. \qquad (4.2.3)
\end{aligned}
$$

Since we are dealing with an N-particle system with no interaction, the full partition function would be expected to be

$$Z = Z(N) = Z_S^N \qquad (4.2.4)$$

and indeed would be true for distinguishable particles. But for identical (classical) particles, we note that Eqs.(4.2.3) and (4.2.4) do not reproduce Eq.(4.2.2) and in this case Eq.(4.2.4) has to be modified to

$$Z = Z(N) = \frac{Z_S^N}{N!}. \qquad \text{(identical classical particles)} \qquad (4.2.5)$$

We are now in a position to make contact with thermodynamics. This comes from Eq.(3.1.18), whence we write the free energy as

$$F = -k_BT \ln Z = -k_BT \left[N \ln V/N + \frac{3N}{2} \ln\left(\frac{2\pi k_BTm}{h^2}\right) + N\right]. \qquad (4.2.6)$$

Thermodynamic quantities follow by differentiation:

- *Equation of state*:

$$P = -\left(\frac{\partial F}{\partial V}\right)_T = \frac{Nk_BT}{V}. \tag{4.2.7}$$

The well known relation for ideal gases is reproduced.

- *Mean energy*: The thermodynamic energy E is found as

$$
\begin{aligned}
E &= \langle E \rangle = \frac{1}{Z}\sum_i \int \frac{p_i^2}{2m} \exp\left[-\sum \frac{p_i^2}{2mk_BT}\right] d\Gamma \\
&= 3Nk_BT.\frac{\Gamma(3/2)}{\Gamma(1/2)} = \frac{3}{2}Nk_BT.
\end{aligned} \tag{4.2.8}
$$

Alternatively, from Eq.(4.2.6)

$$S = -\left(\frac{\partial F}{\partial T}\right)_V = k\left[N \ln \frac{V}{N} + \frac{3N}{2} \ln \frac{2\pi mk_BT}{h^2} + \frac{5}{2}N\right]. \tag{4.2.9}$$

Now, $E = F + TS = \frac{3}{2}Nk_BT$ in agreement with Eq.(4.2.8).

- *Specific heat*: This follows from differentiation of the energy as

$$C = \frac{\partial E}{\partial T} = \frac{3}{2}k_BN. \tag{4.2.10}$$

It should be noted that a constant volume is implicit in the above differentiation and what we have obtained is the specific heat at constant volume.

- *Fluctuation in* $\langle E \rangle$:

$$\langle (\Delta E)^2 \rangle = k_BT^2\frac{\partial E}{\partial T} = \frac{3}{2}N(k_BT)^2. \tag{4.2.11}$$

Consequently,

$$\frac{\langle (\Delta E)^2 \rangle^{1/2}}{\langle E \rangle} = \left(\frac{3}{2}N\right)^{1/2}.\frac{k_BT}{3Nk_BT/2} = \left(\frac{3}{2}\right)^{1/2}\frac{1}{\sqrt{N}}. \tag{4.2.12}$$

As expected, the fractional fluctuation vanishes for $N \to \infty$, and $\langle E \rangle$ is indeed the thermodynamic energy.

It should be noted that in a non interacting set of N particles, the individual energies E_i $(i = 1, 2 \ldots N)$ simply add to give the total energy $E = \sum_i E_i$. It follows that $e^{-E/k_BT} = \Pi_i e^{-E_i/k_BT}$ and hence for such systems $Z_N = Z_S{}^N$.

4.3 Ideal Gas in Grand Canonical Ensemble

The ideal gas is a sufficiently important example to warrant yet another treatment. We now consider the collection of N-particles in the box of volume V to be in contact with a surrounding with which it can exchange both energy and particles. The grand canonical partition function is then

$$
\begin{aligned}
Q &= \sum_N \frac{1}{N!} \int \exp\left(-\sum_i \frac{p_i^2}{2mk_BT}\right) \exp\left(\frac{\mu N}{k_BT}\right) \frac{d^{3N}p\,d^{3N}q}{h^{3N}} \\
&= \sum_N \frac{V^N}{N!} \left(\frac{2\pi mk_BT}{h^2}\right)^{3N/2} e^{\mu N/k_BT} \\
&= \sum_N \frac{Z^N}{N!} \\
&= e^Z
\end{aligned} \tag{4.3.1}
$$

where $Z = V\left(\frac{2\pi mk_BT}{h^2}\right)^{3N/2} e^{\mu/k_BT}$. The thermodynamic connection comes from

$$
PV = k_BT \ln Q = k_BTZ \tag{4.3.2}
$$

while we use Eq.(3.3.14) to obtain the chemical potential as

$$
N = \langle N \rangle = \frac{\partial}{\partial \mu}(PV) = k_BT\frac{\partial Z}{\partial \mu} = z. \tag{4.3.3}
$$

Thus,

$$
e^{\mu/k_BT} = \frac{N}{V}\left(\frac{h^2}{2\pi mk_BT}\right)^{1/2} \tag{4.3.4}
$$

in agreement with Eq.(3.3.4) obtained from the microcanonical picture. Elimination of z between Eqs.(4.3.2) and (4.3.3) yields the equation of state as $PV = Nk_BT$.

We end this section by establishing that the fluctuations in N and E are negligible in the grand canonical ensemble. Using Eq.(3.4.10), Eq.(4.3.3) and the definition of z

$$
\langle(\Delta N)^2\rangle^{1/2} = k_BT\frac{\partial N}{\partial \mu} = k_BT\frac{\partial z}{\partial \mu} = k_BT\frac{z}{k_BT} = z = N
$$

or

$$
\frac{\langle(\Delta N)^2\rangle^{1/2}}{N} = \frac{1}{N^{1/2}}. \tag{4.3.5}
$$

For the fluctuation in E, Eq.(3.4.17) yields the relevant information. First, we calculate the expectation value of E as

$$
\begin{aligned}
\langle E \rangle &= \frac{1}{Q} \sum_N \frac{1}{N!} \int \frac{d^{3N}p \, d^{3N}q}{h^{3N}} \sum_i \left(\frac{p_i^2}{2m} \right) e^{\mu N/k_B T} \exp \left[-\sum_i \frac{p_i^2}{2m k_B T} \right] \\
&= \sum \frac{V^N}{N!} \left(\frac{2\pi m k_B T}{h^2} \right)^{3N/2} e^{\mu N/k_B T} \\
&= \frac{3}{2} k_B T z \\
&= \frac{3}{2} N k_B T
\end{aligned}
$$

in agreement with Eq.(4.2.8). Now, we find

$$
\begin{aligned}
\langle (\Delta E)^2 \rangle &= k_B T^2 \frac{\partial \langle E \rangle}{\partial T} + \mu k_B T \frac{\partial \langle E \rangle}{\partial \mu} \\
&= k_B T^2 \left[\frac{3}{2} N k + \frac{3}{2} k_B T \frac{\partial N}{\partial T} \right] + \frac{3\mu}{2} (k_B T)^2 \frac{\partial N}{\partial \mu} \\
&= \frac{15}{4} N (k_B T)^2
\end{aligned}
$$

after straightforward algebraic manipulation. The fractional fluctuation is

$$
\left[\frac{\langle (\Delta E)^2 \rangle^{1/2}}{\langle E \rangle} \right] = \frac{\sqrt{15}}{2} \frac{2}{3} \frac{1}{\sqrt{N}} = \sqrt{\frac{5}{3N}}
$$

which vanishes in the thermodynamic limit.

4.4   Space-dependent Distribution

4.4.1   In a Gravitational Potential

We consider a collection of non-interacting molecules in a given volume V where the gravitational force acts in the negative z-direction. The whole mass of gas is at a constant temperature T so that equilibrium concepts apply. The energy of a single molecule is

$$
E = \frac{p^2}{2m} + mgz \tag{4.4.1}
$$

and the probability of finding the molecule with energy between E and $E + dE$ is

$$
P(E) = \frac{1}{Z} \exp \left(-\frac{\frac{p^2}{2m} + mgz}{k_B T} \right)
$$

The number of molecules with z-coordinate between z and $z+dz$ is found by integrating over all the other coordinates, i.e. p_x, p_y, p_z, x and y. Thus

$$
\begin{aligned}
P(z)dz &= \frac{1}{Z} \int \frac{d^3pdxdy}{h^3} e^{-p^2/2mk_BT} e^{-mgz/k_BT} dz \\
&= P(0)e^{-mgz/k_BT} dz \qquad (4.4.2)
\end{aligned}
$$

where $P(0)$ is a constant which correctly normalizes the distribution. In the true atmosphere this distribution is not obtained as the temperature does not remain constant throughout.

4.4.2 In a Centrifuge

We consider a collection of macromolecules (large organic molecules) rotated with an angular velocity in an ultra centrifuge. In the rotating frame of reference, the molecules feel a centrifugal force $m\omega^2 r$ radially outward. With $\hat{r}$ representing the unit vector in the radial direction, this force can be obtained from a potential

$$
\vec{F} = m\omega^2 r \hat{r} = -\frac{\partial V}{\partial r}\hat{r}
$$

or

$$
V = -\frac{1}{2}m\omega^2 r^2, \qquad (4.4.3)
$$

so that the total energy of a molecule in the rotating frame is

$$
E = \frac{p^2}{2m} - \frac{1}{2}m\omega^2 r^2 \qquad (4.4.4)
$$

with the probability distribution

$$
P(E) = \frac{1}{Z}\exp\left(-\frac{p^2}{2mk_BT}\right)\exp\left(\frac{m\omega^2 r^2}{2k_BT}\right). \qquad (4.4.5)
$$

Arguments identical to those in subsection 4.3.1 lead to

$$
P(r)dr = P(0)\frac{1}{Z}\exp\left(\frac{m\omega^2 r^2}{2k_BT}\right)dr \qquad (4.4.6)
$$

for the radial distribution of the molecules.

As expected, the molecules crowd towards larger radii. We have taken the rotation speed to be very high and consequently ignored the effect of the Coriolis force. Also note that the number distribution must be the same in rotating and non-rotating frames.

The above distribution leads to a practical method of obtaining the molecular weight m. By determining the concentrations $\rho(r_1)$ and $\rho(r_2)$ at two radii r_1 and r_2 respectively, we find from Eq.(4.4.6)

$$
\frac{\rho(r_1)}{\rho(r_2)} = e^{\frac{1}{2}m\omega^2(r_1^2 - r_2^2)/k_BT}
$$

or

$$m = \frac{2k_BT}{\omega^2}(r_1^2 - r_2^2) \, \ln\left(\frac{\rho(r_1)}{\rho(r_2)}\right). \tag{4.4.7}$$

4.5   Oscillators in Contact with Heat Bath

4.5.1   Classical Oscillators

We now consider a set of N oscillators in three dimensional space in equilibrium with a heat bath characterized by a temperature T. The energy of each oscillator is

$$E = \frac{p^2}{2m} + \frac{1}{2}k_s r^2 \tag{4.5.1}$$

where k_s is the spring constant. The partition function is

$$Z = Z_S{}^{3N} = \left[\int \frac{dpdx}{h} \exp\left(-\frac{p^2}{2mk_BT}\right) . \exp\left(-\frac{k_s x^2}{2k_BT}\right)\right]^{3N} \tag{4.5.2}$$

where Z_S is the single oscillator partition function in one dimension and the average energy

$$
\begin{aligned}
E &= NE_S \\
&= 3N\frac{1}{Z_S}\int \frac{dpdx}{h}\left(\frac{p^2}{2m} + \frac{1}{2}k_s x^2\right)\exp\left(-\frac{p^2/2m + k_s.x^2/2}{k_BT}\right) \\
&= 3Nk_BT.
\end{aligned} \tag{4.5.3}
$$

Note that all our results so far – the ideal gas and the oscillators – give a total energy which corresponds to $\frac{1}{2}k_BT$ for each degree of freedom (i.e. each independent quadratic term in the energy expression). This is a general result known as the classical equipartition theorem. Note that the quadratic form for the energy is essential. The reader should verify that the equipartition theorem is not valid for the quartic oscillator, e.g. where the energy expression (in one dimension) is

$$E = \frac{p^2}{2m} + \frac{1}{2}\lambda x^4.$$

It should be noted that the classical equipartition theorem holds only for classical non-interacting systems and that too where every independent degree of freedom contributes a quadratic term to the expression for energy.

4.5.2   Quantum Oscillators

We now consider a set of one-dimensional oscillators in thermal equilibrium but with the energy levels given by quantum mechanics. The energy levels are

$$E_n = \left(n + \frac{1}{2}\right)\hbar\omega. \tag{4.5.4}$$

We have already considered this example in chapter 2. The single oscillator partition function is

$$Z_S = \sum \exp\left[-\left(n+\frac{1}{2}\right)\frac{\hbar\omega}{k_B T}\right]$$
$$= e^{-\hbar\omega/2k_B T}\left(1 - e^{-\hbar\omega/k_B T}\right)^{-1}. \qquad (4.5.5)$$

The average energy E_s is given by

$$E_S = \frac{\hbar\omega}{Z}\sum\left(n+\frac{1}{2}\right)\exp\left[-\left(n+\frac{1}{2}\right)\frac{\hbar\omega}{k_B T}\right]$$
$$= \hbar\omega\left[\frac{1}{e^{\hbar\omega/k_B T}-1}+\frac{1}{2}\right]. \qquad (4.5.6)$$

For N oscillators in three dimensions

$$E = 3N\hbar\omega\left[\frac{1}{e^{\hbar\omega/k_B T}-1}+\frac{1}{2}\right]. \qquad (4.5.7)$$

The point to note is that for $k_B T \gg \hbar\omega$, the first term within the brackets dominates and we can expand the exponential to find the leading terms giving

$$E = 3Nk_B T \qquad (4.5.8)$$

in agreement with the classical result of Eq.(4.5.3). On the other hand, for $T \to 0$, it is the second term within the brackets in Eq.(4.5.7) that dominates and provides the average energy for $T = 0$. It corresponds to the quantum fluctuations and is the biggest departure from the classical result. The above results are identical to that found in Eqs.(2.1.38) and (2.1.39).

4.6 Specific Heat of Solids

The foregoing discussion of the statistical mechanics of oscillators has a direct application in the evaluation of specific heat of solids. The model of the solids that we assume considers atoms (nucleus together with the tightly bound core electrons constituting an atom) arranged on a regular lattice with the effect of loosely bound electrons ignored in the present picture. The atoms vibrate with small amplitude about their mean positions on the lattice and the energy of the solid comes from the energy associated with these simple harmonic vibrations. Assuming that the oscillators are governed by classical mechanics and considering the molar specific heat ($Nk_B = R$), we see from Eq.(4.5.3)

$$C = \frac{\partial E}{\partial T} = 3R \qquad (4.6.1)$$

independent of temperature and material. This universal result is correct at high temperatures and is known as Dulong and Petit's law. At low temperatures substantial deviation from this result occurs and as $T \to 0$ the specific heat tends to zero in accordance with Nernst's heat theorem.

To treat the low temperature situation, we consider the vibrations to be governed by quantum mechanics and further assume that each oscillator is characterized by the same frequency. This is the Einstein model of solids. The molar specific heat now follows from a differentiation of Eq.(4.5.7) to be

$$C = 3R \left(\frac{\hbar \omega}{k_B T} \right)^2 \frac{e^{\hbar \omega / k_B T}}{(e^{\hbar \omega / k_B T} - 1)^2}. \tag{4.6.2}$$

For $k_B T \gg \hbar \omega$, we recover $C = 3R$ as expected, but for $T \to 0$ we find (with $T_E = \hbar \omega / k$)

$$C \cong 3R \left(\frac{\hbar \omega}{k_B T} \right)^2 e^{-\hbar \omega / k_B T} = 3R \left(\frac{T_E}{T} \right)^2 e^{-T_E / T} \tag{4.6.3}$$

making the specific heat vanish exponentially at low temperatures in exact agreement with Eq.(2.1.40). This removes the difficulty with Nernst's heat theorem but is not borne out experimentally. For $T \ll T_E = \hbar \omega / k$, the decrease of C is not exponential as Eq.(4.6.3) would suggest. Instead C vanishes as an algebraic function of T and the reason for the discrepancy lies in the drastic assumption of all oscillators having the same frequency.

We begin by considering the situation at low frequencies, where the vibrational modes must correspond to the normal modes of oscillation of an elastic medium of the same size and the same elastic constants as the crystal we are considering. At low temperatures, the high frequency modes are frozen out and hence the low frequency modes will give an adequate description. Hence if we correctly enumerate the low frequency vibrational modes, the low temperature behaviour will be correctly obtained.

We consider the solid in the shape of a cube with side L. The low frequency vibrations that we are considering will form standing waves inside the solid with the possible wavelengths in the x, y and z directions as

$$\lambda_i = L/n_i \tag{4.6.4}$$

with $i = \{x, y, z\}$, and n is an integer. If the propagating velocity is c, then the possible frequencies are

$$\nu_i = c.n_i/L$$

or

$$\nu^2 = \frac{c^2}{L^2} (n_x^2 + n_y^2 + n_z^2). \tag{4.6.5}$$

If we now ask how many modes of vibration exist between frequencies ν and $\nu + d\nu$, then the issue is one of finding the number of n_x, n_y and n_z that satisfy Eq.(4.6.5). The different points in n_x, n_y and n_z space satisfying $n_x^2 + n_y^2 + n_z^2 \le \nu^2 L^2/c^2$, can be taken in the limit of very large L to form a continuum and occupy the entire sphere of radius $\nu L/c$. The volume of the annulus between radii corresponding to ν and $\nu + d\nu$, then gives the number of available points n_x, n_y, n_z and thus the number of available modes. This is clearly $4\pi n^2 dn = 4\pi(\nu^2/c^3)L^3 d\nu$. This number corresponds to a given value of c. But there are three possible waves-two transverse with velocity c_T and one longitudinal with velocity c_L. So the total number of modes of vibration $g(\nu)d\nu$ between a frequency ν and $\nu + d\nu$ is

$$
\begin{aligned}
g(\nu)d\nu &= 4\pi\nu^2 V\left(\frac{2}{c_T^3} + \frac{1}{c_L^3}\right)d\nu \\
&= \frac{12V\pi\nu^2}{c^3}.d\nu
\end{aligned} \tag{4.6.6}
$$

where

$$
\frac{2}{c_T^3} + \frac{1}{c_L^3} = \frac{3}{c^3}
$$

The total number of modes cannot exceed $3N$ for N three-dimensional oscillators and hence there must be an upper cut-off ν_m on the available frequencies such that (see Fig. 4.1)

$$
3N = \int_0^\nu g(\nu)d\nu = \frac{12\pi\nu}{c^3}\int_0^{\nu_m}\nu^2 d\nu = \frac{4\pi V}{c^3}\nu_m{}^3. \tag{4.6.7}
$$

This upper limit corresponds to an upper limit on the wave-number and hence a lower limit on the wavelength beyond which the elastic continuum which we have assumed loses meaning. The lower limit on the wavelength corresponds approximately to the lattice spacing.

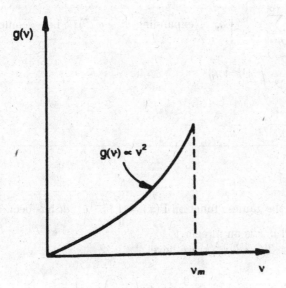

Figure 4.1: The density of states.

We can now evaluate the mean energy (we drop the zero-point term since it cannot give rise to specific heat) and find

$$
\begin{aligned}
E &= \int g(\nu) \frac{h\nu}{e^{h\nu/k_B T} - 1} d\nu \\
&= \frac{12\pi V}{c^3} \int_0^{\nu_m} h\nu \frac{\nu^2 d\nu}{e^{h\nu/k_B T} - 1} \\
&= \frac{12\pi V}{c^3} \frac{(k_B T)^4}{h^3} \int_0^{h\nu_m/k_B T} \frac{x^3 dx}{e^x - 1}.
\end{aligned} \tag{4.6.8}
$$

If we consider very low temperatures ($k_B T \ll h\nu_m$), then the upper limit on the integration can be raised to infinity and we obtain

$$
\begin{aligned}
E &= \frac{12\pi V}{c^3} \cdot \frac{3Nc^3}{4\pi V\nu_m^3} \cdot \frac{(k_B T)^4}{h^3} \int_o^\infty \frac{x^3 dx}{e^x - 1} \\
&= 9Nk_B T \left(\frac{k_B T}{hV_m} \right)^3 I
\end{aligned} \tag{4.6.9}
$$

where

$$
\begin{aligned}
I &= \int_o^\infty x^3 (e^x - 1)^{-1} dx \\
&= \int_o^\infty x^3 e^{-x} \left[\sum_{n=0}^\infty e^{-nx} \right] dx \quad \text{(expanding } (1 - e^{-x})^{-1} \text{ in a geometric series)} \\
&= \sum_{n=0}^\infty \frac{1}{(n+1)^4} \int_o^\infty y^3 e^{-y} dy \\
&= \Gamma(4) \sum_{m=0}^\infty \frac{1}{m^4} \\
&= \frac{\pi^4}{15}
\end{aligned} \tag{4.6.10}
$$

(The definition of the gamma function $\Gamma(x) = \int_0^\infty t^{x-1} e^{-t} dt$ has been used. Note that $\Gamma(x+1) = x!$ if x is an integer.)

Thus for $T \ll T_D = h\nu_m/k$, we have

$$
E = N. \frac{3}{5} \pi^4 k_B T \left(\frac{T}{T_D} \right)^3 \tag{4.6.11}
$$

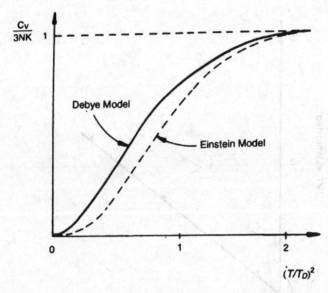

Figure 4.2: Specific heats in the Debye and Einstein models.

and hence the specific heat comes out as

$$C = \frac{\partial E}{\partial T} = N.\frac{12}{5}\pi^4.k.\left(\frac{T}{T_D}\right)^3 \tag{4.6.12}$$

The suppression of density of states at low frequencies (see Eq.(4.6.6) mutes the strong temperature dependence in the specific heat that arises in the Einstein model (Fig 4.2). This model of the specific heat of solids is due to Debye and the T^3-law of Eq.(4.6.12) is borne out very well experimentally (Fig. 4.3). Needless to say, at the higher frequencies this model breaks down as enumeration of the density of states becomes more complicated.

4.7 Model for Paramagnetism

In one of the problems at the end of this chapter we will show that classical physics cannot exhibit paramagnetism or diamagnetism. Yet in this section we will be discussing paramagnetism in the context of classical statistical mechanics. This somewhat paradoxical situation can be understood when we point out that the model assumes the existence of individual magnetic moments (the origin of which lies in quantum mechanics) and what we intend to discuss is how a collection of such magnetic moments (i.e. atoms with magnetic moments) will orient themselves when put in an external magnetic field at a given temperature T. The simplest model, which we begin working with assumes that the individual magnetic moments can only align with or against the direction of the field. If we call the magnetic moment μ, then in the external field H,

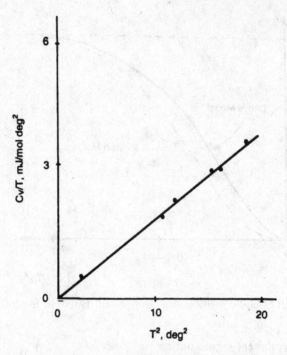

Figure 4.3: Comparison of experimental data with the Debye theory for KCl. It is conventional to plot C/T, so that the electronic contributions (Chapter 7), if any, can show up as non-zero intercepts.

there are only two energy values possible, namely $E_1 = -\mu H$ ($\vec{\mu}$ and $\vec{H}$ parallel) and $E_2 = \mu H$ ($\vec{\mu}$ and $\vec{H}$ antiparallel). The single moment partition function is clearly

$$Z_S = e^{\mu H/k_B T} + e^{-\mu H/k_B T} = 2\cosh \frac{\mu H}{k_B T}. \qquad (4.7.1)$$

For the N-particle system, the free energy is

$$
\begin{aligned}
F &= -k_B T \ \ln \ Z = -N k_B T \ \ln \ Z_S \\
  &= -N k_B T \ \ln \ 2 \ \cosh \frac{\mu H}{k_B T}.
\end{aligned}
\qquad (4.7.2)
$$

The magnetization follows as:

$$M = -\frac{\partial F}{\partial H} = \mu N \ \tanh \frac{\mu H}{k_B T}. \qquad (4.7.3)$$

For $k_B T \gg \mu N$, we have

$$M = \mu^2 N H / k_B T \qquad (4.7.4)$$

leading to the high temperature susceptibility as

$$\chi = \left[\frac{\partial M}{\partial H}\right]_{H \to \infty} = \frac{N\mu^2}{k_B T} \tag{4.7.5}$$

which is the usual Curie law for temperature dependence of susceptibilities. These results are in exact agreement with those found in chapter 2 (see Eq.2.1.30). At very low temperatures, $M \to \mu N$, and the approach of $M \to \mu N$ as $T \to 0$ is given by

$$M \cong \mu N \left[1 - e^{-2\mu H/k_B T}\right]. \tag{4.7.6}$$

[An alternative way of arriving at Eq.(4.7.3) simply requires the evaluation of $\langle \mu \rangle$ using the canonical distribution and this immediately leads to

$$M = N\langle \mu \rangle = N \frac{\mu e^{\mu H/k_B T} - \mu e^{-\mu H/k_B T}}{e^{\mu H/k_B T} + e^{-\mu H/k_B T}} = N\mu \, \tanh \frac{\mu H}{k_B T}$$

where $e^{\mu H/k_B T}$ is the probability for '$+\mu$'].

Having discussed this simple model which has all the essential ingredient, we briefly discuss the more realistic situation where the atom is characterized by a total angular momentum $\vec{J}$ and the magnetic moment $\vec{\mu}$ points in the direction of $\vec{J}$ with the relation

$$\vec{\mu} = g_j \mu_B \vec{J} \tag{4.7.7}$$

where μ_B is the Bohr magneton and g is the Lande g-factor. In a uniform external field, H, in the z-direction, the possible energies of the atom are given by

$$E = -\vec{\mu}.\vec{H} = -g_J \mu_B H J_z \tag{4.7.8}$$

where we know from quantum mechanics that J_z has $2J + 1$ allowed values running from $-J$ to $+J$ in steps of unity. The single atom partition function is clearly

$$\begin{aligned} Z_S &= e^{\beta g_j \mu_B J H} + e^{\beta g_j \mu_B (J-1)H} + \ldots + e^{-\beta g_j \mu_B J H} \\ &= \frac{\sinh\left(g_j \mu_B H \left(J + \frac{1}{2}\right)\right)/k_B T}{\sinh\left(\frac{g_j \mu_B H}{2k_B T}\right)}. \end{aligned} \tag{4.7.9}$$

The free energy for N atoms is $F = -Nk_B T \ln Z_S$ and differentiating with respect to H, leads to the magnetization

$$M = \frac{\partial F}{\partial H} = \frac{N g_j \mu_B}{2} \times B_j \left(\frac{\mu_B g_j H}{2k_B T}\right) \tag{4.7.10}$$

where $B_J(\chi)$ is the Brillouin function

$$B_J(\chi) = (2J + 1) \coth\{(2J + 1)\chi\} - \coth \chi. \tag{4.7.11}$$

At high temperatures $\chi = \frac{\mu_B g_j H}{2k_B T} \ll 1$ and $B_j(\chi) \sim \frac{4}{3}\chi J(J+1)$, so that the Curie susceptibility is given by

$$\chi = \frac{NJ(J+1)g_j^2\mu_B^2}{3k_BT}. \tag{4.7.12}$$

For very low temperatures, $B_J(\chi) \to 2J$ and

$$M \cong Ng_J\mu_B J \tag{4.7.13}$$

i.e. all atoms are oriented parallel to the field.

4.8   Electronic Energy in a Monoatomic Gas

When we dealt with the ideal gas in Sec.4.1, we took the energy as purely translational. However, the electronic motion within each atom leads to definite energies and the total energy of an atom moving with momentum p should be taken to be

$$E = \frac{p^2}{2m} + \epsilon_n \tag{4.8.1}$$

where ϵ_n is the nth eigenvalue of the electronic energy. The total partition function (single atom) is clearly

$$
\begin{aligned}
Z_S &= \sum_n \int d\Gamma e^{-p^2/2mk_BT} . e^{-\epsilon_n/k_BT} \\
&= \left(\frac{2\pi mk_BT}{h^2}\right)^{3/2} V.\bar{Z} = Z_{\text{trans}}.\bar{Z} \tag{4.8.2}
\end{aligned}
$$

where $\bar{Z} = \sum_n e^{-\epsilon_n/k_BT}$ and the remainder Z_{trans} is simply the relevant partition function of the ideal gas obtained in Sec.4.1. Now if the temperature T is much higher than the ionization temperature, then there are no bound electrons and the sum $\bar{Z}$ is meaningless. Thus $\bar{Z}$ is a significant quantity only if $T \leq T_{\text{ion}}$. The ionization energy is of the order of $\epsilon_1 - \epsilon_0$, and hence if $T \leq T_{\text{ion}}$, it is safe to assume that all electrons are in the ground state ϵ_0. Arbitrarily choosing ϵ_0, we have $\bar{Z}=$ degeneracy of the ground state. If the ground state is $L = 0$ (no orbital angular momentum), then the degeneracy is $(2S+1)$, where S is the total spin angular momentum.

For the $L = 0$ ground state there is no fine structure splitting ($\mathbf{L.S} = 0$), but if $L \neq 0$ then the fine structure splittings do exist and the splitting energies ϵ_J may be comparable to k_BT. So one finds the sum $\bar{Z}$ to be

$$\bar{Z} = \sum_J (2J+1)e^{-\epsilon_J/k_BT}. \tag{4.8.3}$$

This is because in the case of fine structure splitting, the degeneracy of a given level is $2J + 1$ and for a given L and S, the J values can range from

$L + S$ to $|L - S|$ in steps of unity. Now if T is high, i.e. $k_B T \gg \epsilon_J$, then all $e^{-\epsilon_J / k_B T} \sim 1$ and

$$\bar{Z} = \sum_{|L-S|}^{|L+S|} (2J + 1) = (2S + 1)(2L + 1) \tag{4.8.4}$$

If, on the other hand, $T \ll \epsilon_J$, then again all the atoms settle for the lowest ϵ_J and $\bar{Z} = (2J_m + 1)$, where the J_m is the J-value for which ϵ_J is lowest.

4.9 Diatomic Molecules

4.9.1 Heteronuclear Molecules

In the previous section we saw that the electronic energy for the monatomic gas is important only if the temperature is much lower than the ionization temperature. Similarly, for a molecule, the binding effects are important only if the temperature is much lower than the dissociation temperature T_{diss} (A typical estimate would be a molecular binding energy between 1 and 2 eV, so that $k_B T_{\text{diss}} \cong 1.5 \ eV$ or $T_{\text{diss}} \sim 1.7 \times 10^4 K$. As before, the gas molecule will be assumed to be in its lowest electronic state. For a diatomic molecule which we are interested in, the total internal energy (the effect of the translational energy has already been taken into account in dealing with the perfect gas: the mass appearing there now stands for the molecular weight of the gas) can be written as

$$E_{\text{int}} = E_{\text{el}} + E_{\text{rot}} + E_{\text{vib}}. \tag{4.9.1}$$

In Eq.(4.9.1) E_{el} is the energy associated with all Coulombic interaction and electronic motion about nuclei. For the electronic ground state $E_{\text{el}} = \epsilon_0$. The energy associated with the rotation E_{rot} is the energy due to the orbital motion of the two nuclei about the centre of mass and can be modelled by a rigid rotator and E_{vib} is the energy associated with the vibrational motion of the nuclei about their equilibrium positions along the line joining the nuclei. The latter can be modelled by a one-dimensional oscillator. Consequently,

$$E_{\text{int}} = \epsilon_0 + \frac{\hbar^2}{2I} K(K + 1) + \hbar \omega \left(n + \frac{1}{2} \right) \tag{4.9.2}$$

where I is the moment of inertia given by $I = \frac{m_1 m_2}{m_1 + m_2} r_0^2$ where m_1 and m_2 are the masses of the two nuclei and r_0 is the equilibrium separation; K is the angular momentum quantum number and ω is the frequency of oscillation of the two-nucleon system along the line joining them. As in Eq.(4.8.2), the total partition function (including translation) can be written as

$$Z = Z_{\text{trans}} . \bar{Z} = Z_{\text{trans}} Z_{\text{el}} . Z_{\text{rot}} . Z_{\text{vib}} \tag{4.9.3}$$

where

$$Z_{el} = e^{-\epsilon_0/k_B T} \tag{4.9.4}$$

$$Z_{rot} = \sum_{K=0}^{\infty} (2K+1) \exp\left[\frac{-\hbar^2 K(K+1)}{2Ik_B T}\right] \tag{4.9.5}$$

and

$$Z_{vib} = \sum_{n=0}^{\infty} \exp\left[\frac{-\hbar\omega}{k_B T}\left(n+\frac{1}{2}\right)\right]. \tag{4.9.6}$$

The free energies, entropies and specific heats calculated from the separate parts add to give the total amounts in each case.

Evaluating Z_{rot} is difficult in general, but we note that for high temperatures, i.e. $k_B T \gg \hbar^2/2I$, the high values of K play an important part in the sum and we can approximate Z_{rot} by

$$Z_{rot} \cong \int_0^{\infty} 2K e^{-\hbar^2 k^2/2Ik_B T} dK = \frac{2Ik_B T}{\hbar^2}. \tag{4.9.7}$$

The contribution to the specific heat in this range of temperature is clearly Nk. For very low temperatures, on the other hand, the first two terms in the sum give a good approximation and

$$Z_{rot} \cong 1 + 3e^{-\hbar^2/Ik_B T} \tag{4.9.8}$$

with the contribution to the rotational specific heat being $C_{rot} \sim 3Nk_B(\hbar^2/Ik_B T)^2 e^{-\hbar^2/Ik_B T}$. As for Z_{vib}, it is exactly identical to the situation in Sec.4.4 and the expression for Z_{vib} is given by Eq.(4.5.5). At high temperatures ($k_B T \gg \hbar\omega$) the contribution to the specific heat is Nk_B, while at low temperatures it falls exponentially. We finally note that, at very high temperatures, the molar specific heat of the diatomic molecule is

$$C_{diatomic} = C_{trans} + C_{rot} + C_{vib} = \left(\frac{3}{2} + 1 + 1\right) R = \frac{7}{2} R. \tag{4.9.9}$$

4.9.2   Homonuclear Molecules

We consider the specific case of the hydrogen molecule (H_2). The new complication is that at low temperatures where quantum effects are important, the exchange of two hydrogen nuclei (protons) should lead to the wavefunction changing sign (for our purpose here the electronic degrees of freedom are considered frozen). The total wave function of the two nuclei can be written as

$$\psi(r_1, r_2) = \psi_{orb}(r_1, r_2) \times \psi_{spin}. \tag{4.9.10}$$

The parity of the orbital part is given by $(-)^K$ and is even for $K = 0, 2, 4$ etc. when interchanging the two nuclei does not change the orbital wave function, while for $K = 1, 3, 5$ etc., interchanging the two nuclei changes the sign of the wave function. As for the χ_{spin} the wave function is symmetric (even under interchange) if $S = 1$ and antisymmetric if $S = 0$. Thus for $S = 1$, the allowed K values are the odd integers while for $S = 0$, the allowed values are the even integers. The degeneracy of the spin wave function is $2S+1$, and hence to one state of $S = 0$, there are three states of $S = 1$ and hence the rotational partition function must now be written as

$$Z_{rot} = \frac{1}{4}\left[\sum_{K_{even}} (2K+1)e^{-\hbar^2 K(K+1)/2Ik_BT} \right.$$

$$\left. + \; 3\sum_{K_{odd}} (2K+1)e^{-\hbar^2 K(K+1)/2Ik_BT} \right]. \tag{4.9.11}$$

The factor of $1/4$ is for normalization. The $S = 1$ state with the larger statistical weight is called ortho hydrogen and $S = 0$ is called para hydrogen. The above formula for Z_{tot} assumes that the ortho and para hydrogen are in thermal equilibrium with each other with the ratio x of the number of ortho molecules to the number or para molecules being given by

$$x = \frac{\sum_{K_{odd}} (2K+1)e^{-K(K+1)\hbar^2/2Ik_BT}}{\sum_{K_{even}} (2K+1)e^{-K(K+1)\hbar^2/2Ik_BT}}. \tag{4.9.12}$$

In practice, this equilibrium is very difficult to set up. The setting up of equilibrium entails collisions which keep changing the states. But collisions between molecules do not change the nuclear spin and thus the ratio x is fixed by observation, rather than by assumption of thermal equilibrium. For all practical purposes x is a constant of motion, having nothing to do with equilibrium. The ortho and para molecules pan be taken to be separately in equilibrium at temperature T with the ratio x fixed by an experimental determination. At room temperatures, $x \cong 3$. Instead of the partition function having the additive form of Eq.(4.9.11), it is the free energies which add and

$$F_{rot} = F_{rot}^{(para)} + F_{rot}^{(ortho)} = -k_BT\left[\ln Z_{rot}^{para} + x \ln Z_{rot}^{ortho}\right] \tag{4.9.13}$$

where

$$Z_{ortho} = \sum_{K_{odd}} (2K+1)e^{-\hbar^2 K(K+1)/(2Ik_BT)} \tag{4.9.14}$$

and

$$Z_{para} = \sum_{K_{even}} (2K+1)e^{-\hbar^2 K(K+1)/(2Ik_BT)}. \tag{4.9.15}$$

The specific heat calculated from the above F_{rot} is in very good agreement with the experiments while the specific heat obtained from Eq.(4.9.11) is not.

4.10   Solid-vapour Equilibrium

We consider an application of the principle of equality of chemical potential to describe the equilibrium between two systems which can exchange particles.

The two systems are a solid and its vapour in equilibrium. The equilibrium is dynamic in the sense that evaporation of atoms continually occurs from the solid surface but there is an equal number of atoms which are condensing on the surface and thus an equilibrium is maintained. Applying the condition of equality of chemical potential for the gas phase and the solid phase we obtain the vapour pressure, p, on the solid surface.

The calculation proceeds by evaluating the chemical potential for the vapour and the solid phases. We consider for simplicity a vapour that is monatomic, i.e. equilibrium between a noble gas and its solid form. The chemical potential can now be written down as [see Eq.(4.3.4)]

$$\mu_{vap} = k_B T \left[\ln\, n - \frac{3}{2} \ln \frac{2\pi m k_B T}{h^2} \right] \tag{4.10.1}$$

where $n = N/V$ in the number density.

Next we need to consider the chemical potential of the solid. To do this we note if we write $\mu_{solid} = G_S/N$, where G_S is the Gibbs potential, then a calculation of G would be necessary, Now writing $G_S = PV_S + E - TS$, we note that solids are characterized by a very high density compared to the vapour and hence the term PV_S be dropped as a first approximation; then $G_S \cong F_S = E - TS$ and we can calculate F_S, within the canonical ensemble. For this, we require a model: we shall take the Einstein model already discussed in connection with the specific heat. The free energy calculated on the basis of this model will not include the cohesive energy E_0 per molecule. The meaning of E_0 is that the potential energy of a solid with all its N atoms at rest at their equilibrium positions is $-NE_0$ compared to the situation where the N atoms are all widely separated from each other. Thus $F_S = -NE_0 +$ contribution from the Einstein model. The contribution of the Einstein model is

$$F_S^{(e)} = -N k_B T \ln Z = -N k_B T \left[\ln\, e^{-T_E/2T} - \ln(1 - e^{-T_E/T}) \right] \tag{4.10.2}$$

where the partition function expression has been taken from Eq.(4.5.5) and T_E is defined in Eq.(4.6.3). We will be interested in temperature $T \gg T_E$ and in this limit, we have

$$F_S^{(e)} = \frac{3}{2} N k_B T_E + 3 N k_B T \ln \frac{T_E}{T} \tag{4.10.3}$$

so that

$$\mu_{solid} \cong F_S/N = -E_0 + \frac{3}{2} k_B T_E + 3 k_B T \ln \frac{T_E}{T}. \tag{4.10.4}$$

The equality $\mu_{vap} = \mu_{solid}$ leads to

$$\ln \left[\left(\frac{T_E}{T} \right)^3 \left(\frac{2\pi m k_B T}{h^2} \right)^{3/2} \frac{1}{n} \right] = \frac{E_0 - \frac{3}{2} k_B T_E}{k_B T}$$

or

$$n \left(\frac{T}{T_E} \right)^3 \left(\frac{h^2}{2\pi m k_B T} \right)^{3/2} = \exp \left(\frac{-E_0 - \frac{3}{2} k_B T_E}{k_B T} \right). \tag{4.10.5}$$

Whence, using $p = n k_B T$,

$$\begin{aligned} p &= k_B T \left(\frac{T_E}{T} \right)^3 \left(\frac{2\pi m k_B T}{h^2} \right)^{3/2} \exp \left(\frac{-E_0 - \frac{3}{2} k_B T_E}{k_B T} \right) \\ &= \frac{(2\pi m)^{3/2}}{(k_B T)^{1/2}} \nu^3 \exp \left(\frac{-E_0 - \frac{3}{2} k_B T_E}{k_B T} \right) \end{aligned} \tag{4.10.6}$$

where $h\nu = k_B T_E$. The temperature dependence of the vapour pressure expressed in Eq.(4.10.6) has been experimentally confirmed.

4.11   Saha Ionization Formula

The calculation of last section can be extended to the equilibrium of reacting gases. It is the chemical potential which is the vital quantity since equilibrium is what we will be concerned with. The specific reaction that we have in mind is of the form.

$$n_1 A + n_2 B = nC \tag{4.11.1}$$

i.e. n_1 moles of material A reacts with n_2 moles of B to produce n moles of C which can dissociate back into A and B. The equilibrium condition is determined by the equality of formation rate and dissociation rate of C. Equilibrium implies that the free energy has to be a minimum. If we denote the concentrations of the species A, B and C by $[A], [B]$ and $[C]$, then the net rate of production of C can be written as

$$\frac{d}{dt}[C] = f_1(T)[A]^{n_1}[B]^{n_2} - f_2(T)[C]^n \tag{4.11.2}$$

In the above $f_1(T)$ describes the rate at which C is produced from A and B and $f_2(T)$ describes the rate at which C dissociates into A and B. The rates depend on the temperature T. At equilibrium $[C]$ is constant and hence,

$$\frac{[C]^n}{[A]^{n_1}[B]^{n_2}} = \frac{f_1(T)}{f_2(T)} = K(T). \tag{4.11.3}$$

As already mentioned, equilibrium implies a minimum in the free energy and hence for small variations about the equilibrium the free energy cannot change. The smallest variation is one step of the reaction Eq.(4.11.1) which means the number N_A of species A changes by a, N_B changes by b and N_c changes by c. The change in N_c is an increment, those in A and B are decrements. We have

$$\Delta F = \left(\frac{\partial F}{\partial N_a} \right)_{V,T} a + \left(\frac{\partial F}{\partial N_B} \right)_{V,T} b - \left(\frac{\partial F}{\partial N_C} \right)_{V,T} c = 0 \tag{4.11.4}$$

Since $dF = dE - TdS - SdT = -PdV + \mu dN - SdT$,

$$\left(\frac{\partial F}{\partial N}\right)_{V,T} = \mu \qquad (4.11.5)$$

and Eq.(4.11.4) leads to

$$a\mu_A + b\mu_B = c\mu_c. \qquad (4.11.6)$$

The partition function for a set of N non interacting molecules can be written as

$$Z = (Z_{\text{trans}} Z_{\text{int}})^N \qquad (4.11.7)$$

where Z_{trans} is the partition function for the translational motion of the centre of mass and Z_{int} is the partition function associated with the internal motion [see Eq.(4.8.2)]. Now, $Z_{\text{trans}} = \left(\frac{V}{\lambda^3}\right)^N \frac{1}{N!}$, where λ is the de Broglie wavelength and hence

$$F = -k_B T \ln Z = -N k_B T \ln \frac{V}{N\lambda^3} - N k_B T \ln Z_{\text{int}}$$

leading to

$$\mu = -k_B T \ln \frac{V}{N\lambda^3} - k_B T \ln Z_{\text{int}} \qquad (4.11.8)$$

in agreement with Eq.(4.3.4). Inserting the above in Eq.(4.11.6)

$$a.\ln \frac{V/N_A}{\lambda_A^3} + b.\ln \frac{V/N_B}{\lambda_B^3} - c.\ln \frac{V/N_C}{\lambda_C^3} = c.\ln Z_{\text{int}}^C - \ln(Z_{\text{int}}^A)^a (Z_{\text{int}}^B)^b.$$

Since $N_A/V = [A]$, the above acquires the form

$$\frac{[C]^C}{[A]^a[B]^b} = \frac{(\lambda_A^3)^a(\lambda_B^3)^b}{(\lambda_C^3)^C} \frac{(Z_{int}^C)^c}{(Z_{int}^A)^a(Z_{int}^B)^b}. \qquad (4.11.9)$$

We would like to apply this formula to the dissociation of the hydrogen atom

$$H = H^+ + e^-. \qquad (4.11.10)$$

Since the mass of H^+ is much greater than the electronic mass, the masses of H and H^+ are nearly equal and hence the de Broglie wavelengths are nearly equal. Neither H^+ nor the electron has internal structure and thus

$$\frac{[H^+][C]}{[H]} = \frac{1}{\lambda_e^3}\left[Z_{int}^H\right]^{-1} = \left(\frac{2\pi m_e k_B T}{h^2}\right)^{3/2} e^{\epsilon_g/k_B T}. \qquad (4.11.11)$$

In the above ϵ_g is the ground state energy and we are working at temperatures such that the ground state population dominates. Since for the hydrogen atom, $\epsilon_* = -13.6 eV = $ - Ionisation energy, we can write Eq.(4.11.11) as

$$\frac{[H^+][e]}{[H]} = \left(\frac{2\pi m_e k_B T}{h^2}\right)^{3/2} \exp\left(-\frac{\text{Ionisation energy}}{k_B T}\right) \qquad (4.11.12)$$

which is the Saha ionization formula.

4.12 Statistical Mechanics of Powders (Edwards' Model)

In this last Section we will tackle a rather unusual problem. This will not be usual statistical mechanics but will use the methods of Chapters 1-3 after certain clarifications regarding the definitions are made. The study was initiated by Edwards and Oakeshott and we will closely follow the steps. It deals with the study of powders. Powders satisfy the tenets of statistical mechanics becaause it can be defined by a small set of parameters and can be constructed in a reproducible manner – in this case repeating the procedure of preparation produces a powder of same density as previously. Our discussion on statistical mechanics has focussed often on the ideal gas – a set of N non-interacting particles moving around in a volume V. The system has a Hamiltonian $H = \sum_{i=1}^{N} [\frac{p_i^2}{2m} + V_{\text{ext}}(\vec{r}_i)]$ where $V_{\text{ext}}(\vec{r}_i)$ is an external potential. A particular value of the Hamiltonian is the energy E of the system and corresponding to a given E and V (macroscopic prescription of the isolated system) there are a large number of microscopic states which are specified by $\{\vec{r}_i, \vec{p}_i\}$, $i = 1, 2, \ldots N$. The number of microstates corresponding to a given E is $\Gamma(E, V, N)$ and an entropy was defined as $S = k_B \ln(E, V, N)$. The temperature is defined by $(\frac{\partial S}{\partial E})_{V,N} = \frac{1}{T}$.

The powder has no energy worth the name – it consists of a large number of grains which are virtually static and external potentials are not relevant in the simplest of situations. What plays the role of E ? It was argued that it was the volume V of the powder. In analogy of the Hamiltonian H, one can introduce a function W of the coordinates of the grains which specifies the volume of the system in terms of positions and orientations of the grains. The form of W will depend on the shape and size of the grains and their configurations. One also needs a function Q which picks out all valid configurations of the grains – configurations that are stable and where the grains do not overlap. In analogy with the corresponding basic postulate of statistical mechanics, one asserts that all configurations W corresponding to a given volume are equally probable. If the number of configurations is $\Gamma(W)$, then one defines the entropy analogously as

$$S = \lambda \ln \Gamma(W) \tag{4.12.1}$$

where λ is the appropriate dimensional constant. What would be the analogue of temperature T ? Edwards and Oakeshott define this as the "frothiness" or "fluffiness" of the powder and denote its measure by X. For $X = 0$ the powder is most compact and for $X \to \infty$, the compactness is least. If a powder has a larger volume than that for $X = \infty$, then it will be unstable and would be characterized by $X < 0$. One has

$$\frac{\partial S}{\partial V} = \frac{1}{X} \tag{4.12.2}$$

in analogy with the definition of the temperature.

In statistical mechanics one has vital quantity called the free energy. This comes from the canonical ensemble when fluctuations in energy are allowed (no longer an isolated system) and the probability of the system having an energy E is proportional to $e^{-E/k_B T}$, giving rise to a partition function

$$Z = \int d\Gamma(E) e^{-E/k_B T} \tag{4.12.3}$$

and the free energy is obtained as

$$F = -k_B T \ln Z \tag{4.12.4}$$

For the powder the corresponding probability of having a volume W can be written as being proportional to $e^{-W/\lambda X}$ and the corresponding "free volume" or effective volume (since "free volume" has a definite meaning in a different context) is denoted by Y and is defined as

$$e^{-Y/(\lambda X)} = \int_{\text{all configurations}} e^{-W/(\lambda X)} d(\text{configurations}) \tag{4.12.5}$$

Clearly

$$Y = V - XS \qquad \text{and} \qquad V = Y - X\frac{\partial Y}{\partial X} \tag{4.12.6}$$

in analogy with $F = E - TS$ and $E = F - T\frac{\partial F}{\partial T}$.

With this set up, Edwards and Oakeshott consider a simple but illustrative model of powder in one dimension (no gravity to compactify it !). The grains are rods of length a each with midpoint located at x_n and $x_n < x_{n-1}$. Clearly for N rods

$$
\begin{aligned}
W &= x_N - x_0 + a \\
&= \sum_1^N (x_n - x_{n-1}) + a
\end{aligned} \tag{4.12.7}
$$

which expresses W as a sum of local volumes. The volume exclusion requires $x_n - x_{n-1} \geq a$. If we put the stability criterion that each rod touches its neighbour, then

$$Q = \prod \delta(x_n - x_{n-1} - a)\Theta(x_n - x_{n-1} - a) \tag{4.12.8}$$

where $\Theta(x) = 1$ if $x > 0$ and is 0 if $x < 0$ and ensures that only configurations with separation greater than a are counted in the δ-function. This obviously allows only one configuration corresponding to $V = Na$.

A more realistic situation would be to try to model a one-dimensional section of a two or three dimensional powder. The grains then do not need to touch on the section but can have a range of separations from a to b. In this case the total number of configurations $\Gamma(V)$ is [the constraint is now $Q = \prod \delta(x_n - x_{n-1} - a)\Theta(a + b - (x_n - x_{n-1}))$]

$$\Gamma(V) = \int \prod dx_n \delta(V - \sum_1^N \delta(x_n - x_{n-1}) - a) \prod_n Q_n \tag{4.12.9}$$

One can alternatively work it out in a canonical ensemble where the volume v of each independent segment can fluctuate, the limit ranging from a to b. The single particle partition function is then $Z_s = e^{-v/(\lambda X)}$ and for N non-interacting "particles",

$$Z = \left[\int_a^{a+b} e^{-v/(\lambda X)} dv \right]^N \tag{4.12.10}$$

The effective Y is $-\lambda \ln Z$ and is given by

$$Y = N \left(\frac{a+b}{2} \right) - N\lambda X \ln \left(2\lambda X \sinh \frac{b}{2\lambda X} \right) \tag{4.12.11}$$

yielding

$$Y = N \left(\frac{a+b}{2} \right) + N\lambda X - \frac{Nb}{2} \coth \left(\frac{b}{2\lambda X} \right). \tag{4.12.12}$$

This yields a nontrivial answer but is not an adequate method of tackling two or three dimensional powders.

The simplest problem in three dimensions would be to study compactivity of a simple powder of uniform grains of the same material in approximately spherical form. Each grain will have neighbours touching it with certain coordination and angular direction. There will also be near neighbours which do not touch. The picture can be quite complicated. To make life simpler, we imagine v_c is the volume occupied by grains with C neighbours, $v_{CC'}$ is a correction which has to do with volume of pairs of neighbouring grains one of which has C neighbours and the other has C' neighbours and so on. The W function is then

$$W = \sum_i v_{C_i} + \sum_{\langle i,j \rangle} v_{C_i C_j} + \dots \tag{4.12.13}$$

where the subscript i indicates that the position of the grain can be affinely deformed to the lattice point $\vec{r}_i$, i.e., v_{C_i} is the volume of the i-th grain with coordination number C_i and $\langle i,j \rangle$ denotes nearest neighbour interactions. To make matters absolutely simple one now imagines a lattice with the grains sitting at the lattice points denoted by i. If at a lattice point i the grain has a coordination number C_i, then the contribution to volume is v_{C_i} and we ignore the second term in Eq.(4.12.13). As a further reduction, we consider only two possible coordination numbers C_0 and C_1 with corresponding volumes v_0 and v_1. This is now a two state system of statistical mechanics and the different sites (N of them) are independent of each other which gives the partition function

$$Z = (e^{-v_0/(\lambda X)} + e^{-v_1/(\lambda X)})^N. \tag{4.12.14}$$

The effective volume is $Y = -\lambda \ln Z$ and the mean volume is obtained from Eq.(4.12.6). Thus

$$Y = N\left(\frac{v_0 + v_1}{2}\right) - \lambda NX \ln \cosh\left(\frac{v_0 - v_1}{\lambda X}\right) \qquad (4.12.15)$$

$$V = N\left(\frac{v_0 + v_1}{2}\right) + N(v_0 - v_1)\tanh\left(\frac{v_0 - v_1}{\lambda X}\right) \qquad (4.12.16)$$

For $X = 0$, i.e., maximum compactness, one has $V = Nv_0$ (the maximum density case) and for $X = \infty$, i.e., minimum compactness or lowest density, one has $V = \frac{N}{2}(v_0 + v_1)$. The model is very simplified but shows that there will be a maximum density which corresponds to the highest coordination number.

The immensely simplifying approximations so far become somewhat more practical if one considers a mixture of powders of type A and B. One assumes that if A is next to A, the contribution to the volume is different from when B is next to B or when A is next to B. The only picture we will retain is one of the nearest neighbour type – we imagine a lattice and there is a contribution to the volume function W only if two neighbouring sites are occupied. We introduce two numbers n_i^A and n_i^B, where i stands for the lattice point and A and B refer to the two species of powders. The value of n can only be 0 or 1, i.e., if $n_i^A = 0$ then there is no A type grain at site i and, if $n_i^A = 1$ there is A type grain at site i. With this in mind we can write the volume as

$$W = \sum_{\langle i,j \rangle} [n_i^A n_j^A v^{AA} + n_i^B n_j^B v^{BB} + n_i^A n_j^B v^{AB}]. \qquad (4.12.17)$$

A few standard manipulations cast this W in the form

$$W = -\frac{1}{2}\sum_{\langle i,j \rangle} J\sigma_i \sigma_j \qquad (4.12.18)$$

where $\sigma_i = \pm 1$ depending whether site i is occupied by type A or type B. The constant J is

$$J = \frac{1}{2}\left[v_{AB} - \frac{1}{2}v_{AA} - \frac{1}{2}v_{BB}\right]. \qquad (4.12.19)$$

As we hall see in Chapter 5 that the model described by Eq.(4.12.18) is the Ising model. For an arbitrary dimensional lattice it cannot be solved exactly but a mean-field approximation (see next Chapter) does yield an answer that is physically meaningful. It turns out that for $\frac{J}{\lambda X} < 1$ (i.e., large X) the two powders are totally miscible but for $\frac{J}{\lambda X} > 1$, the powders begin to separate leading to a complete phase separation as $X \to 0$. This phase separation by somehow controlling the compactivity factor X is a physical effect emerging from an oversimplified model.

4.13 Problems for Chapter 4

1. The nuclei of atoms in a certain crystalline solid have spin 1. According to quantum theory it can exist in three possible quantum states characterized by the quantum number $m=1,0,-1$. Since the charge distribution in a nucleus is not spherically symmetric, the energy depends on the angle between the spin orientation and the local electric field. Accordingly, the energy for the states $m = \pm 1$ is ϵ, and that for $m=0$ is 0. Find the nuclear contribution to the internal energy, entropy and specific heat. By directly calculating the number of accessible states, find the entropy at very high and very low temperatures. Compare with your previous answer.

2. A wire of radius r_o coincides with the axis of a metal cylinder of radius R and length $L(L \gg R)$. The wire is maintained at a positive potential V with respect to the cylinder. The system is at a fairly high temperature T. The electrons emitted from the hot metal form a dilute gas filling the cylinder and in equilibrium with it. The density is low enough to neglect the electrostatic interaction between the electrons. In thermal equilibrium, find the dependence of the electric charge density on the radial distance r.

3. Two atoms interact with the mutual potential energy of interaction of the form

$$U = U_0 \left[\left(\frac{a}{\chi}\right)^{12} - 2 \left(\frac{a}{\chi}\right)^6 \right]$$

where χ is the separation of the particles. The system is in equilibrium at a temperature T, such that $k_B T \ll U_0$, but T is high enough for the applicability of classical statistical mechanics. Calculate the mean separation $\bar{\chi}(T)$ and the expansion coefficient

$$\alpha = \frac{1}{\chi} \frac{\partial \bar{\chi}}{\partial T}$$

(assuming the temperature to the fairly low, set up a perturbative calculation and work to the first non-vanishing order).

4. Consider a collection of N_0 non-interacting magnetic moments in a volume V and at a temperature T. The situation is to be described classically. Consider an external field H applied in the z-direction. Find the mean magnetic moment $\bar{M}$ in the direction of the field.

5. A thin-walled vessel of volume V, kept at constant temperature, contains a gas which slowly leaks out through a small hole of area A. The leakage back into the vessel is negligible. Find the time required for the pressure in the vessel to decrease to half of its initial value. The answer should be in terms of A, V and the mean molecular speed.

6. Helium atoms can be absorbed on the surface of a metal, an amount of work ϕ being then necessary to remove a helium atom from the metal

surface to infinity. The helium atoms are completely free to move on the two-dimensional metal surface. If such a metal surface is in contact with helium gas at a pressure F, and the whole system is in equilibrium at temperature T, what is the mean number of atoms absorbed per unit area of the metal surface?

7. a) Find the free energy of a single quantum oscillator of frequency ω.

b) For a crystal, the above expression needs to be summed over all the frequencies of lattice vibrations. Call this contribution to the free energy $F_0(T)$. If there is a fractional change Δ in the volume of the crystal, the total free energy is

$$F(T; \Delta) = \frac{1}{2}\beta\Delta^2 + F_0(T)$$

where β is the bulk modulus. Due to the volume change, the frequency $\delta\beta$ of a mode changes to $\delta\beta \pm r\omega\Delta$, where r is a constant independent of the mode. Show that minimization of F with respect to Δ, leads to

$$\Delta = rE(T)/\beta$$

where $E(T)$ is the energy density.

8. Consider a classical system of N non-interacting diatomic molecules enclosed in a box of volume V at temperature T. The energy for a single molecule is taken to be

$$E\left(\vec{p}_1, \vec{p}_2, \vec{r}_1, \vec{r}_2\right) = \frac{1}{2m}\left(p_1^2 + p_2^2\right) + \epsilon|\vec{r}_1 - \vec{r}_2|$$

where ϵ and r_0 are given positive constants. Find the specific heat as a function of temperature.

9. Consider an ideal gas contained in a long vertical cylinder of length L and cross section A (gravity has to be included). If the temperature T is uniform throughout, find the specific heat (constant volume) of the gas.

10. For a set of one dimensional anharmonic oscillators with the potential $V(x) = \frac{\lambda}{2n}x^{2N}$, find the average energy and specific heat. If the potential has two terms $V(x) = \frac{1}{2}m\omega^2x^2 + \frac{\lambda}{4}x^4$, find the first correction to the specific heat if λ is small. If the anharmonic term is large, find the first correction induced by the quadratic term. What would you do to write an approximate answer for the specific heat valid for all λ?

Hints for Problem 10:

By a suitable substitution the average energy of the anharmonic oscillator can be written as

$$\langle E \rangle = \frac{\int dx \int dp \left(\frac{p^2}{2m} + \frac{1}{2}m\omega^2x^2 + \frac{\lambda}{4}x^4\right)e^{-E/k_BT}}{\int dx \int dp\, e^{-E/k_BT}}$$

$$= \frac{1}{2}k_BT + \frac{\frac{k_BT}{2}\int dy \left\{y^2 + \frac{\lambda}{2}\left(\frac{k_BT}{m^2\omega^4}\right)y^4\right\}\exp\left[-\left\{\frac{y^2}{2} + \frac{\lambda}{4}\left(\frac{k_BT}{m^2\omega^4}\right)y^4\right\}\right]}{\int dy \exp\left[-\left\{\frac{y^2}{2} + \frac{\lambda}{4}\left(\frac{k_BT}{m^2\omega^4}\right)y^4\right\}\right]}$$

It should be clear that the dimensionless number relevant for the crossover from x^2 to x^4 behaviour (quadratic to quartic) is $\bar{\lambda} = \frac{\lambda k_B T}{(m\omega)^2}$. We could have anticipated this by noting that the length scale "l" at which the two terms in the potential energy become almost equally important is obtained as $l^2 \sim \frac{m\omega^2}{\lambda}$. Consequently the relevant dimensionless number is $\frac{\lambda l^2}{m\omega^2}$. Since, $m\omega^2 l^2$ is a measure of energy at temperature T, we can also estimate $l^2 \sim \frac{k_B T}{m\omega^2}$ and hence the dimensionless number is $\frac{\lambda k_B T}{(m\omega)^2}$.

To order $O(\lambda)$, the above equation becomes,

$$
\begin{aligned}
\langle E \rangle &= \frac{1}{2}k_B T + \frac{1}{2}k_B T . \frac{\int dy \left(y^2 + \frac{\bar{\lambda}}{2}y^4\right) e^{-y^2/2} . \left[1 - \frac{\bar{\lambda}}{4}y^4\right]}{\int dy e^{-y^2/2} . \left[1 - \frac{\bar{\lambda}}{4}y^4\right]} + \dots \\
&= \frac{1}{2}k_B T + \frac{1}{2}k_B T . \frac{\int dy e^{-y^2/2} . \left(y^2 + \frac{\bar{\lambda}}{2}y^4 - \frac{\bar{\lambda}}{4}y^6\right)}{\int dy e^{-y^2/2} . \left[1 - \frac{\bar{\lambda}}{4}y^4\right]} + \dots \\
&= \frac{1}{2}k_B T + \frac{1}{2}k_B T . \frac{1 + \frac{3}{2}\bar{\lambda} - \frac{15}{4}\bar{\lambda}}{1 - \frac{3}{4}\bar{\lambda}} \\
&= \frac{1}{2}k_B T + \frac{1}{2}k_B T \left(1 - \frac{3}{2}\bar{\lambda}\right) + O(\bar{\lambda}^2).
\end{aligned}
$$

The specific heat can be found by differentiation with respect to T. This result should be compared with that obtained from Problem 4 of Chapter 2. The usefulness of the canonical ensemble as a calculational tool should then become obvious.

For $\bar{\lambda} \gg 1$, one needs to rescale y to rewrite $\frac{\bar{\lambda}}{2}y^4 = z^4$ and then

$$
\begin{aligned}
\langle E \rangle &= \frac{1}{2}k_B T + \frac{1}{2}k_B T \frac{\int dz \left(z^4 + z^2\sqrt{\frac{2}{\bar{\lambda}}}\right) \exp\left(-z^2\sqrt{\frac{2}{\bar{\lambda}}} - \frac{z^4}{2}\right)}{\int dz \exp\left(-\frac{z^4}{2}\right) \exp\left(-z^2\sqrt{\frac{2}{\bar{\lambda}}}\right)} \\
&= \frac{1}{2}k_B T + \frac{1}{2}k_B T \frac{\int dz \left(z^4 + z^2\sqrt{\frac{2}{\bar{\lambda}}}\right) e^{-z^4/2} \left(1 - z^2\sqrt{\frac{2}{\bar{\lambda}}} + \dots\right)}{\int dz e^{-z^4/2} \left(1 - z^2\sqrt{\frac{2}{\bar{\lambda}}} + \dots\right)}.
\end{aligned}
$$

Carry out the integrations to find to the lowest nontrivial order

$$
\langle E \rangle = \frac{1}{2}k_B T + \frac{1}{4}k_B T \left[1 + \frac{2\Gamma(3/4)}{\Gamma(1/4)} . \frac{1}{\bar{\lambda}^{1/2}}\right].
$$

Note that if the expansion in $\bar{\lambda}$ is called a weak-coupling expansion then the expansion in inverse powers of $\bar{\lambda}$ is called a strong-coupling expansion.

The technique of constructing a uniformly valid expression for the specific heat uses the method of constructing Pade approximants. For small $\bar{\lambda}$ (weak coupling)

$$\frac{C}{k_B} = 1 - \frac{3}{4}\bar{\lambda} + \dots$$

and for large (strong coupling)

$$\frac{C}{k_B} = \frac{3}{4} + \frac{2\Gamma(3/4)}{\Gamma(1/4)} \cdot \frac{1}{\bar{\lambda}^{1/2}}$$

If $\bar{\lambda}^{1/2} = \mu$, we try

$$\frac{C}{k_B} = \frac{1 + \alpha_1\mu + \alpha_2\mu^2 + \dots}{1 + \beta_1\mu + \beta_2\mu^2 + \dots}$$

We now use all the available information, e.g., $\frac{\alpha_2}{\beta_2} = \frac{3}{4}$, $\alpha_1 = \beta_1$ to infer the coefficients. Find $\alpha_1, \alpha_2, \beta_1$ and β_2.

How good is the approximation? One should numerically obtain $\frac{C}{k_B}$ for different values of μ and find out.

11. Consider a gas of relativistic particles (kinetic energy $= \sqrt{p^2c^2 + m^2c^4} - mc^2$, m being the rest mass) at temperature T. Evaluate the partition function exactly and find the mean energy. There should be a parameter $\lambda = \frac{mc^2}{k_BT}$ in this case which will determine what the average energy is. Carry out weak coupling and strong coupling expansions separately as well as a Pade interpolation and check against the exact answer.

12. Consider an ideal gas enclosed in a cube of sides L. We want to study the first correction coming from the finite size of the container. Using the quantum energy levels for particle in a box, write the partition function and the Euler-Maclaurin formula to obtain the first correction to the thermodynamic limit.

5

Interacting Classical Systems

5.1 The Non-ideal Gas

We now consider N molecules enclosed in a volume V interacting via a pairwise potential $\nu(r_{ij})$, where $r_{ij} = |\vec{r}_i - \vec{r}_j|$, is the magnitude of the separation vector between particles i and j. The potential corresponds to an attractive force at large distances and a very strong repulsion at short distances. A typical $\nu(r)$ is shown in Fig.(5.1).

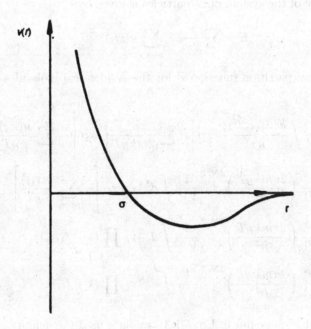

Figure 5.1: $\nu(r)$ vs. r.

For later purposes, we define an auxiliary function

$$f(r) = e^{-\beta \nu(r)} - 1 \tag{5.1.1}$$

shown in Fig.(5.2).

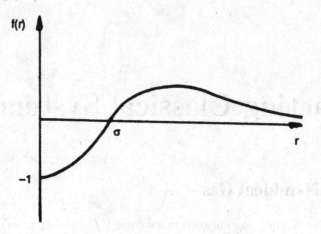

Figure 5.2: $f(r)$ vs. r.

The potential $\nu(r)$ has a zero at $r = \sigma$, which corresponds roughly to the molecular diameter. The function $f(r)$ is zero wherever $\nu(r)$ is so.

The energy of the system of N particles is given by

$$E = \sum \frac{p_i^2}{2m} + \sum_{pairs} \nu(r_{ij}) \tag{5.1.2}$$

and the canonical partition function Z for the N identical molecules is given by

$$
\begin{aligned}
Z &= \frac{1}{N!} \int \frac{d^{3N}p_i d^{3N}r_i}{h^{3N}} \exp\left[-\sum \frac{p_i^2}{2mk_BT}\right] \exp\left[-\sum_{pairs} \frac{\nu(r_{ij})}{k_BT}\right] \\
&= \frac{V^N}{N!} \left(\frac{2\pi mk_BT}{h^2}\right)^{3N/2} \frac{1}{V^N} \int d^{3N}r_i \exp\left[-\sum_{pairs} \frac{\nu(r_{ij})}{k_BT}\right] \\
&= \frac{V^N}{N!} \left(\frac{2\pi mk_BT}{h^2}\right)^{3N/2} \frac{1}{V^N} \int d^{3N}r_i \prod_{pairs}(1 + f(r_{ij})) \\
&= \frac{V^N}{N!} \left(\frac{2\pi mk_BT}{h^2}\right)^{3N/2} \frac{1}{V^N} \int d^{3N}r_i \prod_{pairs}(1 + f_{ij}) \tag{5.1.3}
\end{aligned}
$$

where in the last but one line of Eq.(5.1.3), we have used the definition of $f(r)$ from Eq.(5.1.1) and have introduced the notation $f(r_{ij}) = f_{ij}$ in the last line.

The free energy is consequently

$$
\begin{aligned}
F &= -k_B T \ln Z \\
&= -N k_B T \ln V - \frac{3 N k_B T}{2} \ln \frac{2\pi m k_B T}{h^2} \\
&\quad + N \ln N - N - k_B T \ln \left[\frac{1}{V^N} \int d^3 r_i \prod_{pairs} (1 + f_{ij}) \right].
\end{aligned} \tag{5.1.4}
$$

Stirling's approximation has been used in the equation above and the pressure P follows as

$$
\begin{aligned}
P &= -\left(\frac{\partial F}{\partial V} \right)_T \\
&= \frac{N k_B T}{V} + k_B T \frac{\partial}{\partial V} \ln \frac{1}{V^N} \int d^3 r_i \prod_{pairs} (1 + f_{ij}) \\
&= \frac{N k_B T}{V} \left[1 + \frac{V}{N} \frac{\partial}{\partial V} \ln \frac{1}{V^N} \int d^3 r_i \prod_{pairs} (1 + r_{ij}) \right] \\
&= \frac{N k_B T}{V} \left[1 + V \frac{\partial \bar{F}}{\partial V} \right]
\end{aligned} \tag{5.1.5}
$$

where

$$
\bar{F} = \frac{1}{N} \ln \frac{1}{V^N} \int d^3 r_i \prod_{pairs} (1 + f_{ij}). \tag{5.1.6}
$$

At this stage, we have an exact result. If $\nu_{ij} = 0$, $f_{ij} = 0$ and hence $\bar{F} = 0$, we get back the ideal gas law. Since $\bar{F}$ cannot be exactly obtained for any non-zero ν_{ij} we try to get a handle on the problem by assuming that ν_{ij} is small and hence f_{ij} is small. Accordingly, we retain only the $O(f)$ terms in Eq.(5.1.6) and find

$$
\begin{aligned}
\bar{F} &= \frac{1}{N} \ln \frac{1}{V^N} \int d^3 r_i (1 + \prod_{pairs} f_{ij}) \\
&= \frac{1}{N} \ln \frac{1}{V^N} \left[V^N + \frac{N^2 V^{N-1}}{2} \int f(r) d^3 r \right] \\
&= \frac{1}{N} \ln \left[1 + \frac{N^2}{2V} \int f(r) d^3 r \right]
\end{aligned} \tag{5.1.7}
$$

where we have approximated the number of pairs $N(N-1)/2$ by $N^2/2$ for large N. Evaluating the derivative $\partial \bar{F}/\partial V$ to $O(f)$, we find

$$
\frac{\partial \bar{F}}{\partial V} = -\frac{N}{2V^2} \int f(r) d^3 r. \tag{5.1.8}
$$

The equation of state [Eq.(5.1.5)] now becomes

$$\frac{PV}{Nk_BT} = 1 - \frac{N}{2V} \int f(r)d^3r \qquad (5.1.9)$$

A phenomenological equation of state for the real gas is Van der Waals equation which reads

$$\left(P + \frac{a}{V^2}\right)(V - b) = Nk_BT \qquad (5.1.10)$$

where a and b are constants for a given gas. We can rewrite this equation of state as

$$\begin{aligned} P &= \frac{Nk_BT}{V - b} - \frac{a}{V^2} \\ &= \frac{Nk_BT}{V}\left(1 + \frac{b}{V}\right) - \frac{1}{V^2} + O\left(\frac{1}{V^3}\right) \\ &= \frac{Nk_BT}{V} + \frac{bNk_BT - a}{V^2} + \dots \end{aligned}$$

or

$$\frac{PV}{Nk_BT} = 1 + \left(b - \frac{a}{Nk_BT}\right)\frac{1}{V} + \dots. \qquad (5.1.11)$$

• To see how this compares with Eq.(5.1.9), we need to have an expression for $f(r)$. We make the approximation

$$\begin{aligned} f(r) &= -1, \qquad \text{for } r < \sigma \\ \text{and} \qquad f(r) &= -(1 - e^{-\beta\nu(r)}), \qquad \text{for } r > \sigma \qquad (5.1.12) \end{aligned}$$

leading to

$$\int f(r)d^3r = -\frac{4}{3}\pi\sigma^3 - 4\pi \int_\sigma^\infty r^2(1 - e^{-\beta\nu(r)})dr$$

and

$$\begin{aligned} \frac{PV}{Nk_BT} &= 1 + \frac{2\pi}{3}\frac{N}{V}\sigma^3 - \frac{2\pi N}{V}\int_\sigma^\infty r^2(e^{-\beta\nu(r)} - 1)dr \\ &\cong 1 + \frac{2\pi}{3}\frac{N}{\nu}\sigma^3 - \frac{2\pi n}{Vk_BT}\int_\sigma^\infty (1 - \nu(r))r^2dr \qquad (5.1.13) \end{aligned}$$

for weak potentials, i.e. $\nu(r) \ll k_BT$. Note that $\int_\sigma^\infty (-\nu(r)r^2)dr$ is positive definite since beyond σ, $\nu(r)$ is attractive. Thus as an expansion in inverse powers of V, Eq.(5.1.13) has the same structure as Van der Waals equation to the lowest order, when we identify

$$b = \frac{2}{3}\pi N\sigma^3 \qquad (5.1.14)$$

and

$$a = 2\pi N^2 \int\limits_{\sigma}^{\infty} \{-\nu(r)\}r^2 dr \qquad (5.1.15)$$

If we could retain all powers of f, i.e., considered arbitrarily strong potentials, we would generate an equation of state of the form

$$\frac{P}{nk_B T} = 1 + A_1 n + A_2 n^2 + \ldots \qquad (5.1.16)$$

where $n = N/V$ is the number density. The various A_1, $A_2, \ldots$ etc. are known as virial coefficients. What we have generated is the first virial coefficient A_1. The interesting feature is that A_1 has a zero at a finite temperature. This fact has been experimentally verified.

5.2 Linked Cluster Expansion

This is a systematic way of obtaining the equation of state of a gas. The technique is useful as it shares many features of the diagrammatic techniques of many-body physics, but the end result is in a sense disappointing. It is incapable of providing an equation of state which correctly describes the high density behaviour and hence cannot describe a phase transition. In that sense, the phenomenological equation of state of Van der Waals is superior.

We return to Eq.(5.1.3) and write it as

$$Z = \frac{1}{N!} \left(\frac{2\pi m k_B T}{h^2} \right)^{3N/2} Z_N = \frac{V}{N!} \frac{1}{\lambda^{3N}} Z_N \qquad (5.2.1)$$

where $\lambda = h/\sqrt{2\pi m k_B T}$ is the De Broglie wavelength, and

$$\begin{aligned} Z_N &= \int d^3 r_1 d^3 r_2 \ldots d^3 r_n \prod_{pairs} (1 + f_{ij}) \\ &= \int d^3 r_1 d^3 r_2 \ldots d^3 r_n \, (1 + f_{12} + f_{13} + f_{23} + \ldots \\ &\quad + \; f_{12}f_{13} + f_{12}f_{23} + \ldots + f_{12}f_{23}f_{31} + \ldots) . \end{aligned} \qquad (5.2.2)$$

The various terms in the brackets in Eq.(5.2.2) can be given a diagrammatic representation by setting up the following rules:

1. Each particle is represented by a circle. So an N-particle situation will involve N circles.

2. We represent f_{pq} by a line linking the pth circle (i.e., circle representing the pth particle) to the qth circle.

3. In any given diagram some circles may be linked and some may be free. Every linkage contributes to f while each unlinked circle contributes unity.

As an example consider $N = 3$, for which Eq.(5.2.2) can be explicitly written as,

$$
\begin{aligned}
Z_3 &= \int d^3r_1 d^3r_2 d^3r_3 \left[1 + f_{12} + f_{23} + f_{13} \right. \\
&\quad + \left. f_{12}f_{13} + f_{12}f_{23} + f_{12}f_{23} + f_{13}f_{23}f_{13}\right] \\
&= Z_{30} + Z_{31} + Z_{32} + Z_{33}
\end{aligned}
\tag{5.2.3}
$$

where

$$
\begin{aligned}
Z_{30} &= \int d^3r_1 d^3r_2 d^3r_3 \\[2mm]
Z_{31} &= \int d^3r_1 d^3r_2 d^3 [f_{12} + f_{23} + f_{13}] \\[2mm]
Z_{32} &= \int d^3r_1 d^3r_2 d^3r_3[f_{12}f_{23} + f_{13}f_{23} + f_{12}f_{23}] \\[2mm]
Z_{33} &= \int d^3r_1 d^3r_2 d^3r_3 f_{12}f_{23}f_{13}
\end{aligned}
$$

and the diagrammatic representations are shown in Fig. 5.3.

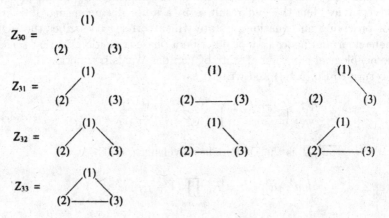

Figure 5.3: The three particle graphs.

A graph with N circles is called a N-particle graph. From the above, it is clear that for $N = 3$, Z_3 [of Eq.(5.2.3)] = sum of all *distinct* three-particle graphs. An example of graphs which are not distinct is shown in Fig. 5.4.

For arbitrary N,

$$
Z_N = \text{Sum of all distinct } N-\text{particle graphs}
\tag{5.2.4}
$$

Figure 5.4: Non-distinct three particle graphs.

For a given N, any graph, in general, can be viewed as a product of several subgraphs. Two examples for $N = 8$ is shown in (Fig.5.5).

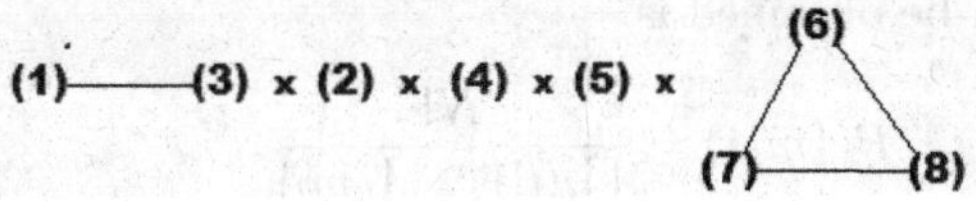

(1)———(3) × **(2)** × **(4)** × **(5)** ×

Figure 5.5: Factoring of graphs.

The factoring shown in Fig.5.5 leads to the contribution

$$\left[\int d^3r_1 d^3r_3 f_{13}\right] \left[\int d^3r_2\right] \left[\int d^3r_4\right] \left[\int d^3r_5\right]$$
$$\times \left[\int d^3r_6 d^3r_7 d^3r_8 f_{67} f_{78} f_{68}\right]$$

from that particular graph. It should be noted that each factor corresponds to a connected graph-connected in the sense that each circle is connected to a line and thus directly or indirectly connected to every other circle in the graph.

If l circles are involved in a connected graph, then it is called an l-cluster. We first define a dimensionless number b_l associated with each cluster integral by

$$b_l = \frac{1}{\lambda^{(3l-3)}} \cdot \frac{1}{V} \cdot \frac{1}{l!} \times [\text{Sum of all possible } l-\text{clusters}]. \qquad (5.2.5)$$

This definition ensures that

1. b_l is dimensionless.

2. $\lim_{(V\to\infty)} b_l$ is finite. This is true because all the f_{ij} have a finite range and only the integration over the centre of gravity produces a factor of V.

A general N-particle graph will be composed of l-clusters where l may range from 1 to N. In a given situation, an l-cluster occurs m_l times and we must have $\sum m_l.l = N$. Now, the set of numbers $\{m_l\}$ is not unique, e.g. if $N = 3$, we can have $l = 1$, $m_l = 3$; or $l = l$, $m_l = 1$ and $l = 2$, $m_l = 1$; or $l = 3$, $m_l = 1$. Thus, there are three ways of satisfying $\sum m_l.l = 3$. These three ways are the three types of nontrivial graphs (Z_{31}, Z_{32}, Z_{33}) in Fig. 5.3.

Consider an arbitrary N-particle graph with a given distribution of m_l. The magnitude of the integral ($\{m_l\}$) associated with this graph is

$$I = \prod_{l=1}^{N} \{b_l \lambda^{3l-3} V l!\}^{m_l}. \qquad (5.2.6)$$

We now need to know in how many different ways these l-clusters with the distribution $\{m_l\}$ can be formed. Clearly all N particles can be permuted among themselves and thus there are $N!$ possibilities. But not all these permutations can generate distinct graphs. Clearly the permutation of the m_l number of l-clusters among themselves cannot generate new graphs. That cuts the number down by $\prod_l m_l!$. Further, within the same cluster, the l-particles can be permuted among themselves in $l!$ ways without producing a distinct graph and since there are m_l number of l-clusters, the number of such redundant permutations is $(l!)^{m_l}$. Thus the total number of distinct ways P in which a given distribution $\{m_l\}$ can be obtained is

$$P(\{m_l\}) = \frac{N!}{\prod_l (l!)^{m_l} \prod_l m_l!}. \tag{5.2.7}$$

The contribution of this distribution $\{m_l\}$ to Z_N is PI and can be written as

$$
\begin{aligned}
Z(\{m_l\}) &= \frac{N!}{\prod_l (l!)^{m_l} \prod_l m_l!} \cdot \prod_l \left(b_l \lambda^{3l-3} V l! \right)^{m_l} \\
&= \prod_l \frac{N!}{m_l!} \left(\frac{V}{\lambda^3} \right)^{m_l} \lambda^{3N} b_l{}^{m_l}. \tag{5.2.8}
\end{aligned}
$$

The partition function is

$$Z_N = \sum_{\{m_l\}} Z_N(\{m_l\}) \tag{5.2.9}$$

with the constraint $\sum l m_l = N$.

It is this constraint which makes the evaluation of the sum impossible. Hence we resort to the grand canonical ensemble where the partition function can be written as

$$
\begin{aligned}
Q &= \sum_{\epsilon,N} e^{-\beta\epsilon} e^{\mu N \beta} \\
&= \sum_N e^{\mu N \beta} Z(N) \\
&= \sum_N e^{\mu N \beta} \frac{1}{N!} \frac{V}{\lambda^{3N}} \cdot Z_N \\
&= \sum_N e^{\mu N \beta} \sum_{\{m_l\}} \prod_l \frac{1}{m_l!} \left(\frac{V}{\lambda^3} \right)^{m_l} \cdot b_l{}^{m_l} \\
&= \sum_{N=0}^{\infty} \sum_{\{m_l\}} \prod_{l=1}^{N} \frac{1}{m_l!} \left(\frac{V b_l e^{\beta \mu l}}{\lambda^3} \right)^{m_l} \tag{5.2.10}
\end{aligned}
$$

where we have made use of Eq.(5.2.1) for $Z(N)$ (which is just the Z of that equation) and Eqs.(5.2.8) and (5.2.9). Now the various sums over the set $\{m_l\}$

are freed from the constraint of fixed N and consequently we can perform the unrestricted sum over each m_l for a given l and finally perform the product over l. Thus

$$Q = \prod_l \sum_{m_1} \frac{1}{m_1!} \left(\frac{V b_1 e^{\beta \mu l}}{\lambda^3} \right)^{m_1} \cdot \sum_{m_2} \frac{1}{m_2!} \left(\frac{V b_2 e^{\beta \mu l}}{\lambda^3} \right)^{m_2} \cdots$$

$$= \prod_l \exp \left[\frac{V b_l e^{\mu \beta l}}{\lambda^3} \right]. \tag{5.2.11}$$

The equation of state follows from the relation,

$$PV = k_B T \ \ln \ Q = k_B T \sum_l \frac{V}{\lambda^3} b_l e^{\mu l / k_B T}. \tag{5.2.12}$$

Once μ is obtained from

$$N = \left[\frac{\partial (PV)}{\partial \mu} \right]_T = \frac{V}{\lambda^3} \sum l b_l e^{\mu l / k_B T}, \tag{5.2.13}$$

elimination of μ between Eqs.(5.2.12) and (5.2.13) yields the equation of state – a relation between P, V, N and T. In principle, one needs to obtain the relation between P, N and T for arbitrary N and V and then take the thermodynamic limit. Needless to say, this is virtually impossible. If we take the limit of $N \to \infty$ and $V \to \infty$ with $n = N/V$ finite, then Eqs.(5.2.12) and (5.2.13) become

$$\frac{P}{k_B T} = \frac{1}{\lambda^3} \sum b_l e^{\mu l / k_B T} \tag{5.2.14}$$

and

$$n = \frac{1}{\lambda^3} \sum l b_l e^{l \beta \mu} \tag{5.2.15}$$

where the b_ls are finite numbers in this limit. We can obtain a series solution for last two equations as

$$\frac{P}{n k_B T} = 1 + A_1 n + A_2 n^2 + \ldots \tag{5.2.16}$$

which is the virial expansion introduced in Sec.5.1. This expansion essentially treats n as small and hence from Eq.(5.2.15), we get

$$n = \frac{1}{\lambda^3} (b_1 e^{\beta \mu} + 2 b_2 e^{2 \beta \mu} + \ldots)$$

$$= \frac{e^{\beta \mu} b_1}{\lambda^3} \left(1 + \frac{2 b_2}{b_1} e^{\beta \mu} + \frac{3 b_3}{b_1} e^{2 \beta \mu} \ldots \right) \tag{5.2.17}$$

and a perturbative solution for the chemical potential yields as the first approximation

$$\frac{e^{\beta \mu} b_1}{\lambda^3} = n \tag{5.2.18}$$

and

$$\frac{e^{\beta\mu}b_1}{\lambda^3} = n\left(1 - 2\frac{b_2}{b_1^2}\lambda^3 n\right) + O(n^3) \tag{5.2.19}$$

as the second approximation. Now, the pressure can be expressed from Eq.(5.2.14) as

$$\begin{aligned}
\frac{P}{k_B T} &= \frac{b_1}{\lambda^3}e^{\beta\mu} + \frac{b_2}{\lambda^3}e^{2\beta\mu} \\
&= n - \frac{2b_2}{b_1^2}\lambda^3 n^2 + \frac{b_2}{b_1^2}\lambda^3 n^2 + O(n^3) \\
&= n\left[1 - \frac{b_2}{b_1^2}\lambda^3 n + O(n^2)\right]
\end{aligned}$$

or

$$\frac{P}{nk_B T} = \left(1 - \frac{b_2}{b_1^2}\lambda^3 n + O(n^2)\right). \tag{5.2.20}$$

Thus the first virial coefficient is given by $(b_2/b_1^2)\lambda^3$. Referring to the definition of b_l [see Eq.(5.2.5)], it is evident that $b_1 = 1$ and

$$b_2 = \frac{1}{\lambda^3}\cdot\frac{1}{V}\cdot\frac{1}{2} \times (\text{All possible 2−clusters}). \tag{5.2.21}$$

But there can be only one kind of 2−cluster and that has the contribution $\int d^3 r_1 d^3 r_2 f(r_{12}) = V \int f(r)d^3 r$, so that Eq.(5.2.20), becomes

$$\frac{P}{nk_B T} = 1 - \frac{n}{2}\int f(r)d^3 r + O(n^2) \tag{5.2.22}$$

exactly identical to Eq.(5.1.9). Thus, what we have generated now is a systematic way of evaluating the different terms of the virial expansion of Eq.(5.1.16). However, this constraint of going to the thermodynamic limit before obtaining the equation of state makes the virial expansion useless at high densities and hence for studying phase transitions. Phenomenological equations of state such as that due to Van der Waals are consequently very useful.

5.3   Van der Waals Equation of State

For one mole of a real gas, the equation of state proposed by Van der Waals [see Eq.(5.1.10)] can be written as

$$P = \frac{RT}{V - b} - \frac{a}{V^2}. \tag{5.3.1}$$

Clearly $P \to \infty$ as $V \to b$ and $P \to 0$ as $V \to \infty$, for a fixed value of T. The intermediate behaviour is not monotonic as extrema exist at

$$\frac{RT}{(V - b)^2} = \frac{2a}{V^3}$$

or

$$V^3 - \frac{2a}{RT}V^2 + \frac{4ab}{RT}V - \frac{2ab^2}{RT} = 0 \qquad (5.3.2)$$

This gives rise to three roots for V, one of which is smaller than b and hence irrelevant for our purpose. Of the two relevant ones, one is a maximum, the other a minimum. As the temperature is raised, the separation between the maximum and the minimum decreases and if we consider the isotherm at $T = T_c$, the maximum and the minimum merge and there is only one point at which $\partial P/\partial V = 0$. It is obvious that this limiting situation must be a point of inflection and $\partial^2 P/\partial V^2$ must be zero. If the molar volume is V_c at $T = T_c$, then the second derivative of Eq.(5.3.1) yields

$$\left(\frac{\partial^2 P}{\partial V^2}\right)_{T=T_c} = 0 \quad \text{or} \quad \frac{RT_c}{(V_c - b)^3} = \frac{3a}{V_c^4} \qquad (5.3.3)$$

and together with Eq.(5.3.2) at $T = T_c$ and $V = V_c$ (the critical point) we obtain

$$V_c = 3b, \qquad RT_c = \frac{8a}{27b}. \qquad (5.3.4)$$

The above result could also be obtained in a different way. We could have asked for the condition where the two roots of the cubic come together, i.e. Eq.(5.3.2) has the structure $(V - \alpha)^2(V - \beta) = 0$. This yields Eq.(5.3.4) once more. Yet another way of arriving at Eq.(5.3.4) is to consider how many values of V yield the same pressure on a given isotherm, i.e., we cast Eq.(5.3.1) itself as a cubic in V for a fixed T and P and determine the three roots.

All three roots are real for $T < T_c$, as it must according to our previous discussion of the maximum, minima and the behaviour at $V = b$ and $V = \infty$. For $T < T_c$, the relation between P and V has got to be monotonic and at $T = T_c$, when there is an extremum as a point of inflection, the three roots of V must come together and we should have $(V - V_c)^3 = 0$. The condition that the cubic in V has this structure again yields Eq.(5.3.4). In the light of the above discussion it is fairly obvious what the different isotherms should look like and they are shown in Fig.5.6.

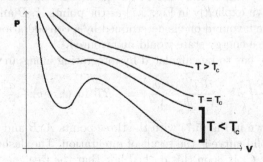

Figure 5.6: Isotherms for the Van der Waals equation.

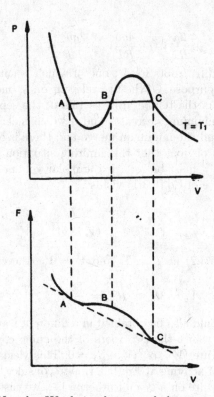

Figure 5.7: Van der Waals isotherm and the two-phase region.

For $T > T_c$, the isotherms are like a standard vapour with $\partial P/\partial V < 0$ everywhere. It is for $T < T_c$, in the region between the minimum and maximum that Van der Waals equation produces an unphysical situation by yielding $\partial P/\partial V > 0$. For the equation of state to make sense for $T < T_c$, it can no longer represent a homogeneous substance as we have assumed so far – it must represent two phases in equilibrium. To see whether this is the case we must construct the free energies of phases which are in thermal and mechanical equilibrium, i.e., same values of T and P. As discussed before there are three such candidates shown explicitly in Fig.(5.7) as the points A, B and C. Which of them would be the favoured one is determined by a consideration of the free energy; the lowest free energy state would be favoured.

The free energy has to be calculated by integrating along an isotherm

$$F = -\int PdV = -\int dV \left[\frac{RT}{V-b} - \frac{a}{V^2}\right] = -RT \ln (V-b) - \frac{a}{V} + F_0 \quad (5.3.5)$$

and the result is shown in Fig.(5.7) with the three points A, B and C indicated on the curve. The solid curve is the result of integration. The dashed line joins the tangents at A at C. It is on this dashed line that the free energy of mixed phase composed of A and C will be lying and as can be seen from Fig.(5.7), the mixed phase has a lower free energy than the homogeneous phase by the solid line. It should be noted that the area above the line ABC in the P vs. V curve is equal to the area below it. This forms the basis of the Maxwell construction.

Thus, between A and C, there is a coexistence of two phases and Van der Waals equation of state shows a phase transition as T is lowered below T_c. For $T < T_c$, between the points A and C, the pressure will remain constant as the molar volume of phase A changes to the molar volume of phase C. This is a first-order transition and the latent heat involved can be calculated from the Clausius-Clapeyron relation. The limiting transition occurs at $T = T_c$. Here there is no latent heat involved, but as one passes T_c from above, one passes from a one-phase (vapour) region to a two-phase coexistence state. The compressibility is infinite at $T = T_c$. The transition at $T = T_c$, is a second-order transition characterized by the absence of latent heat in the transition and singularities in thermodynamic quantities. At a temperature $T > T_c$, we determine the isothermal compressibility as

$$\frac{1}{\chi_T} = -\left(\frac{\partial P}{\partial V}\right)_T = \frac{RT}{(V-b)^2} - \frac{2a}{V^3}. \tag{5.3.6}$$

Along the line of critical volume $V = V_c = 3b$, if we change the temperature, then

$$\begin{aligned} \frac{1}{\chi_T} &= \frac{RT}{4b^2} - \frac{2a}{27b^3} \\ &= \frac{R}{4b^2}\left[T - \frac{8a}{27b}\right] \\ &= \frac{R}{4b^2}(T - T_c) \end{aligned}$$

or

$$\chi_T \propto (T - T_c)^{-1}. \tag{5.3.7}$$

Thus the susceptibility diverges in a characteristic fashion as T approaches T_c and while the absolute magnitude of the susceptibility is dependent on the nature of the substance (reflected in b and T_c), the nature of the singularity at T_c, i.e., $(T - T_c)^{-1}$, is universal. As the temperature is lowered towards the critical temperature along the critical isochore, the susceptibilities of all gases approach infinity, the behaviour being described by $(T-T_c)^{-1}$. This universality shown by the Van der Waals gas is quite remarkable and, as we will see later, is a standard feature of all second-order phase transitions.

We can explore yet another singular behavior pertaining to the Van der Waals gas close to T_c but below it. As we remarked before, at T_c the maximum and the minimum come together, causing a merging of two of the roots of Eq.(5.3.2). If $T < T_c$, then the two roots will be displaced slightly from V_c. Thus for $T < T_c$ if one writes $T = T_c(1 - \epsilon)$ then clearly the structure of Eq.(5.3.2) to $O(\epsilon)$ will be

$$V^3 - \frac{2a}{RT_c}V^2 + \frac{4abV}{RT_c} - \frac{2ab^2}{RT_c} - \epsilon\left(\frac{2aV_c^2}{RT_c} - \frac{4abV_c}{RT_c} + \frac{2ab^2}{RT_c}\right) = 0$$

or

$$V^3 - \frac{2a}{RT_c}V^2 + \frac{4abV}{RT_c} - \frac{2ab^2}{RT_c} = \epsilon 27b^3 \tag{5.3.8}$$

The two roots of V which were at $V = V_c = 3b$, are now at $V_{1,2} = V_c(1+\delta)$. Using this in Eq.(5.3.8) and working to $O(\delta^2)$

$$V_c^3(1 + 3\delta + 3\delta^2) - \frac{2a}{RT_c}V_c^2(1 + 2\delta + \delta^2) + \frac{4abV_c}{RT_c}(1 + \delta) - \frac{2ab^2}{RT_c} = 27b^3\epsilon$$

or

$$\delta\left[3V_c^3 - \frac{4aV_c^2}{RT_c} + \frac{4abV_c}{RT_c}\right] + \delta^2\left(3V_c^3 - \frac{2a}{RT_c}V_c^2\right) = 27b^3\epsilon. \tag{5.3.9}$$

But the coefficient of δ evaluates to

$$3V_c^3 - \frac{4aV_c^2}{RT_c} + \frac{4abV_c}{RT_c} = b^3\left(81 - \frac{27 \times 9}{2} + \frac{27 \times 3}{2}\right) = 0 \tag{5.3.10}$$

while the coefficient of δ^2 becomes

$$3V_c^3 - \frac{2a}{RT_c}.V_c^2 = b^3\left[81 - \frac{27 \times 9}{4}\right] = \frac{81}{4}b^3$$

or

$$\delta^2 = \frac{4}{3}\epsilon. \tag{5.3.11}$$

Consequently, for $T \cong T_c$ but below it, the difference between the two molar volumes is given by

$$\Delta V = V_1 - V_2 = 2V_c\delta = V_c\frac{4}{\sqrt{3}}\epsilon^{1/2}. \tag{5.3.12}$$

The difference in densities between the two phases is

$$\Delta\rho \propto \Delta V \sim V_c\epsilon^{1/2} \propto (T_c - T)^{1/2}. \tag{5.3.13}$$

The singular behavior [i.e., $(T_c - T)^{1/2}$] is once again universal, while the actual $\Delta\rho$ will depend on the material properties. It should be noted that $\Delta\rho = 0$ in the one-phase region $(T > T_c)$ and is non-zero only in the two- phase region $(T > T_c)$ and hence is an important characteristic of this second-order phase transition taking place at $T = T_c$. It is an example of what in general is called an order parameter for the second-order transition – a quantity which is zero above the transition and non-zero below. The analogue is the spontaneous magnetization in the paramagnetic-ferromagnetic transition.

5.4 Existence of Phase Transitions

The analysis of the Van der Waals gas in the last section showed that phase transitions lead to singularities in the thermodynamic functions. It is natural to ask how the partition function which appears to be an analytic function of its arguments develops a singularity. The answer lies in the limit $V \to \infty$. As we take the limit $V \to \infty$, singularities may develop as the limit function of a sequence of analytic functions need not be analytic.

Considering the grand partition function Q, we have

$$\frac{P}{k_B T} = \frac{1}{V} \ln Q \tag{5.4.1}$$

and

$$n = \frac{1}{V} z \frac{\partial}{\partial z} \ln Q \tag{5.4.2}$$

where $z = e^{\mu/k_B T}$. To have the possibility of singularities, we must take the thermodynamic limit $V \to \infty$ and write

$$\frac{P}{k_B T} = \lim_{V \to \infty} \frac{1}{V} \ln Q \tag{5.4.3}$$

$$n = \lim_{V \to \infty} \frac{1}{V} z \frac{\partial}{\partial z} \ln Q \tag{5.4.4}$$

with n fixed. In Eq.(5.4.4) the limit and the differentiation can be interchanged only if the convergence is uniform.

It was shown by Yang and Lee that the phase transitions are controlled by the zeros of the grand partition function in the complex z-plane. A transition occurs if a zero approaches the real axis in the limit $V \to \infty$. The two theorems governing the transition are:

Theorem 1: The limit $F_\infty(z) = \lim_{V \to \infty} \frac{1}{V} \ln Q$ exists for all $z > 0$ and is a continuous non-decreasing function of z and independent of the shape of the confining volume so long as the area of the surface that encloses the volume increases no faster than $V^{2/3}$ as the volume is increased.

Theorem 2: If R is a region of the complex z-plane containing a segment of the real axis where there are no zeros of Q, then in this region the quantity $\frac{1}{V} \ln Q$ converges uniformly to its limit as $V \to \infty$. The limit is analytic for all z in R.

A thermodynamic phase is defined by the values of z described in Theorem 2. In this region uniform convergence implies that limits can be interchanged in Eq.(5.4.4) and we have

$$\frac{P(z)}{k_B T} = F_\infty(z) \tag{5.4.5}$$

and

$$n = z\frac{\partial}{\partial z}F_\infty(z) \tag{5.4.6}$$

The equation of state can be obtained by graphical elimination of z (see Fig.5.8).

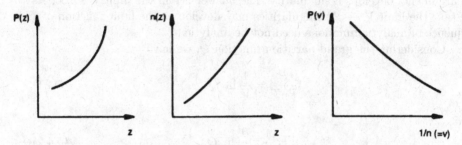

Figure 5.8: No zero of the partition function on the real axis and the corresponding single phase equation of state.

If the entire real axis is contained in R, then there is only one thermodynamic phase possible. Note that the fact that $n(z)$ is a non-decreasing function of z follows from

$$\begin{aligned}
z\frac{\partial}{\partial z}\{n(z)\} &= z\frac{\partial}{\partial z}\left(z\frac{\partial}{\partial z}\frac{1}{V}\ln Q\right)\\
&= \langle n^2\rangle - \langle n\rangle^2\\
&\geq 0 \tag{5.4.7}
\end{aligned}$$

If we now consider a situation where a zero of the grand partition function approaches a point z_0 on the real axis, then there are two thermodynamic phases, corresponding to distinct regions R_1 and R_2 (separated by z_0) where the conditions of Theorem 2 hold separately. At $z = z_0$, $F_\infty(z)$ and hence $P(z)$ are continuous by Theorem 1, but the derivative may be discontinuous and that leads to the two different phases for $z < z_0$ and $z > z_0$. Now since $n(z)$ is a non-decreasing function of z, it will have to make a finite jump at the point where $\partial P/\partial z$ is discontinuous, corresponding to a first-order transition. The situation is shown in Fig.5.9. The similarity of the P vs $1/n$ plot with the isotherm of the Van der Waals gas is obvious.

5.5   Correlation Functions

For dense gases and liquids a particularly important quantity is the structure function or the correlation function or what is also known as the distribution function. The distinguishing feature of these systems is the interaction between particles and if $V(\vec{r}_1, \vec{r}_2, \ldots, \vec{r}_N)$ be the general potential energy of the N-particle system, the probability of finding a particle at $\vec{r}_1$, another at $\vec{r}_2$, a third

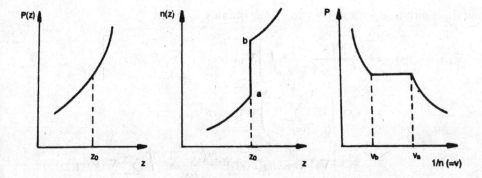

Figure 5.9: Van der Waals isotherm and the two-phase region.

at $\vec{r}_3$, and so on for all the N particles is given by the canonical distribution function $(1/Z_N)e^{-V(\vec{r}_1,\vec{r}_2,\ldots,\vec{r}_N)/k_BT}$.

To obtain the probability of finding a particle at the point $\vec{r}_1$, we need to integrate over all the other coordinates $(\vec{r}_2,\vec{r}_3,\ldots,\vec{r}_N)$ and thus the one-particle distribution function $n_1(\vec{r}_1)$ is defined as

$$n_1(r_1) = \frac{N}{Z_N}\int d^3r_2\ldots d^3r_N \exp\left[-\frac{V(\vec{r}_1,\vec{r}_2,\ldots,\vec{r}_N)}{k_BT}\right]. \qquad (5.5.1)$$

The two-particle distribution function or the two-point correlation function $n_2(\vec{r}_1,\vec{r}_2)$ is similarly defined by integrating out over the $(N-2)$ remaining coordinates as

$$\begin{aligned}n_2(\vec{r}_1,\vec{r}_2) &= (2!)\cdot\frac{N!}{2!(N-2)!}\cdot\frac{1}{Z_N}\int d^3r_3\ldots d^3r_N \\[2mm] &\times \exp\left[-\frac{V(\vec{r}_1,\vec{r}_2,\ldots,\vec{r}_N)}{k_BT}\right].\end{aligned} \qquad (5.5.2)$$

If we expect translational invariance and rotational invariance, then $n_2(\vec{r}_1,\vec{r}_2) = n_2(r_{12})$ where $r_{12} = |\vec{r}_1 - \vec{r}_2|$. As $r_{12} \to \infty$, $n_2 \to n_1^2$ since at large distances the particles are essentially independent of each other.

The immediate question is whether one can find a differential equation for $n_2(\vec{r}_1,\vec{r}_2)$ which would lead to a procedure for finding $n_2(\vec{r}_1,\vec{r}_2)$, if the potential is known.

Accordingly, one writes (assuming a pairwise interaction for the remainder of the section),

$$\begin{aligned}n_2(\vec{r}_1,\vec{r}_2) &= \frac{N(N-1)}{Z_N}\int \exp\left[-\beta\left\{\sum_{i\neq 1}V(r_{i1}) + \sum_{i\neq j\neq 1}V(r_{ij})\right\}\right] \\[2mm] &\times\ d^3r_3 d^3r_4\ldots d^3r_N\end{aligned} \qquad (5.5.3)$$

and differentiating with respect to $\vec{r}_1$ one obtains,

$$
\nabla_1 n_2(\vec{r}_1, \vec{r}_2) = -\frac{N(N-1)}{Z_N} \beta \int \left[\nabla_1 V(r_{12}) + \sum_{i \neq 1,2} \nabla_1 V(r_{i1}) \right]
$$

$$
\times \quad \exp\left(-\beta \sum_{pairs} V \right) d^3 r_3 \ldots d^3 r_n
$$

$$
= -\beta \left[\nabla_1 V(r_{12}) n_2(r_{12}) + \frac{N(N-1)}{Z_N} \int \sum_{i \neq 1,2} \nabla_1 V(r_{li}) \right.
$$

$$
\times \quad \left. \exp\left(-\beta \sum_{pairs} V \right) d^3 r_3 \ldots d^3 r_n \right]. \qquad (5.5.4)
$$

We now define the three-point correlation function or the three-particle distribution function as

$$
n_3(\vec{r}_1, \vec{r}_2, \vec{r}_3) = -\frac{N!}{(N-3)!} \frac{1}{Z_N} \int d^3 r_4 \ldots d^3 r_N \exp\left[-\beta \sum_{pairs} V(r_{ij}) \right] \qquad (5.5.5)
$$

and recognizing that the second term on the right hand side of Eq.(5.5.4) can be written as

$$
\frac{N(N-1)(N-2)}{Z_N} \int \nabla_1 V(r_{13}) d^3 r_3 \int d^3 r_4 \ldots d^3 r_N \cdot e^{-\beta \sum_{pairs} V}
$$

we arrive at

$$
\nabla_1 n_{12}(r_{12}) = -\beta \left[(\nabla_1 V(r_{12})) n_2(r_{12}) \right.
$$

$$
+ \quad \left. \int d^3 r_3 \nabla_1 V(r_{13}) n_3(\vec{r}_1, \vec{r}_2, \vec{r}_3) \right]. \qquad (5.5.6)
$$

This is the Born-Green equation and it illustrates a problem typical of all non-linear systems. The two-point correlation function n_2 is related to the three-point correlation function n_3. If we now want to find an equation for n_3, we would find that it would be involving n_4 and so on. This is the closure problem which is found in turbulence, or in any system described by a nonlinear Langevin equation. An approximate solution is to truncate the hierarchy at some point by invoking a relation between a higher correlation and lower correlations. One such approximation is that of Kirkwood, which assumes that

$$
n_3(\vec{r}_1, \vec{r}_2, \vec{r}_3) = n_2(\vec{r}_1, \vec{r}_2) \times n_2(\vec{r}_1, \vec{r}_3) \times n_2(\vec{r}_2, \vec{r}_3)/n_1^3. \qquad (5.5.7)
$$

Substituting for $n_3(\vec{r}_1, \vec{r}_2, \vec{r}_3)$ from Eq.(5.5.7) in Eq.(5.5.6), we may solve for n_2 – still a formidable task but nevertheless solvable.

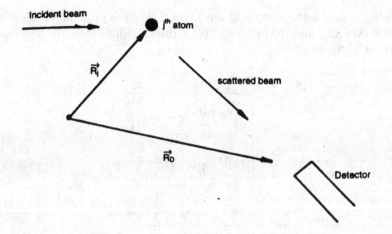

Figure 5.10: The scattering geometry.

The two-point correlation function can be determined fairly accurately experimentally by neutron or X-ray scattering from the gas or liquid. The typical experimental arrangement is shown in Fig.5.10.

The scattering amplitude from the jth atom (position vector $\vec{R}_j$)is

$$S_j = a(\theta)e^{i\vec{k}_{in}.\vec{R}_j}\frac{e^{ikR_{jD}}}{R_{jD}}. \tag{5.5.8}$$

In the above $\vec{k}_{in}$, is the momentum of the incoming beam and R_{jD} is the distance between the jth atom and the detector and

$$\vec{R}_{jD} = \vec{R}_D - \vec{R}_j \tag{5.5.9}$$

The angle θ is the angle between the directions of the incident and scattered beams.

The collision is elastic and hence the magnitude k of the momentum of the incoming beam is the same as that of the outgoing beam. Thus, (since $\vec{k}_{out}$ and $\vec{R}_{jD}$ are in the same direction)

$$kR_{jD} = \vec{k}.\vec{R}_{jD} = \vec{k}_{out}.(\vec{R}_D - \vec{R}_j) \tag{5.5.10}$$

and since the detector is very far away the magnitudes of $\vec{R}_{jD}$ and $\vec{R}_D$ are approximately equal, i.e.,

$$R_{jD} \cong R_D. \tag{5.5.11}$$

With the above results, the scattering amplitude may be written as

$$S_j \cong a(\theta)\frac{e^{i\vec{k}_{out}.\vec{R}_D}}{R_D}e^{i\vec{R}_j.(\vec{k}_{in}-\vec{k}_{out})}$$

$$= a(\theta)\frac{e^{i\vec{k}_{out}.\vec{R}_D}}{R_D}e^{i\vec{q}.\vec{R}_j} \tag{5.5.12}$$

where $\vec{q} = \vec{k}_{in} - \vec{k}_{out}$ is the scattering wave vector. For the total amplitude, we have to sum over all j and the intensity $I(\theta)$ is then obtained from the absolute magnitude of $\sum S_j$, i.e.,

$$
\begin{aligned}
I(\theta) &= \left| \sum S_j(\theta) \right|^2 \\
&= \left| a(\theta) \frac{e^{i\vec{k}.\vec{R}_D}}{R_D} \sum e^{i\vec{q}.\vec{R}_j} \right|^2 \\
&= |a(\theta)|^2 \cdot \frac{1}{R_D^2} \cdot |G|^2
\end{aligned}
\tag{5.5.13}
$$

where

$$
|G|^2 = \sum_i e^{i\vec{q}.\vec{R}_i} \sum_j e^{-i\vec{q}.\vec{R}_j} = \sum_{ij} e^{i\vec{q}.\vec{R}_{ij}}.
$$

For a liquid or gas, we need to average this $|G|^2$ over the probability of finding scatterers at $\vec{R}_i$ and $\vec{R}_j$ and hence the measured quantity is

$$
\begin{aligned}
\langle |G|^2 \rangle &= \sum_{pairs} \int d^3R_1 \ldots d^3R_N \frac{1}{Z_N} \exp\left[-\beta \sum_{pairs} V(\vec{R}_{ij}) + i\vec{q}.\vec{R}_{ij} \right] \\
&= \frac{N(N-1)}{2} \int d^3R_1 \ldots d^3R_N \frac{e^{i\vec{q}.\vec{R}_{12}}}{Z_N} \cdot e^{-\beta \sum_{pairs} V(\vec{R}_{ij})} \\
&= \int d^3R_1 d^3R_2 e^{i\vec{q}.\vec{R}_{12}} n_2(\vec{R}_1 - \vec{R}_2) \\
&= \int d^3R_{cm} d^3R_{12} e^{i\vec{q}.\vec{R}_{12}} n_2(\vec{R}_{12}) \\
&= \frac{4\pi V}{q} \int r n_2(r) \sin qr \, dr
\end{aligned}
\tag{5.5.14}
$$

where r is the magnitude of R_{12}, $\vec{R}_{cm} = (\vec{R}_1 + \vec{R}_2)/2$ and $q = 2k \sin \frac{\theta}{2}$.

In arriving at the final form we have assumed translational and rotational invariance which makes $n_2(\vec{R}_1, \vec{R}_2)$ a function of R_{12} alone. Thus, measuring the scattering intensity at different values of θ (i.e. different q) would allow one to invert the relation expressed in Eq.(5.5.14) and obtain the correlation function $n_2(r)$.

5.6 Ising Model

This is a model of interacting magnetic moments at various points in space. For convenience, we imagine a definite lattice in space (linear lattice in one dimension, square lattice in two dimensions, cubic in three dimensions and so on) and put classical spins (essentially magnetic moment vectors) on each site. Each spin has the same magnitude but can have different orientations. In the

Ising model, the spins can only point up or down. The interaction is short ranged – specifically we shall take them only to extend to nearest neighbours and it is such that the parallel alignment of spins is favoured. The energy function of this interacting system is taken to be

$$E = -\sum_{nn} J_{ij} S_i S_j \tag{5.6.1}$$

where S_i is the spin at site i and can have two values $+1$ and -1 and nn is an abbreviation for "nearest neighbour". J_{ij} is positive and exists only if i and j are two nearest neighbour positions. A simplifying assumption is to take $J_{ij} = J$, independent of the sites and then we have the standard Ising model energy expression

$$E = -J \sum_{nn} S_i S_j \tag{5.6.2}$$

If the whole system is put in an uniform external magnetic field h, then an additional term in the energy expression is required depicting the interaction of the spins with the magnetic field and we have

$$E = -J \sum_{nn} S_i S_j - h \sum S_i. \tag{5.6.3}$$

Exact solution of the model with $h = 0$ is straightforward in one dimension and difficult in two dimensions. In three dimensions, the model cannot be exactly solved. With $h \neq 0$, the solution in one dimension is again straightforward and in two dimensions somewhat more involved. We will present the exact solution in one dimension and present an approximate method for solving the problem in any dimensions.

The important question that one has to address is whether the model would show spontaneous magnetization (i.e., magnetization in the limit of $h \to 0$) at a finite temperature. Certainly at high temperatures the average magnetic moment would be zero, because the free energy would be dominated by the entropy term. But as the temperature is lowered, the energy term begins to dominate and it becomes increasingly difficult to turn the spin, so that it is possible that at low temperatures the spins will be aligned causing a phase transition from a state of zero average magnetization (paramagnet) to a state of non-zero average magnetization (ferromagnet).

5.6.1 Exact Solution in One Dimension

We need to evaluate the partition function

$$Z = \sum_{\{S_i\}} e^{\beta J \sum_i S_i S_{i+1}} \tag{5.6.4}$$

in the case of zero external field. This can be done by a trick. The product $S_i S_{i+1}$ can only acquire two values $+1$ or -1. So we define a new set of variables η, which are defined by

$$
\begin{aligned}
\eta_i &= +1 \text{ if } S_i S_{i+1} = 1 \\
\eta_i &= -1 \text{ if } S_i S_{i-1} = 1.
\end{aligned} \tag{5.6.5}
$$

The partition function can be written in terms of these variables as

$$
Z = \sum_{\eta_i = \pm 1} e^{\beta J \sum \eta_i} = \prod_i \sum_{\eta_i = \pm 1} e^{\beta J \eta_i} = (2 \cosh \beta J)^N. \tag{5.6.6}
$$

The free energy is $F = -k_B T \ln Z = -N k_B T \ln \cosh(\beta J)$

In the presence of an external magnetic field, the above trick fails and we resort to a technique which has very general applicability. We consider the problem with N sites and evaluate the partition function without summing over the spin S_N. Then,

$$
Z(S_N) = \sum_{S_1, S_2, \ldots, S_{N-1}} e^{\sum \beta J S_i S_{i+1} + h\beta \sum S_i}. \tag{5.6.7}
$$

We put another site $(N+1)$ and perform the sum over the spin at site N to write

$$
Z(S_{N+1}) = \sum_{S_N} e^{\beta J S_N S_{N+1}} Z_N(S_N) e^{\beta h (S_N + S_{N+1})}. \tag{5.6.8}
$$

Now, S_N can acquire two values $+1$ and -1 and hence we can write Z_N as a two-component vector $Z_N(+)$ and $Z_N(-)$. Clearly the relation between Z_{N+1} and Z_N is given by

$$
\begin{aligned}
Z_{N+1}(+) &= e^{\beta J + 2\beta h} Z_N(+) + e^{-\beta J} Z_N(-) \\
Z_{N+1}(-) &= e^{-\beta J} Z_N(+) + e^{\beta J - 2\beta h} Z_N(-)
\end{aligned} \tag{5.6.9}
$$

or in matrix form

$$
Z_{N+1}^\alpha = \sum_\beta T_{\alpha\beta} Z_N^\beta \tag{5.6.10}
$$

where $T_{\alpha\beta}$, called the transfer matrix for obvious reasons, is given by

$$
T = \begin{pmatrix} e^{\beta J + 2\beta h} & e^{-\beta J} \\ e^{-\beta J} & e^{\beta J - 2\beta h} \end{pmatrix} \tag{5.6.11}
$$

The matrix T is a real and symmetric and hence can be diagonalized and its eigenvectors will span the two-dimensional vector space we are dealing with. If we call the two eigenvalues λ_1 and λ_2 and the associated eigenvectors $\begin{pmatrix} \alpha_1 \\ \alpha_2 \end{pmatrix}$ and $\begin{pmatrix} \beta_1 \\ \beta_2 \end{pmatrix}$ then clearly we can write

$$
\begin{pmatrix} Z_1^+ \\ Z_1^- \end{pmatrix} = c_1 \begin{pmatrix} \alpha_1 \\ \alpha_2 \end{pmatrix} + c_2 \begin{pmatrix} \beta_1 \\ \beta_2 \end{pmatrix}. \tag{5.6.12}
$$

From Eq.(5.6.10) it is obvious that

$$\begin{pmatrix} Z_{N+1}^+ \\ Z_{N+1}^- \end{pmatrix} = (T)^N Z_1 = \lambda_1^N c_1 \begin{pmatrix} \alpha_1 \\ \alpha_2 \end{pmatrix} + \lambda_2^N c_2 \begin{pmatrix} \beta_1 \\ \beta_2 \end{pmatrix}. \tag{5.6.13}$$

and the partition function

$$Z_{N+1} = Z_{N+1}^+ + Z_{N+1}^- = \lambda_1^N(c_1\alpha_1 + c_1\alpha_2) + \lambda_2^N(c_2\beta_1 + c_2\beta_2). \tag{5.6.14}$$

In the thermodynamic limit only the larger eigenvalue survives and if $\lambda_1 > \lambda_2$, we have for $N \gg 1$

$$Z_{N+1} = \lambda_1^N(c_1\alpha_1 + c_1\alpha_2) \tag{5.6.15}$$

and the free energy

$$F = -Nk_BT \ln \lambda_1. \tag{5.6.16}$$

In the present case the two eigenvalues of T of Eq.(5.6.11) are

$$\lambda_{1,2} = e^{\beta J} \cosh 2\beta h \pm [e^{-2\beta J} + e^{2\beta J} \sinh^2 2\beta h]^{1/2} \tag{5.6.17}$$

and hence

$$F = -\frac{N}{\beta} \ln \left[e^{\beta J} \cosh 2\beta h + \sqrt{e^{-2\beta J} + e^{2\beta J} \sinh^2 2\beta h} \right]. \tag{5.6.18}$$

The magnetization M is given by

$$M = -\frac{\partial F}{\partial h} = N \frac{2e^{\beta J} \sinh 2\beta h}{\sqrt{e^{-2\beta J} + e^{2\beta J} \sinh^2 2\beta h}}. \tag{5.6.19}$$

As $h \to 0$,

$$M \cong 4\beta N h e^{2\beta J} \tag{5.6.20}$$

showing that there can be no spontaneous magnetization as $h \to 0$ unless $T \to 0$. The susceptibility is given by

$$\chi = \left(\frac{\partial M}{\partial h}\right)_{h=0} = \frac{4N}{(k_BT)} e^{2J/k_BT} \tag{5.6.21}$$

Thus as $T \to 0$, the susceptibility shows a singularity. Hence a phase transition occurs in the one dimensional Ising model at $T = 0$. Later we will show that in two dimensions a phase transition occurs at a finite temperature.

5.6.2 Approximate Method for Any Dimensions

We will present the technique for $h = 0$, the generalization to $h \neq 0$ being straightforward. In this approximation, we first write the energy expression as $E = -J \sum S_i S_j = -\sum_i h_i S_i$, where $h_i = J \sum_{j \neq i}$ is the field at the site i due to all the other spins. At this point there is no approximation – it comes when we say that h_i is the same at every site i equal to some average $h = \langle h_i \rangle = J\alpha\langle m \rangle$, m being the average magnetization and α a constant. The constant α is $2D$ where D is the dimensionality of the lattice. The energy expression is now

$$E = -h \sum s_i \tag{5.6.22}$$

and the partition function can be trivially evaluated. In the above, h is an unknown quantity. So it has to be consistently fixed by determining the magnetization m from the partition function. We now have an explicit equation for m and the approximation is operationally complete. This is called the 'mean field approximation' and has a wide range of applicability as a first crack at a difficult problem involving interactions. For the energy of Eq.(5.6.22), the mean magnetization m is

$$m = \frac{e^{\beta h} - e^{-\beta h}}{e^{\beta h} + e^{-\beta h}} = \tanh\ \beta h = \tanh\ (\alpha\beta Jm). \tag{5.6.23}$$

The desired equation for m has been obtained.

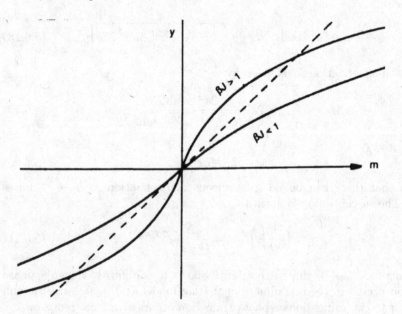

Figure 5.11: Solution of the mean field equation.

The solution has to be obtained graphically as shown in Fig.5.11. The dashed line is the plot $y = m$ and the solid curves correspond to $y = \tanh \alpha\beta Jm$ for various values of βJ. Clearly for $\alpha\beta J < 1$, there is only one solution, namely $m = 0$. However for $\alpha\beta J > 1$ there is a second solution with $m \neq 0$.

It m is very small, in other words $\alpha\beta J$ greater than but very close to unity, the hyperbolic tangent in Eq.(5.6.23) can be expanded to give

$$m = \alpha\beta Jm - \frac{1}{3}\alpha^3(\beta Jm)^3 m^3 + \ldots$$

or

$$\frac{\alpha^3(\beta J)^3}{3}m^3 = m(\alpha\beta J - 1). \qquad (5.6.24)$$

For $\beta J < 1$, clearly the only solution is $m = 0$, but for $\beta J > 1$, one has the non-zero value of m given by

$$
\begin{aligned}
m^2 &= \frac{3}{\alpha^3(\beta J)^3}(\alpha\beta J - 1) \\
&= 3\left(\frac{\beta}{\beta_c} - 1\right) \\
&= 3\left(\frac{T_c - T}{T}\right) \\
&\cong 3\left(\frac{T_c - T}{T_c}\right).
\end{aligned}
\qquad (5.6.25)
$$

where $\alpha\beta_c J = 1$ defines the critical temperature, below which a solution with $m \neq 0$ exists.

That this is the favoured phase for $T < T_c$, can be established only after the free energy has been constructed. This will be dealt with in one of the problems at the end of the chapter. For the moment, we accept that when it exists, the solution with $m \neq 0$ is the stable solution and hence we have a second-order phase transition (no latent heat) at $T - T_c$, with $m = 0$ (paramagnet) above T_c, and $m \neq 0$ (ferromagnet) below T_c. Thus the magnetization is the order parameter for the transition and its temperature dependence just near the transition is given by

$$m \propto (T_c - T)^{1/2}. \qquad (5.6.26)$$

Once again a singular behaviour and a universal behaviour independent of the material that is getting magnetized are seen. In fact, the even wider universality may be noticed by comparing with Eq.(5.3.13) describing the order parameter for the Van der Waals gas near the critical point. There too the behaviour is $(T_c - T)^{1/2}$.

The approximation described above is independent of spatial dimensionality. Hence this approximation would lead to a phase transition at finite temperature even in one dimension which we have seen above, does not happen. Thus the approximation can be quite faulty. However, if a transition is known to occur, then this approximation is a good way of describing it. As a rule of thumb, it is a bad approximation in low dimensions but good in higher dimensions.

5.7   Problems for Chapter 5

1. The pair correlation function $D(\vec{r}_1, \vec{r}_2)$ of the system of particles is defined as follows: $D(\vec{r}_1, \vec{r}_2)d^3 r_1 d^3 r_2 = $ probability of simultaneously finding a particle the volume element $d^3 r_1$ about $\vec{r}_1$ and a particle in the volume element $d^3 r_2$ about $\vec{r}_2$. For a gas of classical particles with pairwise interaction (repulsive at very short range and attractive but decaying fast at long range), find the correlation function to the lowest non-zero order in the interaction. The density of the gas is very low.

2. An ensemble of 10^{10} weakly interacting spinless particles, each with the mass of an electron, are identical in appearance but obey classical statistics. They are confined in a critical box which is 10^{-6}cm on an edge. Each particle undergoes a potential interaction with the box which is of two sorts. One is attractive and leads to a bound state well localized near the centre of the box and having an energy of -1eV. The other interaction is a strong repulsive one which prevents the particles from escaping through the walls of the box. Find at what temperature the pressure in the box is 1atm.

3. Long molecules, each of length l are constrained to lie at equal separations along a one-dimensional chain. The molecules are capable of two orientations: parallel or perpendicular to the chain. The nearest neighbour interaction energy is $-\epsilon$ if one is parallel and the other perpendicular and $+\epsilon$ if both are perpendicular to the chain length. If the total number of molecules is N, find the average length of the chain as a function of temperature.

4. Consider the one-dimensional Ising model at temperatures close to $T = 0$. At $T = 0$, all the spins are aligned. For very low temperatures, there will be a few breaks in the aligned state. The following figure shows a situation with two breaks.

Figure 5.12: Figure for problem 4.

Each break costs an energy J. The statistical mechanics of the Ising model for $T = 0$ can be described as the statistical mechanics of a dilute gas of breaks (non-interacting). Use this model to determine the correlation function of the Ising model as $T \to 0$.

5. Consider the approximate treatment for the magnetic transition carried out in Sec.5.6. For $T < T_c$ show by explicit calculation of the free energy that the state with $m \neq 0$ will be favoured.

6. If the Ising spins are put in a amount external magnetic field H, then Eq.(5.6.22) becomes $E = -h + H \sum S_i$. Write down the equivalent of

Eq.(5.6.23). For both $T > T_c$ and $T < T_c$, show that the Isothermal zero field susceptibility diverges as $|T - T_c|^{-1}$. The amplitudes of the divergent temperature dependence are different in the two cases. Find the ratio of the two amplitudes. At $T = T_c$ find the ratio between m and H as $H \to 0$.

7. For the weakly interacting monatomic gas of Sec.5.1, write down the expression for total energy. From that find the first correction to the constant volume specific heat of the ideal gas.

6

Quantum Statistics

6.1 Introduction

Our discussion so far has ignored quantum correlations among particles. This is valid if the temperature is relatively high so that the De Broglie wavelength is much smaller than the relevant length scales or the density is low enough so that inter-particle separation, a_0, is much greater than the atomic size.

We now remove the restriction of low density and high temperatures which means that the De Broglie wavelength $(= h/\sqrt{mkT})$ becomes of the order of $a_0 \approx n^{-1/3}$, $(n = N/V)$. The quantum nature of the particles is now going to become effective and the necessary restrictions have to be taken into account. For a system of identical particles this restriction amounts to the symmetry under interchange of particles in the quantum mechanical many-body wave function. Since two successive interchanges of any two particles (say in single particle states i and j) reproduces the same many-body state (so far as probabilities go), a particular exchange either keeps the state unaltered (symmetric wave function) or introduces a change of sign (antisymmetric wave function). If the wave function changes sign, then clearly there cannot be two particles in the same quantum state i, since under exchange the same wavefunction would be reproduced but with changed sign. This can only happen if the wave function vanishes identically. Thus Pauli's exclusion principle is obeyed if the wavefunction is antisymmetric and particles with such wavefunctions are called fermions. If the wavefunction is symmetric, then there is no restriction on the number of particles in a particular state and such particles are called bosons. The spin statistics theorem relates the intrinsic spin of a particle to its symmetry property and thus half integer spin particles are fermions while integer spin ones are bosons. Electrons and nucleons are fermions having spin 1/2, while pions (spin 0), photons (spin 1) etc. are bosons.

6.2   Quantum Distribution Functions

We consider a quantum particle capable of existing in any one of a series of single particle quantum states labelled by i ($i = 1, 2, 3, \ldots$). The energy of the i-th state is ϵ_i.

A collection of such non-interacting particles in equilibrium at a temperature T is considered and we want to find the average number $\langle n_i \rangle$ of particles in the i-th state. If the number of particles is n_i in any given configuration, then the total number of particles is

$$N = \sum n_i \tag{6.2.1}$$

and the total energy is

$$E = \sum \epsilon_i n_i \tag{6.2.2}$$

(In this energy expression, we are making use of the vital assumption of the absence of interaction among the particles). To find the average value of n_i, it is best to consider one of the standard ensembles, evaluate the partition function and thence obtain $\langle n_i \rangle$. The grand canonical ensemble is best suited for our purpose since in that case neither E nor N need to be held constant. Unconstrained sums are easiest to do and hence we evaluate the grand canonical partition function

$$Q = \sum_{E,N} e^{-\beta(E-\mu N)} \tag{6.2.3}$$

where μ is the chemical potential. Using Eqs.(6.2.1) and (6.2.2)

$$Q = \sum_{\epsilon_i, n_i} e^{-\beta \sum_i (\epsilon_i n_i - \mu n_i)}. \tag{6.2.4}$$

It is convenient to do the sum over n_i first and write

$$Q = \prod_{\epsilon_i} Q^{(i)} = \prod \left[\sum_{n_i} e^{-\beta n_i (\epsilon_i - \mu)} \right] \tag{6.2.5}$$

If the particles are fermions, then each n is restricted to be zero or one and

$$Q_F^{(i)} = \sum_{n_i} e^{-\beta(\epsilon_i - \mu)n_i} = 1 + e^{-\beta(\epsilon_i - \mu)}. \tag{6.2.6}$$

For bosons on the other hand, n_i can range from zero to infinity and we have

$$Q_B^{(i)} = \sum_{n_i=0}^{\infty} e^{-\beta(\epsilon_i - \mu)n_i} = \left[1 - e^{-\beta(\epsilon_i - \mu)} \right]^{-1}. \tag{6.2.7}$$

Subsequently, the average number of particles with energy ϵ_i per quantum state is obtained using an expression similar to the last equation of Section 3.3.

6.2.1 Fermions

For fermions

$$\langle n_i \rangle = k_B T \frac{\partial}{\partial \mu} \ln Q_F^{(i)}$$

$$= \frac{1}{e^{\beta(\epsilon_i - \mu)} + 1}. \tag{6.2.8}$$

It is instructive to picture the average occupation per state for different temperatures. The most important case is $T = 0$, i.e., $\beta = \infty$. In this case for $\epsilon_i < \mu$, we have $e^{\beta(\epsilon_i - \mu)} \to 0$ and hence there is on the average one particle state in this energy range. On the other hand, for $\epsilon_i > \mu$, yields $e^{\beta(\epsilon_i - \mu)} \to \infty$ and hence $\langle n_i \rangle \to 0$. Thus all the particles are accommodated in the states below the chemical potential μ and the distribution is pictured in Fig.6.1.

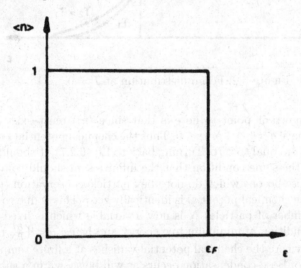

Figure 6.1: Fermi distribution at $T = 0$ given by a plot of $\langle n \rangle$ vs. ϵ.

The chemical potential μ at $T = 0$ is called the fermi energy ϵ_F. For temperatures $T > 0$, the distribution develops a tail for $\epsilon_i > \mu$ with $\langle n_j \rangle = 1/2$ at $\epsilon_i = \mu$. As the temperature increases, the tail becomes more and more pronounced as shown in Fig.6.2.

6.2.2 Bosons

In the case of bosons

$$\langle n_i \rangle = k_B T \frac{\partial}{\partial \mu} \ln Q_B^{(i)}$$

$$= \frac{1}{e^{\beta(\epsilon_i - \mu)} - 1}. \tag{6.2.9}$$

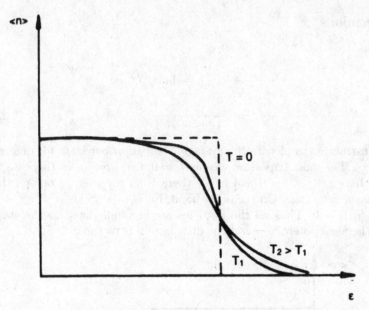

Figure 6.2: Fermi distribution at $T \neq 0$.

Here the important point to note is that since $\langle n_i \rangle$ must exist even for $\epsilon_i = 0$, one must have $e^{-\beta \mu} \geq 1$ or $\mu \leq 0$. Thus the chemical potential for bosons must be less than or equal to zero. Turning back to Eq.(6.2.7), it should now be clear that this is the same condition that the infinite sum should converge for all $\epsilon_i \geq 0$. For those bosons which do not obey particle conservation (photons, phonons, etc.), the chemical potential is identically zero. This is due to the fact that the total number of particles N is now a variable which is fixed in equilibrium by the condition of minimum free energy and hence $\mu = \left(\frac{\partial F}{\partial N} \right)_{V,T} = 0$. For those bosons where the chemical potential vanishes at a finite temperature, the phenomenon of Bose-condensation occurs as will be shown in a subsequent chapter.

We end this section by writing down the number of particles between energy ϵ and $\epsilon + d\epsilon$ in three dimensional space. The particles are taken to be free (i.e., no interaction and no confining potential) and hence the number of quantum states per unit energy interval denoted by $\frac{dg}{d\epsilon}$ is

$$\frac{dg}{d\epsilon} = (2s+1) \frac{2\pi V}{h^3} (2m)^{3/2} \frac{\epsilon^{1/2}}{e^{\beta(\epsilon-\mu)} \mp 1} \qquad (6.2.10)$$

where s is the spin of the particle and hence number of particles $n(\epsilon)$ in the energy range $d\epsilon$ is

$$n(\epsilon)d\epsilon = (2s+1) \frac{2\pi V}{h^3} (2m)^{3/2} \frac{\epsilon^{1/2}}{e^{\beta(\epsilon-\mu)} \mp 1} d\epsilon \qquad (6.2.11)$$

with the upper sign holding for fermions and the lower sign for bosons.

6.3 Equation of State

Before establishing the equation of state, we will prove a relationship which is independent of statistics – this relates the pressure volume product to the total energy. Using the standard result of grand canonical ensemble, we have

$$PV = k_B T \sum_i \ln Q_{F,B}^{(i)}$$

$$= \pm k_B T (2s+1) \frac{2\pi V (2m)^{3/2}}{h^3}$$

$$\times \int_0^\infty \ln\left(1 \pm e^{-(\epsilon-\mu)/k_B T}\right) \epsilon^{1/2} d\epsilon \qquad (6.3.1)$$

where the upper sign holds for fermions while the lower sign holds for bosons. Integrating by parts the integral on the right hand side of Eq.(6.3.1), we arrive at

$$PV = \pm k_B T (2s+1) \frac{2\pi V (2m)^{3/2}}{h^3} \left\{ \left[\frac{2}{3}\epsilon^{3/2} \ln\left(1 \pm e^{-(\epsilon-\mu)/k_B T}\right)\right]_0^\infty \right.$$

$$+ \frac{2}{3} \int_0^\infty \left[\frac{\pm e^{(\epsilon-\mu)/k_B T}}{1 \pm e^{-(\epsilon-\mu)/k_B T}} \cdot \frac{\epsilon^{3/2} d\epsilon}{k_B T} \right] \right\}$$

$$= \frac{2}{3}(2s+1) \frac{2\pi V (2m)^{3/2}}{h^3} \int_0^\infty \frac{\epsilon^{3/2} d\epsilon}{e^{(\epsilon-\mu)/k_B T} \pm 1}$$

$$= \frac{2}{3} \int_0^\infty \epsilon n(\epsilon) d\epsilon$$

$$\Rightarrow PV = \frac{2}{3} E \qquad (6.3.2)$$

where E is the average energy of the system. The above result, as should be clear from the derivation, holds irrespective of whether the particles are fermions or bosons – in fact by comparing with Eqs.(4.1.7) and (4.1.8) it can be seen that it holds for the classical ideal gas as well. This is expected from Eq.(6.3.2) since it is a result true regardless of values of T and μ, and hence must be valid at large negative values of μ, which corresponds to the classical ideal gas. To obtain the equation of state, we need in addition to Eq.(6.3.1), an equation for the total number of particles N,

$$N = \frac{\partial}{\partial \mu}(PV)$$

$$= \pm(2s+1) \frac{2\pi V}{h^3} (2m)^{3/2} \int_0^\infty \frac{(\pm)\epsilon^{3/2} d\epsilon}{e^{(\epsilon-\mu)/k_B T} \pm 1}. \qquad (6.3.3)$$

Our purpose will be to eliminate the chemical potential between Eqs.(6.3.1) and (6.3.3) and obtain a relation of the form

$$\frac{PV}{Nk_BT} = 1 + \sum_{l=1}^{\infty} A_l n^l \qquad (6.3.4)$$

where $n = N/V$ is the number density.

We proceed by assuming that μ is negative and hence a binomial expansion of $(1 \pm e^{-(\epsilon-\mu)/k_BT})^{-1}$ is possible. This allows us to write Eq.(6.3.3) as

$$
\begin{aligned}
N &= (2s+1)\frac{2\pi V}{h^3}\cdot(2m)^{3/2} \int_0^{\infty} \frac{\epsilon^{1/2}d\epsilon}{e^{(\epsilon-\mu)/k_BT} \pm 1} \\
&= (2s+1)\frac{2\pi V}{h^3}\cdot(2m)^{3/2} \int_0^{\infty} e^{-(\epsilon-\mu)/k_BT} \\
&\times \left[1 \mp e^{(\mu-\epsilon)/k_BT} + e^{2(\mu-\epsilon)/k_BT} \mp \dots\right] \epsilon^{1/2}d\epsilon \\
&= (2s+1)\frac{2\pi V}{h^3}\cdot(2m)^{3/2} \int_0^{\infty} e^{(\mu-\epsilon)/k_BT} \sum_{j=0}^{\infty} (\mp)^j e^{j(\mu-\epsilon)/k_BT} \epsilon^{1/2}d\epsilon \\
&= (2s+1)\frac{2\pi V}{h^3}\cdot(2mk_BT)^{3/2} e^{\mu/k_BT} \sum_{j=0}^{\infty} (\mp)^j \frac{e^{j\mu/k_BT}}{(j+1)^{3/2}}\Gamma\left(\frac{3}{2}\right) \\
&= (2s+1)V\left(\frac{2\pi mk_BT}{h^2}\right)^{3/2} e^{\mu/k_BT} \sum_{j=0}^{\infty} \frac{(\mp)^j e^{j\mu/k_BT}}{(j+1)^{3/2}}. \qquad (6.3.5)
\end{aligned}
$$

Identifying the De Broglie wavelength as $\lambda_d = \frac{h}{(2\pi mk_BT)^{1/2}}$, we can write Eq.(6.3.5) as

$$e^{\mu/k_BT} \sum_{j=0}^{\infty} \frac{(\mp)^j e^{j\mu/k_BT}}{(j+1)^{3/2}} = \frac{n}{2s+1}\lambda_d^3 = n\lambda^3 \qquad (6.3.6)$$

(we have written $\lambda_d = (2s+1)^{1/3}\lambda$).

Now, if $n\lambda^3 \ll 1$, which is attained if the density is very low (small n) or λ is very small (high T), then the left hand side of Eq.(6.3.6) must be much smaller than unity and hence we must have $e^{\mu/k_BT} \ll 1$, which requires μ to be large and negative. We proceed by writing Eq.(6.3.6) as

$$e^{\mu/k_BT}\left\{1 \mp \frac{e^{\mu/k_BT}}{2^{3/2}} + \frac{e^{2\mu/k_BT}}{3^{3/2}} \mp \dots\right\} = n\lambda^3. \qquad (6.3.7)$$

Expanding

$$e^{\mu/k_BT} = n\lambda^3 + \sum_{l=2}^{\infty} B_l(n\lambda^3)^l \tag{6.3.8}$$

and inserting in Eq.(6.3.7) we can write,

$$
\begin{aligned}
n\lambda^3 &= n\lambda^3[1 + B_2(n\lambda^3) + \ldots] \\
&\times \left\{ 1 \mp \frac{n\lambda^3}{2^{3/2}}(1 + B_2 n\lambda^3 + \ldots) + \frac{1}{3^{3/2}}(n\lambda^3)^2 + \ldots \right\}. \tag{6.3.9}
\end{aligned}
$$

Matching terms of equal power of $n\lambda^3$ on either side of Eq.(6.3.9), we find

$$B_2 = \pm\frac{1}{2^{3/2}}$$

by working to $O(n\lambda^3)^2$. Higher-order matchings would lead to a systematic determination of the higher-order coefficients. To this order, we have

$$e^{\mu/k_BT} = n\lambda^3 \left[1 \pm \frac{n\lambda^3}{2^{3/2}} + \ldots \right]. \tag{6.3.10}$$

To find the equation of state, we make use of Eq.(6.3.2), and putting $x = e^{-\epsilon/k_BT}$ as we go along, we obtain

$$
\begin{aligned}
\frac{PV}{Nk_BT} &= \frac{2}{3}\frac{E}{Nk_BT} \\
&= \frac{2}{3}(2s+1)\frac{2\pi V(2m)^{3/2}}{h^3 Nk_BT} \int_0^{\infty} \frac{\epsilon^{3/2}d\epsilon}{e^{(\epsilon-\mu)/k_BT} \pm 1} \\
&= \frac{2}{3}(2s+1)\frac{2\pi V(2m)^{3/2}}{h^3 Nk_BT} \int_0^{\infty} e^{-(\epsilon-\mu)/k_BT} \\
&\times \left[1 \mp e^{-(\epsilon-\mu)/k_BT} + \ldots \right] \epsilon^{3/2}d\epsilon \\
&= \frac{2}{3}(2s+1)\frac{2\pi V(2mk_BT)^{3/2}}{Nh^3}.e^{\mu/k_BT} \int_0^{\infty} \sum_{j=0}^{\infty} \frac{(\mp)^j e^{\mu j/k_BT}}{(j+1)^{5/2}} x^{3/2}e^{-x}dx \\
&= (2s+1)\frac{V}{N}\left(\frac{2\pi mk_BT}{h^2}\right)^{3/2} e^{\mu/k_BT} \sum_{j=0}^{\infty} \frac{(\mp)^j e^{\mu j/k_BT}}{(j+1)^{5/2}} \\
&= \frac{e^{\mu/k_BT}}{n\lambda^3} \sum_{j=0}^{\infty} (\mp)^j \frac{e^{\mu j/k_BT}}{(j+1)^{5/2}} \\
&= \frac{e^{\mu/k_BT}}{n\lambda^3}\left(1 \mp \frac{e^{\mu/k_BT}}{2^{5/2}} + \ldots \right). \tag{6.3.11}
\end{aligned}
$$

Using the fugacity expansion from Eq.(6.3.10), we arrive at the equation of state

$$\begin{aligned}
\frac{PV}{Nk_BT} &= \left(1 \pm \frac{n\lambda^3}{2^{3/2}} + \dots\right)\left(1 \pm \frac{n\lambda^3}{2^{5/2}} + \dots\right) \\
&= 1 \pm \frac{n\lambda^3}{2^{3/2}} \pm \frac{n\lambda^3}{2^{5/2}} - O\left((n\lambda^3)^2\right) \\
&= 1 \pm \frac{3n\lambda^3}{2^{5/2}} + O\left[(n\lambda^3)^2\right].
\end{aligned} \tag{6.3.12}$$

Clearly, we have obtained an equation of state in the desired form of Eq.(6.3.4), with

$$A_1 = \pm\frac{3\lambda^3}{2^{5/2}} \tag{6.3.13}$$

The correction to the classical ideal gas now is, as expected, quantum mechanical in nature and determined by the De Broglie wavelength. For high temperatures and low densities, when quantum effects are expected to be unimportant, the classical equation of state is obtained. Also, as expected the pressure of fermions increases, and that of bosons decreases.

6.4   Entropy of the Quantum Ideal Gas

We will carry out the calculation of the entropy in the microcanonical ensemble. This will illustrate the use of Boltzmann's equation $S = k \ln \Gamma(E, N)$ in the quantum mechanical case. The energy E is now almost fixed (allowing for only a very small spread) and the total number of particles N is fixed as well. As usual, the non-interacting particles are enclosed in an extremely large box which causes the momentum to be quantized and thus the energy levels ϵ_i, (labelled by momentum) are very closely spaced. If there are n_i particles in the level ϵ_i, then

$$\sum n_i = N \quad \text{(fixed)} \tag{6.4.1}$$

$$\sum \epsilon_i n_i = E \quad \text{(fixed)} \tag{6.4.2}$$

First, we need to find out the number of ways in which the total number of particles can be arranged to give n_i, particles in the ith level. We call this $W(\{n_i\})$. Clearly,

$$\Gamma(E, N) = \sum_{\{n_i\}} W(\{n_i\}). \tag{6.4.3}$$

To find $W(\{n_i\})$, we need to follow a standard trick. The energy ϵ_i is taken to be the mean energy of a bunch of levels centred around ϵ_i and g_i in number. This sounds quite semiclassical and supports the point of view that statistical mechanics is the quantum mechanics of macroscopic systems. These g_i levels are clustered together forming a cell whose occupation number is the number n_i introduced. We now consider $W(n_i, g_i)$, the number of ways in which n_i particles are distributed among the g_i states in the cell. Each arrangement within a cell is independent of the arrangements within the other cells and hence

$$W(\{n_i\}) = \prod_i W(n_i, g_i). \tag{6.4.4}$$

The determination of $W(n_i, g_i)$ is statistics dependent.

i) Fermions: To each level g_i, there can be at most one particle. So n_i of the g_i levels will be occupied in a given arrangement. The total number of arrangements are obviously equal to the number of ways in which n_i things can be chosen from a total g_i and hence

$$W^F(n_i, g_i) = \frac{g_i!}{n_i!(g_i - n_i)!} \tag{6.4.5}$$

ii) Bosons: To each level there can be upto n_i, particles in this case. We consider the $(g_i - 1)$ partitions of the g_i levels and together with the n_i particles, we have a collection of $(n_i + g_i - 1)$ objects. The $(g_i - 1)$ partitions can be placed anywhere interspersed between the n_i particles and the total number of configurations obtained therefore will give $W^B(g_i, n_i)$. The total number of arrangements is obviously the total number of permutations of the $(n_i + g_i - 1)$ objects, which is equal to $(n_i + g_i - 1)!$. But of these the n_i identical particles and the $(g_i - 1)$ identical partitions can arrange themselves in $n_i!$ and $(g_i - 1)!$ ways respectively without giving a new arrangement. So the number of independent configurations is

$$W^B(g_i, n_i) = \frac{(n_i + g_i - 1)!}{n_i!(g_i - 1)!} \tag{6.4.6}$$

The calculation of Γ is through the relation of Eq.(6.4.3) which yields

$$\begin{aligned} \Gamma^F &= \sum_{\{n_i\}} \prod_i W^F \\ &= \sum_{\{n_i\}} \prod_i \frac{g_i!}{n_i!(g_i - n_i)!} \end{aligned} \tag{6.4.7}$$

and similarly

$$\Gamma^B = \sum_{\{n_i\}} \prod_i \frac{(n_i + g_i - 1)!}{n_i!(g_i - 1)!}. \tag{6.4.8}$$

The summation over $\{n_i\}$ keeping the constraints of Eqs.(6.4.1) and (6.4.2) is impossible and hence we replace the sum by the single term $\{n_i\} = \{\bar{n}_i\}$, where $\{\bar{n}_i\}$ is that particular set which maximizes W under the given constraints. This clearly follows from the fast that the equilibrium distribution which is the one that maximizes Γ is overwhelmingly probable. It is more convenient to maximize $\ln W$ and hence, for fermions, we have using Stirling's approximation, viz., $\ln n! = n \ln n - n$,

$$
\begin{aligned}
\ln W^F &= \sum_i [\ln g_i! - \ln n_i! - \ln (g_i - n_i)!] \\
&= \sum_i [g_i \ln g_i - g_i - n_i \ln n_i + n_i - (g_i - n_i) \ln (g_i - n_i) + g_i - n_i] \\
&= \sum_i \{g_i [\ln g_i - \ln (g_i - n_i)] - n_i [\ln n_i - \ln (g_i - n_i)]\}.
\end{aligned}
$$

Differentiation with respect to n_i leads to

$$
\begin{aligned}
\delta(\ln W^F) &= \sum_i \left[\frac{(g_i - n_i)}{g_i - n_i} \delta n_i + \ln (g_i - n_i) \delta n_i - \ln n_i \delta n_i \right] \\
&= \sum_i \ln \left(\frac{g_i}{n_i} - 1 \right) \delta n_i.
\end{aligned}
\tag{6.4.9}
$$

Similarly,

$$
\begin{aligned}
\ln W^B &= \sum_i [(n_i + g_i - 1) \ln (n_i + g_i - 1) - (n_i + g_i - 1) \\
&\quad - n_i \ln n_i + n_i - (g_i - 1) \ln (g_i - 1) + g_i - 1] \\
&= \sum_i [(n_i + g_i - 1) \ln (n_i + g_i - 1) \\
&\quad - n_i \ln n_i - (g_i - 1) \ln (g_i - 1)]
\end{aligned}
\tag{6.4.10}
$$

leading to

$$
\begin{aligned}
\delta(\ln W^B) &= \sum_i [\ln (n_i + g_i - 1) \delta n_i - \ln n_i \delta n_i] \\
&= \sum_i \ln \left(1 + \frac{g_i - 1}{n_i} \right) \delta n_i.
\end{aligned}
\tag{6.4.11}
$$

To take into account the constraints $\sum n_i = N$ and $\sum \epsilon_i n_i = E$ (with N and E fixed), we need to write them in the differential form

$$
\sum_i \delta n_i = 0
\tag{6.4.12}
$$

$$
\sum_i \epsilon_i \delta n_i = 0.
\tag{6.4.13}
$$

Introducing Lagrange multipliers λ_1 and λ_2 we have for fermions

$$
\sum_i \left(\ln \left(\frac{g_i}{n_i} - 1 \right) - \lambda_1 - \lambda_2 \epsilon_i \right) \delta n_i = 0
\tag{6.4.14}
$$

and for bosons

$$\sum_i \left(\ln \left(1 + \frac{g_i - 1}{n_i} \right) - \lambda_1 - \lambda_2 \epsilon_i \right) \delta n_1 = 0. \tag{6.4.15}$$

The maximizing set $\{\bar{n}_i\}$ for fermions satisfy

$$\frac{g_i}{n_i} = 1 + e^{\lambda_1 + \lambda_2 \epsilon_1}$$

or

$$\bar{n}_i = \frac{g_i}{e^{\lambda_1 + \lambda_2 \epsilon_i} + 1} \tag{6.4.16}$$

while for bosons

$$\bar{n}_i = \frac{g_i}{e^{\lambda_1 + \lambda_2 \epsilon_i} - 1} \tag{6.4.17}$$

where we have assumed $g_i \gg 1$. Eqs.(6.4.16) and (6.4.17) are exactly identical to the grand canonical results expressed in Eq.(6.2.11), once we have identified $\lambda_2 = \beta$, $\lambda_1 = \beta\mu$ and $g_i = g(\epsilon)d\epsilon = (2s + 1)\frac{2\pi V}{h^3}(2m^{3/2}\epsilon^{1/2}d\epsilon)$. As noted after Eq.(6.4.8), our strategy is to write

$$\Gamma^{F,B} = W^{F,B}(\{\bar{n}_i)\} \tag{6.4.18}$$

which can be safely done if the fluctuation in $\bar{n}_i$ is small. Hence for self consistency, we will need to demonstrate that $\langle \bar{n}_i^2 \rangle - \langle \bar{n}_i \rangle^2$ vanishes in the thermodynamic limit.

Before doing so we complete the calculation of the entropy by writing

$$\Gamma^F = \prod_i \frac{g_i!}{\bar{n}_i!(g_i - \bar{n}_i)!} \tag{6.4.19}$$

and

$$\Gamma^B = \prod_i \frac{(\bar{n}_i + g_i - 1)!}{\bar{n}_i(g_i - 1)!} \tag{6.4.20}$$

which lead to

$$
\begin{aligned}
S^F &= k \left[\sum_i g_i \ln g_i - \bar{n}_i \ln \bar{n}_i - (g_i - \bar{n}_i) \ln (g_i - \bar{n}_i) \right] \\
&= k \sum_i \{ g_i[\ln g_i - \ln (g_i - \bar{n}_i)] - \bar{n}_i[\ln \bar{n}_i - \ln (g_i - \bar{n}_i)] \} \\
&= k \sum_i \left\{ \bar{n}_i \ln \left(\frac{g_i}{\bar{n}_i} - 1 \right) - g_i \ln \left(1 - \frac{\bar{n}_i}{g_i} \right) \right\}
\end{aligned}
\tag{6.4.21}
$$

and

$$S^B = k \sum_i \left\{ \bar{n}_i \ln \left(1 + \frac{g_i}{\bar{n}_i} \right) + g_i \ln \left(1 + \frac{\bar{n}_i}{g_i} \right) \right\}. \tag{6.4.22}$$

To conclude our argument, we need to show that the fluctuations in n_i, are negligible. We proceed by noting that [see Eqs.(6.2.8) and (6.2.9)]

$$\langle n_i \rangle = -\frac{\partial}{\partial \epsilon_i} k_B T \ln \, Q^{(i)}. \tag{6.4.23}$$

Differentiating,

$$\frac{\partial}{\partial \epsilon_i} \langle n_i \rangle = -k_B T \frac{\partial}{\partial \epsilon_i} \left(\frac{1}{Q^{(i)}} \frac{\partial Q^{(i)}}{\partial \epsilon_i} \right)$$

$$= -k_B T \left[\frac{1}{Q^{(i)}} \frac{\partial^2 Q^{(i)}}{\partial \epsilon_i^2} - \left(\frac{1}{Q^i} \frac{\partial^2 Q^{(i)}}{\partial \epsilon_i} \right)^2 \right]. \tag{6.4.24}$$

We note that the average occupation can be written as

$$\langle n_i \rangle = \frac{1}{Q} \sum n_i . e^{\mu \beta \sum n_i - \beta \sum \epsilon_i n_i} = -k_B T \frac{\partial}{\partial \epsilon_i} \ln \, Q^{(i)}$$

while

$$\langle n_i^2 \rangle = \frac{1}{Q} \sum n_i^2 . e^{\mu \beta \sum n_i - \beta \sum \epsilon_i n_i}$$

$$= \frac{(k_B T)^2}{Q} \frac{\partial^2}{\partial \epsilon_i^2} Q$$

$$= \frac{(k_B T)^2}{Q^{(i)}} \frac{\partial^2}{\partial \epsilon_i^2} Q^{(i)}. \tag{6.4.25}$$

Using the above result in Eq.(6.4.24), we obtain,

$$\langle (\delta n_i)^2 \rangle = \langle n_i \rangle^2 - \langle n_i \rangle^2$$

$$= -k_B T \frac{\partial}{\partial \epsilon_i} \langle n_i \rangle$$

$$= -k_B T \frac{\partial}{\partial \epsilon_i} \left[e^{\beta(\epsilon_i - \mu)} \pm 1 \right]^{-1}$$

$$= \frac{e^{\beta(\epsilon_i - \mu)}}{(e^{\beta(\epsilon_i - \mu)} \pm 1)^2}$$

$$= \langle n_i \rangle \mp \langle n_i \rangle^2 \tag{6.4.26}$$

(once again, the upper sign is for fermions and the lower sign for bosons).

In the above form, the fluctuations in n_i, are not necessarily small. However, as we consider the infinite volume limit, the level distribution becomes continuous and instead of a single level i we consider a group of levels clustered around the level i so that

$$n_i = \sum_k n_k$$

leading to,

$$\langle n_i^2 \rangle - \langle n_i \rangle^2 = \left\langle \left(\sum n_k \right)^2 \right\rangle - \left\langle \sum n_k \right\rangle^2$$

$$= \left\langle \sum_{m,l} n_l n_m \right\rangle - \left\langle \sum_l n_l \right\rangle \left\langle \sum_m n_m \right\rangle . \quad (6.4.27)$$

If the suffixes m and l are unequal, the right hand side of Eq.(6.4.27) vanishes which we prove by noting the following:

$$\frac{\partial}{\partial \epsilon_i} \langle n_m \rangle = -k_B T \frac{\partial}{\partial \epsilon_i} \left(\frac{1}{Q^{(m)}} \frac{\partial Q^{(m)}}{\partial \epsilon_m} \right)$$

$$= -k_B T \frac{\partial^2}{\partial \epsilon_i \partial \epsilon_m} \ln Q$$

$$= -k_B T \frac{\partial}{\partial \epsilon_i} \left[\frac{1}{Q} \frac{\partial Q}{\partial \epsilon_m} \right]$$

$$= -k_B T \left[\frac{1}{Q} \frac{\partial^2 Q}{\partial \epsilon_i \partial \epsilon_m} - \frac{1}{Q^2} \frac{\partial Q}{\partial \epsilon_i} \frac{\partial Q}{\partial \epsilon_m} \right]$$

$$= -\beta \left[\langle n_i n_m \rangle - \langle n_i \rangle \langle n_m \rangle \right].$$

For $m \neq i$, $\frac{\partial}{\partial \epsilon_i} \langle n_m \rangle = 0$ and hence the result. Using this result, we can write Eq.(6.4.27) as

$$\langle n_i^2 \rangle - \langle n_i \rangle^2 = \sum_k \left(\langle n_k^2 \rangle - \langle n_k \rangle^2 \right). \quad (6.4.28)$$

From Eq.(6.4.26), we can write

$$\sum_k \left(\langle n_k^2 \rangle - \langle n_k \rangle^2 \right) = \sum_k \left(\langle n_k \rangle \mp \langle n_k \rangle^2 \right)$$

$$= \langle n_i \rangle \mp \sum_k \langle n_k \rangle^2$$

so that,

$$\langle n_k^2 \rangle - \langle n_k \rangle^2 = \langle n_i \rangle \mp \sum_k \langle n_k \rangle^2 \quad (6.4.29)$$

which is the desired result. The right hand side is proportional to the volume V. While the first term on the right hand side is automatically so, the second term too is proportional to the volume as well which appears in the measure of integration as $V d^3 k / (2\pi)^3$. Thus the relative fluctuation $\frac{(\langle n_i^2 \rangle - \langle n_i \rangle^2)}{\langle n_i \rangle^2}$ is $O\left(\frac{1}{V}\right)$ and in the infinite volume limit, the fluctuations vanish and hence the entropy expressions of Eqs.(6.4.21) and (6.4.22) are perfectly valid.

6.5   Chemical Potential in Two Dimensions

In this section we point out that the chemical potential of both the Fermi and Bose gases can be obtained exactly in $D = 2$. To do this, we first need the density of states in two dimensions for the single particle. This is

$$\frac{d^2r d^2p}{h^2} = \frac{d^2r 2\pi p dp}{h^2} = \frac{2m\pi d^2r}{h^2} d\left(\frac{p^2}{2m}\right) = \frac{2m\pi}{h^2} d^2r d\epsilon. \tag{6.5.1}$$

The total number of particles N is

$$N = \int n(\epsilon)\frac{d^2r d^2p}{h^2} = \frac{2\pi m A}{h^2} \int \frac{d\epsilon}{e^{(\epsilon-\mu)\beta} \pm 1}. \tag{6.5.2}$$

In the above, A is the confining area, so that $N/A = n$ is the areal density of the particles. The integration over ϵ can be carried out exactly to yield

$$n\lambda^2 = \pm \ln\ (1 \pm e^{\beta\mu}) \tag{6.5.3}$$

For fermions, we have

$$e^{\beta\mu} = e^{n\lambda^2} - 1 \tag{6.5.4}$$

As $T \to 0$, $n\lambda^2 \gg 1$ and $\beta\mu \gg 1$, so that

$$\mu_{(T=0)} = \frac{nh^2}{2\pi m} \tag{6.5.5}$$

which is the zero temperature value of the chemical potential for fermions. At high temperature, $n\lambda^2 \ll 1$ and expanding the exponential in Eq.(6.5.4), we have

$$e^{\beta\mu} \approx n\lambda^2 \tag{6.5.6}$$

which is the correct classical limit.

For bosons on the other hand,

$$e^{\beta\mu} = 1 - e^{-n\lambda^2} \tag{6.5.7}$$

at high temperature, when $n\lambda^2 \ll 1$ this reduces to Eq.(6.5.6), the classical limit. We saw in Sec.6.1 that the chemical potential for free bosons cannot be positive. This is borne out by Eq.(6.5.7). The final question that we can ask is at what temperature does μ become zero for a fixed value of n. As can be seen from Eq.(6.5.7), this can only occur at $T = 0$. When the magnitude of μ is very small, Eq.(6.5.7) shows

$$\mu \approx -e^{-n\lambda^2}. \tag{6.5.8}$$

The fact that μ is finite (however small) at all finite temperatures according to Eq.(6.5.8) implies that there can be no Bose-Einstein condensation in two dimensions.

6.6 Problems for Chapter 6

1. Find the number of energy states per unit energy interval in arbitrary spatial dimension D (analogue of Eq. (6.2.10) which is for $D = 3$). Use it to find the relation beteween E, P and V in D dimensions.

2. Prove that in 2-dimensions the specific heat $C_V(N, T)$ of an ideal nonrelativistic gas of spinless fermions is identical to the specific heat of a corresponding gas of bosons for all values of N and T.

7

Fermi Distribution: Examples

7.1 Degenerate Fermi Gas

A collection of independent fermions is called degenerate if it is at very low temperatures or if its density is very high. The characteristic of the degenerate Fermi gas is that the distribution function $f(\epsilon) = 1$ if $\epsilon < \mu$ and zero if $\epsilon > \mu$. This is because the temperature of the gas can be considered to be absolute zero and the limiting form of the Fermi distribution discussed in Sec. 6.2 obtains. We will consider spin half fermions so that the factor $2s + 1 = 2$ and we have

$$
\begin{aligned}
n(\epsilon)d\epsilon &= 2.2\pi V \left(\frac{2m}{h^2}\right)^{3/2} \epsilon^{1/2} d\epsilon, \text{ for } \epsilon < \mu \\
&= 0 \quad \text{for } \epsilon > \mu.
\end{aligned}
\tag{7.1.1}
$$

If the total number of fermions is N, then

$$
\begin{aligned}
N &= \int_o^\mu n(\epsilon)d\epsilon = 4\pi V \left(\frac{2m}{h^2}\right)^{3/2} \frac{2}{3} \cdot \mu^{3/2} \\
&= \frac{8\pi}{3} V \left(\frac{2m}{h^2}\right)^{3/2} \epsilon_F^{3/2}
\end{aligned}
\tag{7.1.2}
$$

where we identify the chemical potential μ at absolute zero as the Fermi energy ϵ_F, as defined in Chapter 6. In terms of the number density $n = N/V$, the Fermi energy can be written as

$$
\epsilon_F = \left(\frac{3n}{8\pi}\right)^{2/3} \frac{h^2}{2m}.
\tag{7.1.3}
$$

The Fermi temperature T_F is defined as $\epsilon_F = k_B T_F$, while the Fermi velocity v_F is defined as $v_F = (2\epsilon_F/m)^{1/2}$ and the Fermi momentum as $p_F =$

mv_F. Clearly the Fermi temperature is proportional to the two thirds power of density and hence a high density Fermi gas has a high Fermi temperature. The Fermi gas is degenerate if its temperature T is much lower than T_F - this is generally the criterion for determining whether the limiting form of the distribution for the degenerate gas is applicable in a given situation or not. The total energy of the degenerate Fermi gas (non-relativistic) is

$$
\begin{aligned}
E &= \int_0^{\epsilon_F} \epsilon n(\epsilon) d\epsilon = 4\pi V \left(\frac{2m}{h^2}\right)^{3/2} \int_0^{\epsilon_F} \epsilon^{3/2} d\epsilon \\
&= \frac{8}{3}\pi V \left(\frac{2m}{h^2}\right)^{3/2} \epsilon_F{}^{3/2} \frac{3}{5}\epsilon_F \\
&= \frac{3}{5} N\epsilon_F.
\end{aligned}
\tag{7.1.4}
$$

The pressure exerted by the Fermi gas is found from the general expression $P = 2E/3V$ and hence

$$
P = \frac{2}{3} n\epsilon_F = \frac{2}{3} nk_B T_F.
\tag{7.1.5}
$$

Thus, one pictures the degenerate Fermi gas as being at absolute zero with all the levels below $\epsilon = \epsilon_F$ occupied, the occupancy factor being $(2s+1)$, where s is the spin of the fermions. The number of fermions and the zero point kinetic energy coming from the uncertainty principle allows the gas to exert pressure, which is given by Eq.(7.1.5). The intense pressure which the degenerate Fermi gas can exert manifests itself in the form of the white dwarf stars, which we deal with in the next section.

7.2   White Dwarfs

The brightness of a star diminishes as the colour becomes redder and in a brightness against colour plot (known as Russell-Hertzprung diagram (Fig. 7.1)) most of the stars lie in a strip.

Certain stars are brighter than their red colour warrants-these are the red giants, while certain others are dimmer than their white colour would suggest-these are the white dwarfs. The reason for their dimness is the absence of hydrogen. The mass of the star is provided by the helium nuclei while the stability is provided by the ionized electrons. The mass of the star would try to drive it towards gravitational collapse-this is prevented by the zero point pressure exerted by the electrons which can be considered a degenerate Fermi gas. We consider the Fermi gas of electrons to consist of N particles. The number of nucleons is then $2N$ and the mass of the star is approximately $2Nm_n$ where m_n is the nucleonic mass. Knowing the approximate mass of the star, we can estimate N and from the approximate size estimate the volume V. Thus the density $n = N/V$ is known and the Fermi temperature of the star can be calculated. It is roughly $10^{11}K \gg 10^7K$, the temperature of the star. Hence

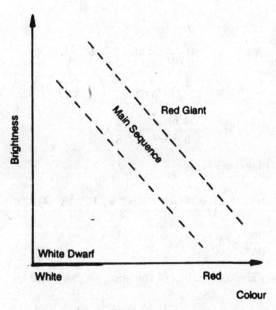

Figure 7.1: Russel-Hertzprung diagram

the electron gas can be taken to be at zero temperature and thus completely degenerate.

As explained above, the role of the nucleons is to provide the mass while the role of the electrons is to exert the pressure. The typical energy of the electrons is much larger than the rest mass of the electrons and hence we need to use the relativistic expression for energy in calculating the pressure of the degenerate electron gas:

$$
\begin{aligned}
E &= V.\frac{8\pi}{h^3} \int_0^{p_F} \sqrt{(p^2c^2 + m^2c^4)}\, p^2\, dp \\
&\approx V.\frac{8\pi}{h^3}.c \int_0^{p_F} \left(1 + \frac{m^2c^2}{2p^2}\right).p^3\, dp \\
&= V.\frac{8\pi}{h^3}.c \left(\frac{p_F^4}{4} + \frac{m^2c^2}{4}p_F^2 + \dots\right) \\
&= V.\frac{8\pi}{h^3}.\frac{m^4c^5}{4}(X_F^4 + X_F^2 + \dots)
\end{aligned}
\qquad (7.2.1)
$$

where $X_F = p_F/mc$. The pressure can be obtained by differentiating E with respect to V as

$$
P = -\frac{\partial E}{\partial V} = -\frac{2\pi}{h^3}m^4c^5\left(X_F^4 + X_F^2 + V\frac{\partial X_F^2}{\partial V}(2X_F^2 + 1) + \dots\right). \quad (7.2.2)
$$

Now,

$$V.\frac{\partial}{\partial V}X_F{}^2 = 2V\frac{\partial}{\partial V}\frac{\epsilon_F}{mc^2} = \frac{2}{mc^2}V.\frac{\partial\epsilon_F}{\partial V}$$

$$= \frac{2}{mc^2}V.\frac{\partial}{\partial V}\left(\frac{h^2}{2m}\left(\frac{3}{8\pi}\frac{N}{V}\right)^{2/3}\right)$$

$$= -\frac{4}{3}\frac{m\epsilon_F}{m^2c^2} = -\frac{2}{3}X_F^2 \qquad (7.2.3)$$

and hence Eq.(7.2.2) becomes

$$P = -\frac{2\pi m^4 c^5}{h^3}\left(X_F{}^4 + X_F{}^2 - \frac{4}{3}X_F{}^4 - \frac{2}{3}X_F{}^2 + \ldots\right)$$

$$= \frac{2\pi m^4 c^5}{3h^3}(X_F{}^4 - X_F{}^2 + \ldots). \qquad (7.2.4)$$

In terms of the mass and radius of the star, we can write

$$M = N.m_n \quad \text{and} \quad V = \frac{4}{3}\pi R^3$$

leading to

$$2m_n.n = \frac{3M}{4\pi R^3} \quad \text{or} \quad n = \frac{3M}{8\pi m_n}.R^3$$

and

$$X_F = \frac{\bar{M}^{1/3}}{\bar{R}} \qquad (7.2.5)$$

where

$$\bar{M} = \frac{9\pi}{8}\frac{M}{m_n} \quad \text{and} \quad \bar{R} = R.\frac{mc}{h}. \qquad (7.2.6)$$

In terms of $\bar{M}$ and $\bar{R}$ we can write the pressure as

$$P = K\left(\frac{\bar{M}^{4/3}}{\bar{R}^4} - \frac{\bar{M}^{2/3}}{\bar{R}^2}\right) \qquad (7.2.7)$$

where

$$K = \frac{2\pi}{3}mc^2\left(\frac{mc}{h}\right)^3. \qquad (7.2.8)$$

The equilibrium radius is found by equating the work done against the pressure to the gravitational potential energy. The gravitational potential energy of a system of total mass M and radius R is

$$V(R) = -\frac{\alpha GM^2}{R} = -\alpha G\left(\frac{8m_n}{9\pi}\right)^2\frac{mc}{h^3}\frac{\bar{M}^2}{\bar{R}} \qquad (7.2.9)$$

where α is a number of order unity and can be calculated only when the density is known as a function of the radial distance. The work done against the pressure is

$$W = \int_{\infty}^{R} P.4\pi r^2 dr = 4\pi \left(\frac{h}{mc}\right)^3 \int_{\infty}^{\bar{R}} P(\bar{r})4\pi \bar{r}^2 d\bar{r} \qquad (7.2.10)$$

which must be equated to the potential energy $V(R)$. Differentiating the right hand sides of Eqs.(7.2.9) and (7.2.10),

$$4\pi \bar{R}^2 P(\bar{R}) = \alpha G \left(\frac{8m_n}{9\pi}\right)^2 \left(\frac{mc}{h}\right)^4 \frac{\bar{M}^2}{\bar{R}^2} \qquad (7.2.11)$$

or

$$K \left(\frac{\bar{M}^{4/3}}{\bar{R}^4} - \frac{\bar{M}^{2/3}}{\bar{R}^2}\right) = K' \frac{\bar{M}^2}{\bar{R}^4}$$

where we have substituted for $P(\bar{R})$ from Eq.(7.2.7) and defined

$$K' = \frac{\alpha G}{4\pi} \cdot \left(\frac{8m_n}{9\pi}\right)^2 \left(\frac{mc}{h}\right)^4.$$

Rearranging,

$$\frac{1}{\bar{R}^4}\left(\frac{K}{K'}\bar{M}^{4/3} - \bar{M}^2\right) = \frac{K}{K'} \cdot \frac{\bar{M}^{2/3}}{\bar{R}^2}$$

or

$$\bar{R}^2 = \bar{M}^{2/3}\left(1 - \frac{\bar{M}^{2/3}}{\bar{M}_0^{2/3}}\right) \qquad (7.2.12)$$

where

$$\begin{aligned}
\bar{M}_0^{2/3} &= \frac{K}{K'} = \left[\frac{2\pi}{3}mc^2\left(\frac{mc}{h}\right)^3\right] \bigg/ \left[\frac{\alpha G}{4\pi}\left(\frac{8m_n}{9\pi}\right)^2\left(\frac{mc}{h}\right)^4\right] \\
&= \frac{2\pi}{3}.4\pi\left(\frac{1}{2\pi}\right)^3 \cdot \left(\frac{9\pi}{8}\right)^2 \frac{1}{\alpha}\frac{hc}{Gm_n^2} \\
&= \frac{27\pi}{64\alpha}\frac{hc}{Gm_n^2} \\
&= \frac{27\pi}{64\alpha} \cdot \left(\frac{M_p}{m_n}\right)^2 \qquad (7.2.13)
\end{aligned}$$

where $M_p = hc/G$ is the Planck mass which appears naturally, since, the phenomenon that we are considering is a combination of quantum mechanics, relativity and gravitation. Clearly for $\bar{R}$ as given by Eq.(7.2.12) to exist, we must have $\bar{M} < \bar{M}_0$. Thus $\bar{M}_0$ is the upper limit on the mass of a white dwarf. The dimensionless number $hc/Gm_n^2 \approx 10^{39}$. The dimensional mass M_0 corresponding to $\bar{M}_0$ is (for $\alpha \approx 1$)

$$M_0 = \frac{8}{9\pi} m_n . \bar{M}_0 \approx 10^{33} \text{grams} \approx M_{sun}$$

For an accurate determination of M_0, one must have a proper estimate of α and when this is taken into account $M_0 \approx 1.4 M_{sun}$-an upper limit on the white dwarf mass known as the Chandrasekhar limit.

7.3   Specific Heat of an Electron Gas

In the last two sections we dealt with a gas of fermions at $T = 0$. We now consider the case of finite temperature when it is possible for the fermions to populate states with energy above the chemical potential μ and consequently the chemical potential is now going to be a function of the total number of particles N and the temperature T. It is this temperature-dependent chemical potential which we want to determine. The most convenient way to proceed is to use Eq.(6.3.2) and write

$$PV = \frac{2}{3}E = \frac{2}{3} . \frac{2V}{4\pi^2} \left(\frac{2m}{h^2}\right)^{3/2} \int\limits_{0}^{\infty} \frac{\epsilon^{3/2} d\epsilon}{e^{\beta(\epsilon-\mu)} + 1} . \qquad (7.3.1)$$

[Note that for a spin s fermion, the factor 2 associated with V in the above expression will be changed to $(2s + 1)$]. To evaluate the integral note that

$$\int\limits_{0}^{\infty} \frac{\epsilon^{3/2} d\epsilon}{e^{\beta(\epsilon-\mu)} + 1} = (k_B T)^{5/2} \int\limits_{-\beta\mu}^{\infty} \frac{(\chi + \mu\beta)^{3/2} d\chi}{e^{\chi} + 1} = (k_B T)^{5/2} I(\mu\beta) \qquad (7.3.2)$$

where

$$I(\alpha) = \int\limits_{-\alpha}^{\infty} \frac{(\chi + \alpha)^{3/2}}{e^{\chi} + 1} d\chi. \qquad (7.3.3)$$

We first note that an exact evaluation of $I(\alpha)$ is impossible and one has to settle for a perturbative calculation. Having done the $T = 0$ calculation in the previous sections, we now wish to extend those results to $T \neq 0$ but small nonetheless. This means $\alpha = \mu/k_B T \gg 1$ and our approximate evaluation of

$I(\alpha)$ will deal with the range where $\alpha \to \infty$. Thus,

$$
\begin{aligned}
I(\alpha) &= \int\limits_{-\alpha}^{\infty} \frac{(\chi+\alpha)^{3/2}}{e^{\chi}+1} d\chi = \int\limits_{-\alpha}^{0} \frac{(\chi+\alpha)^{3/2}}{e^{\chi}+1} d\chi + \int\limits_{0}^{\infty} \frac{(\chi+\alpha)^{3/2}}{e^{\chi}+1} d\chi \\[2mm]
&= \int\limits_{0}^{\alpha} \frac{(\alpha-\chi)^{3/2}}{1+e^{-\chi}} d\chi + \int\limits_{0}^{\infty} \frac{(\chi+\alpha)^{3/2}}{e^{\chi}+1} d\chi \\[2mm]
&= \int\limits_{0}^{\alpha} \frac{e^{\chi}}{1+e^{\chi}}(\alpha-\chi)^{3/2} d\chi + \int\limits_{0}^{\infty} \frac{(\alpha+\chi)^{3/2}}{e^{\chi}+1} d\chi \\[2mm]
&= \int\limits_{0}^{\alpha} (\alpha-\chi)^{3/2} d\chi - \int\limits_{0}^{\alpha} \frac{\alpha-\chi^{3/2}}{1+e^{\chi}} d\chi + \int\limits_{0}^{\infty} \frac{(\alpha-\chi)^{3/2}}{e^{\chi}+1} d\chi \\[2mm]
&= \frac{2}{5}\alpha^{5/2} + \int\limits_{0}^{\infty} \frac{(\alpha+\chi)^{3/2} - |(\alpha-\chi)|^{3/2}}{e^{\chi}+1} + \int\limits_{\alpha}^{\infty} \frac{|(\alpha-\chi)|^{3/2}}{e^{\chi}+1} d\chi .
\end{aligned}
$$

$$(7.3.4)$$

The last term on the right hand side in $O(e^{-\alpha})$ and hence negligible. The second term can be expanded as

$$
\begin{aligned}
(\alpha+\chi)^{3/2} - (\alpha-\chi)^{3/2} &= \alpha^{3/2} + \frac{3}{2}\alpha^{1/2}\chi + \cdots - \left(\alpha^{3/2} = \frac{3}{2}\alpha^{1/2}\chi + \cdots\right) \\[2mm]
&= 3\alpha^{1/2}\chi + O\left(\frac{1}{\alpha^{1/2}}\right). \quad (7.3.5)
\end{aligned}
$$

Evaluation of $I(\alpha)$ to two-term accuracy ($\alpha^{5/2}$ and $\alpha^{1/2}$), then requires the evaluation of the definite integral $I_1 = \int\limits_{0}^{\infty} \frac{\chi d\chi}{e^{\chi}+1}$, which we can write as,

$$
\begin{aligned}
I_1 &= \int\limits_{0}^{\infty} \chi e^{-\chi} \left[1+e^{-\chi}\right]^{-1} d\chi \\[2mm]
&= \int_{0}^{\infty} \chi e^{-\chi} \sum_{n=0}^{\infty} (-1)^n e^{-n\chi} d\chi \\[2mm]
&= \sum_{n=0}^{\infty} \int\limits_{0}^{\infty} \chi e^{-(n+1)\chi}(-1)^n d\chi \\[2mm]
&= \sum_{n=0}^{\infty} (-1)^n \frac{1}{(n+1)^2} \int\limits_{0}^{\infty} y e^{-y} dy \\[2mm]
&= \sum_{n=0}^{\infty} \frac{(-1)^n}{(n+1)^2} = \frac{1}{1^2} + \frac{1}{3^2} + \frac{1}{5^2} - \left(\frac{1}{2^2} + \frac{1}{4^2} + \frac{1}{6^2}\right)
\end{aligned}
$$

$$= \frac{1}{1^2} + \frac{1}{2^2} + \frac{1}{3^2} + \ldots - 2\left(\frac{1}{2^2} + \frac{1}{4^2} + \frac{1}{6^2} + \ldots\right)$$

$$= \zeta(2) - 2 \cdot \frac{1}{2^2}\left(\frac{1}{1^2} + \frac{1}{2^2} + \frac{1}{3^2} + \ldots\right)$$

$$= \zeta(2) - \frac{1}{2}\zeta(2)$$

$$= \frac{1}{2}\zeta(2) \tag{7.3.6}$$

where

$$\zeta(n) = \frac{1}{1^n} + \frac{1}{2^n} + \frac{1}{3^n} + \ldots \quad . \tag{7.3.7}$$

is the Riemann zeta-function. From any text on mathematical physics it can be verified that $\zeta(2) = \pi^2/6$ and hence the required integral $I_1 = \pi^2/12$. Now using Eqs.(7.3.4) and (7.3.5), we have

$$I(\alpha) = \frac{2}{5}\alpha^{5/2} + \frac{\pi^2}{4}\alpha^{1/2} + O\left(\frac{1}{\alpha^{1/2}}\right) \tag{7.3.8}$$

and from Eqs.(7.3.1) and (7.3.2), we obtain,

$$PV = \frac{V}{3\pi^2}\left(\frac{2m}{h^2}\right)^{3/2}\left[\frac{2}{5}\mu^{5/2} + \frac{\pi^2}{4}(k_BT)^2\mu^{1/2}\right]. \tag{7.3.9}$$

Differentiating with respect to μ yields N

$$N = \frac{V}{4\pi^2}\left(\frac{2m}{h^2}\right)^{3/2}\left[\mu^{3/2} + \frac{\pi^2}{8}(k_BT)^2\frac{1}{\mu^{1/2}} + \ldots\right]. \tag{7.3.10}$$

As $T \to 0$, the first term correctly reproduces the zero temperature result of Eq.(7.1.2). The Fermi energy ϵ_F is given by Eq.(7.1.3) and eliminating N/V in favour of ϵ_F, we find

$$\mu = \epsilon_F\left[1 + \frac{\pi^2}{8}\left(\frac{k_BT}{\mu}\right)^2 + \ldots\right]^{-2/3}$$

$$\cong \epsilon_F\left[1 - \frac{\pi^2}{12}\left(\frac{k_BT}{\epsilon_F}\right)^2 + \ldots\right] \tag{7.3.11}$$

to the leading order in the temperature T. Note that this result does go in the expected direction in that, for very high temperatures, the chemical potential is negative (see Chapter 6) and thus the first correction to the positive zero temperature chemical potential is to decrease it.

To find the specific heat, we first need to find the energy. This is best done by noting PV is $2/3E$ and thus from Eq.(7.3.9)

$$
\begin{aligned}
E &= \frac{V}{2\pi^2}\left(\frac{2m}{h^2}\right)^{3/2}\left[\frac{2}{5}\mu^{5/2} + \frac{\pi^2}{4}(k_BT)^2\mu^{1/2} + \ldots\right] \\
&= \frac{V}{2\pi^2}\left(\frac{2m}{h^2}\right)^{3/2}\left[\frac{2}{5}\epsilon_F^{5/2}\left(1 - \frac{5}{2}\cdot\frac{\pi^2}{12}\left(\frac{k_BT}{\epsilon_F}\right)^2 + \ldots\right)\right. \\
&\quad \left. + \frac{\pi^2}{4}(k_BT)^2\epsilon_F^{1/2} + \ldots\right] \\
&= \frac{V}{2}\left(\frac{2m}{h^2}\right)^{3/2}\left[\frac{2}{5}\epsilon_F^{5/2} + \frac{\pi^2}{6}(k_BT)^2\epsilon_F^{12} + \ldots\right]. \quad (7.3.12)
\end{aligned}
$$

Differentiating with respect to T at constant volume

$$
\begin{aligned}
C &= \frac{\partial E}{\partial T} = \frac{V}{2\pi^2}\cdot\left(\frac{2m}{h^2}\right)^{3/2}\cdot\frac{\pi^2}{3}.k^2\epsilon_F^{1/2}T + \ldots \\
&= N.3\pi^2\left(\frac{h^2}{2m}\right)^{3/2}\frac{1}{\epsilon_F^{3/2}}\frac{1}{2\pi^2}\left(\frac{2m}{h^2}\right)^{3/2}\frac{\pi^2}{3}k(k_BT)\epsilon_F^{1/2} + \ldots \\
&= \frac{\pi^2}{2}Nk.\frac{k_BT}{\epsilon_F} + \ldots \quad . \quad (7.3.13)
\end{aligned}
$$

Thus, as $T \to 0$, the specific heat of the electron gas is given by $C = \frac{\pi^2}{2}Nk\frac{k_BT}{\epsilon_F}$, which vanishes linearly with T as T tends to zero. This is to be contrasted with the T^3-dependence of the contribution of the vibrations of the nuclei forming the crystal lattice. The total specific heat for a metal will be a combination of that obtained above in Eq.(7.3.13) and the lattice part obtained in Eq.(4.6.12). At very low temperatures the linear part dominates while at higher temperatures the T^3 effect is more prominent.

The results are in surprisingly good agreement [Fig.(7.2)] with experiments for a wide variety of metals in spite of the fact that effects like presence of impurities, periodic structure of the lattice, etc. have been ignored. In the next section, the effect of the periodic structure of the lattice will be discussed.

7.4 One Dimensional Metal: Effect of Periodic Lattice Structure

7.4.1 Energy Levels for a Periodic Potential

To begin with we would like to find out the allowed energy levels ϵ of an electron moving in a weak periodic potential $U(x)$. We assume the periodicity parameter to be a, so that

$$U(x+a) = U(x). \quad (7.4.1)$$

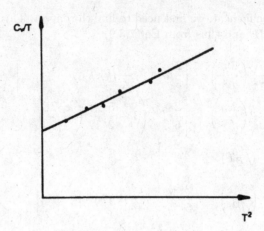

Figure 7.2: Measured specific of Copper at low temperatures showing the linear behaviour of the electron gas and the additional T^3 correction of the lattice vibrations.

Further, we assume that the length L of the sample is such that

$$\int_0^L U(x)dx = 0. \qquad (7.4.2)$$

The energy levels ϵ are to be obtained by a perturbative treatment of the Schrodinger equation

$$H\psi = \left[-\frac{\hbar^2}{2m}\frac{\partial^2}{\partial x^2} + U(x) \right]\psi(x) = \epsilon\psi(x), \qquad (7.4.3)$$

where the potential $U(x)$ is the perturbation. The unperturbed energies are those of the free particle,

$$\epsilon = \frac{p^2}{2m} \qquad (7.4.4)$$

where p is the momentum and the wavefunctions are the suitably normalized plane wave solutions $\psi_p = \frac{1}{\sqrt{L}}e^{ipx/h}$. The first-order energy shift E_1 is given by the standard formula

$$E_1 = \int_0^L \psi_p U(x)\psi_p dx = \frac{1}{L}\int_0^L U(x)dx = 0. \qquad (7.4.5)$$

However, the above formula for the first-order energy shift is true for non-degenerate states alone and clearly there are two wavefunctions $\psi_{\pm p} = (1/\sqrt{L})e^{\pm ipx/h}$ belonging to the same energy level $\epsilon = p^2/2m$ and accordingly we must deal with the degenerate state perturbation theory which will yield non-trivial results whenever the potential can cause transitions between the two degenerate states, i.e. whenever the matrix element

$$M_{12} = \int \psi^*_{-p} U(x) \psi_p dx \tag{7.4.6}$$

will be non-zero. Expanding the periodic potential in a Fourier series as

$$U(x) = \sum_{n=0}^{\infty} U_n e^{inqx} + c.c \tag{7.4.7}$$

where

$$q = \frac{2\pi}{a} \tag{7.4.8}$$

(*a* being the lattice periodicity introduced in Eq.(7.4.1)) it is easy to see that $M_{12} \neq 0$ whenever

$$2p = \pm nq\hbar \quad \text{or} \quad p = \pm n\frac{\pi}{a}\hbar. \tag{7.4.9}$$

We will concentrate on the $n = 1$ case for the present. The energy levels when p is in the vicinity of $(\pi/a)\hbar$, i.e. $p = \left(\frac{\pi}{a} + k\right)\hbar$ can be found by diagonalizing the 2 x 2 matrix whose diagonal elements are $\int \psi^*_p H \psi_p dx$ and $\int \psi^*_{-p} H \psi_{-p} dx$ (which are clearly equal to $\left[\left(\frac{\pi}{a} \pm k\right)^2\right]\hbar^2$) and the off diagonal elements are $\int \psi^*_p H \psi_{-p} dx$ and $\int \psi^*_{-p} H \psi_p dx$ (which from Eq.(7.4.6) are clearly seen to be U_1 and U_1^*). Thus the energy levels ϵ are found from

$$\text{Det}\left[\begin{array}{cc} \frac{1}{2m}\left(\frac{\pi}{a} + k\right)^2 \hbar^2 - \epsilon & U_1 \\ U_1^* & \frac{1}{2m}\left(\frac{\pi}{a} - k\right)^2 \hbar^2 - \epsilon \end{array} \right] = 0$$

or

$$\epsilon^2 - \epsilon\frac{\hbar^2}{m}\left(\frac{\pi^2}{a^2} + k^2\right) + \frac{\hbar^4}{4m^2}\left(\frac{\pi^2}{a^2} - k^2\right)^2 - |U_1|^2 = 0$$

giving

$$\epsilon_{1,2} = \frac{\hbar^2}{2m}\left(\frac{\pi^2}{a^2} + k^2\right) \pm \left(\frac{\pi^2 k^2 \hbar^4}{a^2 m^2} + |U_1|^2\right)^{1/2}. \tag{7.4.10}$$

At $k = 0$, i.e. $p = \pm(\pi/a)\hbar$ there is an energy gap given by $\Delta\epsilon = \epsilon_1 - \epsilon_2 = 2U_1$. The situation is shown in Fig.7.3.

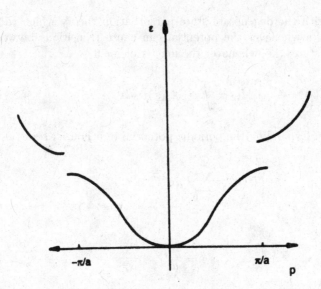

Figure 7.3: The discontinuity in the ϵ vs. p plot.

Far from $p = \pm(\pi/a)\hbar$ the free particle result in valid to first order in U, while at $p = \pm(\pi/a)\hbar$ there is a discontinuity in ϵ with the approach to the two limiting values being quadratic in nature as is obvious from Eq.(7.4.10).

While we have restricted the above discussion to $n = 1$, the same phenomenon occurs for all n and whenever $\epsilon = n^2\pi^2\hbar^2/2ma^2$, we have a discontinuity in the ϵ vs. p plot.

7.4.2   Fermi Level and Density of States

We now address the question regarding the position of the Fermi level in this one dimensional system. If the total number of electrons is N, then we have for a system of length L,

$$N = \frac{2L}{h} \int_{-p_F}^{p_F} dp = \frac{4Lp_F}{h} = \frac{2L}{\pi\hbar}p_F. \qquad (7.4.11)$$

Now, if there is one electron per site (sites come at separations of a), then $N = L/a$ and

$$p_F = \frac{\pi}{2a}.\hbar. \qquad (7.4.12)$$

The Fermi level is far away from the energy gap and we have the usual metallic properties associated with the electron gas, which is to say that it is possible to excite electrons across the Fermi level with a small amount of thermal energy and thus electrical and thermal conduction is feasible.

On the other hand, if there are two electrons per site (the maximum allowable), then $N = 2L/a$ and

$$p_F = \frac{\pi}{a}\hbar. \tag{7.4.13}$$

The Fermi level is now right in the middle of the energy discontinuity and this can have very significant consequences.

One of the quantities that is strongly affected is the density of states. There are no states at $p = p_F$ for ϵ lying between $\epsilon_F + |U_1|$ and $\epsilon_F - |U_1|$. We want the density of states as ϵ approaches $\epsilon_F - |U_1|$. The general formula for the density of states $g(\epsilon)$ in one dimension can be written as

$$\frac{dg(\epsilon)}{d\epsilon} = \frac{2L}{h} \int dp \delta(\epsilon - \epsilon_p) \tag{7.4.14}$$

which is the trivial statement that if we are to convert the basic density of states $2L dp/h$ to the energy variable, then we must integrate over all momenta which give the same energy. In using this relation in the present case, we note that it is most convenient to set the zero of the energy at ϵ_F and work in terms of small deviations Δ from ϵ_F, i.e.,

$$\epsilon = \epsilon_F + \Delta. \tag{7.4.15}$$

For the momentum, we work in terms of small deviation from $\pi\hbar/a$. Further, remembering $\Delta \ll \epsilon_F$, Eq.(7.4.14) leads to

$$g(\epsilon) \cong \frac{2L}{h} \frac{|\Delta|}{\sqrt{\Delta^2 - |U_1|^2}} \tag{7.4.16}$$

for $\Delta > U_1$.

7.4.3 Specific Heat

We can now determine the energy difference $E(T) - E(0)$ for $T \ll T_F$ by the following argument. At $T = 0$, all the electrons occupy the states with $\epsilon < \epsilon_F$, while for $T > 0$, there are some electrons in states $\epsilon > \epsilon_F$. For every electron raised above the Fermi level, there is a hole created in the erstwhile filled Fermi sea. If we measure the energy from the Fermi level ϵ_F, i.e. $\epsilon' = \epsilon_F + \Delta$ [Eq.(7.4.15)] then the absence of a particle with energy ϵ' from the Fermi sea corresponds to a hole of energy $\Delta' = \epsilon_F - \epsilon'$. Thus for $T > 0$, we have a gas of holes with energy lying between 0 and ϵ_F and a gas of particles (electrons) with energy greater than ϵ_F. At $T = 0$, this gas disappears. For $T \ll T_F$, this gas will have vanishingly small energy and will be determined by the distribution in the vicinity of the Fermi level. The probability of finding a hole with energy $\Delta' = \epsilon_F - \epsilon'$ is clearly $1 - f(\epsilon') = [e^{\beta(\epsilon_F - \epsilon')} + 1]^{-1} = (e^{\beta\Delta'} + 1)^{-1}$ which is the same as the distribution of particles. For a given energy Δ $(= \epsilon - \epsilon_F)$, the distribution of holes is $(e^{\beta\Delta'} + 1)^{-1}$ and the distribution of particles is also

$(e^{\beta\Delta} + 1)^{-1}$ for $T \ll T_F$ and the energy of this gas of particles and holes is determined to a first approximation by the density of states at the Fermi level. The density of states is the same for electrons and holes and hence

$$E(T) - E(0) = \int_o^\infty \Delta(e^{\beta\Delta} + 1)^{-1} n(\epsilon_F) d\Delta. \qquad (7.4.17)$$

The density of states near $\epsilon = \epsilon_F$ is given by Eq.(7.4.16) and hence

$$E(T) - E(0) = \frac{2L}{h}\left(\frac{2m}{\epsilon_F}\right)^{1/2} 0\int_0^\infty \frac{\Delta^2}{\sqrt{\Delta^2 - |U_1|^2}} \frac{d\Delta}{e^{\beta\Delta} + 1}$$

$$\approx \frac{2L}{h}\left(\frac{2m}{\epsilon_F}\right)^{1/2}\left(\frac{\pi}{\Delta^2}\right)^{1/2}$$

$$\times U_1^{3/2}(k_BT)^{1/2}e^{-U_1/k_BT}. \qquad (7.4.18)$$

The leading term in the specific heat will accordingly be

$$C(T) \approx \frac{2L}{h}\sqrt{2m/\epsilon_F}\sqrt{\pi/\Delta^2}e^{-U_1/k_BT}U_1^{5/2}(k_BT)^{-3/2}k_B. \qquad (7.4.19)$$

This is a very different specific heat from the normal situation where $C \propto T$. In this case, where there is a gap at the Fermi level, the specific heat is much smaller (it approaches zero exponentially as $T \to 0$) and this is as it should be because there is a barrier of height $2|U_1|$ formed at the point where the particle-hole gas is created and thus temperatures of the order $k_BT \cong |U_1|$ are required for the formation of particle-hole pairs. The conductivity is determined by this gas of particles and holes. For $\Delta \gg k_BT$, the formation of this gas is strongly suppressed and one has an insulator, while for smaller Δ the behaviour is that of a semiconductor.

7.4.4 Peierls' Instability

We now compare the ground state energy $(T = 0)$ of the system with and without the energy gap. Clearly,

$$E(U_1) - E(U_1 = 0) = \int_{-\epsilon_F}^{U_F} [n(U_1) - n(U_1 = 0)]\Delta d\Delta \qquad (7.4.20)$$

where $n(U_1)$ is the function exhibited in Eq.(7.4.17).

Note that $n(U_1)$ differs from $n(U_1 = 0)$ only near $\Delta = 0$ and thus while $n(U_1)$ is expressed by Eq.(7.4.17), $n(U_1 = 0)$ is to be evaluated at the Fermi surface.

Thus to the lowest order of accuracy, we can write,

$$E(U_1) - E(0) \cong \frac{2L}{h}\left(\frac{2m}{\epsilon_F}\right)\int_{-\epsilon_F}^{U_1}\left(\frac{\Delta^2}{\sqrt{\Delta^2 - |U_1|^2}} - \Delta\right)d\Delta$$

$$= \frac{2L}{h}\left(\frac{2m}{\epsilon_F}\right)^{1/2}\left[\int_{-\epsilon_F}^{U_1}\left[\sqrt{\Delta^2 - |U_1|^2} - \Delta\right]d\Delta\right.$$

$$+ \left.|U_1|^2\int_{-\epsilon_F}^{U_1}\frac{d\Delta}{\sqrt{\Delta^2 - |U_1|^2}}\right]$$

$$= \frac{2L}{h}\left(\frac{2m}{\epsilon_F}\right)^{1/2}|U_1|^2\left[\int_{-\epsilon_F}^{U_1}\frac{d\Delta}{\sqrt{\Delta^2 - |U_1|^2}} - \frac{1}{2}\int_{-\epsilon_F}^{U_1}\frac{d\Delta}{\Delta}\right]$$

$$= -\frac{1}{h}\left(\frac{2m}{\epsilon_F}\right)^{1/2}|U_1|^2\ln\left(\epsilon_F\right)/|U_1|). \qquad (7.4.21)$$

Hence for $N = 2L/a$, when there is a gap at the Fermi surface, the system with the gap has a substantially lower energy.

We now return to a lattice where $N = L/a$. The Fermi level in this case is far removed from the energy gap and the usual metallic properties will be found, But the interaction between the electron and lattice can now deform the lattice. Accordingly, we assume that every other atom in the lattice suffers a displacement b, so that the periodicity of the lattice is now $2a$ instead of a as shown in Fig.(7.4).

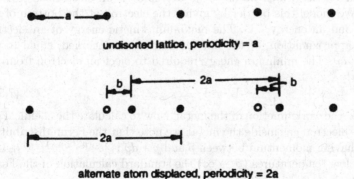

undisorted lattice, periodicity = a

alternate atom displaced, periodicity = 2a

Figure 7.4: Lattice distortions.

We note that there is a weak energy discontinuity at $p = \pi/2a$ which is the position of the Fermi level with $N = L/a$. This discontinuity is weak since the lattice is still almost periodic with period 'a' but the small deformation 'b'

induces a period $2a$, i.e. we can write the potential $U(x, 0) + \dots$

$$U(x, b) = U(x, 0) + bU'(x, 0) + \dots$$

where $U(x, 0)$ is our previous potential with period 'a' while $U'(x, 0)$ is the potential with period $2a$. The strength of the potential term causing the discontinuity at the Fermi surface is bU_1'. If we consider the energetics, then to achieve a deformation b will require an energy $kb^2/2$, where k is an elastic constant. Opening up a gap at the Fermi level, however, lowers the energy by the amount shown in Eq.(7.4.21) where we now have to replace U_1 by bU_1'. So the change in energy achieved by distorting the lattice is

$$\Delta E(b) = \frac{1}{2}Kb^2 - \frac{L}{h}\sqrt{2m/\epsilon_F b^2}|U_1|^2 \ \ln \ \frac{\epsilon_F}{b|U_1|}. \tag{7.4.22}$$

Clearly for small b, the second term will dominate over the first and it is energetically favourable for the lattice to deform in such a manner that a gap opens up at the Fermi surface. Thus electron-lattice interaction in one dimension causes the lattice to deform structurally in the above manner and the one-dimensional metal loses its metallic property because of the gap at the Fermi surface. This instability is known as the Peierl's instability. It does not occur in two or three dimensions where the effect of the periodicity of the lattice on the electrons shows up in the form of charge density waves.

7.5   Emission of Electrons from a Metal Surface

We consider a metal in equilibrium with a dilute electron gas at very low temperatures. The periodic distribution of the ions which produces the periodic potential in which the electrons move is interrupted at the surface giving rise to a potential barrier which needs to be overcome if an electron is to leave the surface. We model this barrier by giving the electron at the bottom of the conduction band an energy $-V$. The maximum kinetic energy of an electron is to a good approximation, at the low temperatures concerned, equal to the Fermi energy ϵ_F. The minimum energy required to eject an electron from the metal is clearly

$$W = V - \epsilon_F \tag{7.5.1}$$

which is called the work function of the metal. Now to calculate the chemical potential of this electron gas inside the metal, we note that the Fermi distribution for particles having momentum between $\vec{p}$ and $\vec{p} + d\vec{p}$ is $[e^{\beta(p^2/2m - V - \mu)} + 1]^{-1}$ and hence at low temperatures $(\beta \to \infty)$ the standard calculation of the Fermi energy yields

$$\epsilon_F = V + \mu. \tag{7.5.2}$$

Comparing Eqs.(7.5.1) and (7.5.2), we have

$$\mu = -W \tag{7.5.3}$$

inside the metal.

Outside the metal, the density of the electron gas is very low and hence this corresponds to a situation where the Fermi distribution will reduce to the classical distribution and thus the number of electrons between $\vec{p}$ and $\vec{p} + d\vec{p}$ will be given by $e^{-\beta(p^2/2m - \mu_0)}$, where μ_0 is the chemical potential outside the metal. If the total number of electrons is N and the volume V, then outside the metal

$$N = 2.V \frac{4\pi}{h^3} \int_0^\infty p^2 e^{-p^2/2mk_BT} e^{\mu_0/k_BT} dp \qquad (7.5.4)$$

(spin $1/2$ of the electron has been taken into account), giving rise to the electron density

$$n = 2. \left(\frac{2\pi m k_B T}{h^2} \right)^{3/2} e^{\mu_0/k_BT}. \qquad (7.5.5)$$

For equilibrium between the metal and the electron gas, the chemical potentials must be equal and hence

$$\mu_0 = \mu = -W. \qquad (7.5.6)$$

Thus the number density of the electrons outside the metal is given by

$$n = 2. \left(\frac{2\pi m k_B T}{h^2} \right)^{3/2} e^{-W/k_BT} = \frac{e^{-W/k_BT}}{\lambda^3}. \qquad (7.5.7)$$

The equilibrium that is achieved here is a dynamical one in that in every unit time interval, a certain number of electrons strike the metal surface and a fraction of them are absorbed. To keep the number of electrons fixed, an equal number of electrons must emerge from the metal surface per unit area per unit time. This gives the emission current from the metal surface. The number of electrons with momentum between p_z and $p_z + dp_z$ (the surface is $z = 0$) and density given by Eq.(7.5.7) striking a unit area of the metal in unit time, is clearly $(2n\lambda^3/h^2) \int_{-\infty}^\infty dp_x \int_{-\infty}^\infty dp_y \int_0^\infty dp_z v_z e^{-p^2/2mk_BT}$ and hence the total number of electrons, N, striking the metal surface per unit area per unit time is

$$N = 2 \frac{e^{-W/k_BT}}{h^3} \int_0^\infty \frac{p_z}{m} \exp\left(\frac{-p_z^2}{2mk_BT} \right) dp_z$$

$$\times \int_{-\infty}^\infty \frac{p_x}{m} \exp\left(\frac{-p_x^2}{2mk_BT} \right) dp_x \int_{-\infty}^\infty \frac{p_y}{m} \exp\left(\frac{-p_y^2}{2mk_BT} \right) dp_y$$

$$= 2. \frac{2mk_BT\pi}{h^2} . \frac{k_BT}{h} e^{-W/k_BT}. \qquad (7.5.8)$$

If the reflection coefficient at the metal surface is r, then the number of electrons absorbed on the metal surface per unit area per unit time is $(1 -$

$r)4\pi m k_B T/h^2.(k_B T/h)e^{-W/k_B T}$ and this must equal the emission rate. The emission current I is consequently given by

$$I = \frac{4\pi m (k_B T)^2}{h^3}(1-r)e^{-W/k_B T}. \tag{7.5.9}$$

This thermionic current is associated with the work of Richardson and Dushman. The factor $e^{-W/k_B T}$ reflects the depletion of the current due to the effect of the potential barrier.

7.6   Correlations in a Fermi Gas

The two-point distribution function $n(r_1, r_2)$ was defined in Chapter 5 and evaluated in the case when the classical particles have a pairwise interaction. For independently moving particles it should be easy to see that

$$n(r_1, r_2) = n^2 + n\delta(r_1 - r_2). \tag{7.6.1}$$

For fermions which are otherwise free, the correlation function $n(r_1, r_2) = n(r_{12})$ does not have the trivial form of Eq.(7.6.1) since the two point function has to keep track of the fact that fermions obey the exclusion principle and hence there cannot be two fermions in the same state. This restriction is best implemented in momentum space where we define the Fourier transform of $n(r_{12})$ as $C(k)$ given by

$$C(k) = \int d^3 r_{12} e^{-ik.r_{12}} n(r_{12}). \tag{7.6.2}$$

Now $C(k)$ represents the particle scattering amplitude with momentum transfer k as we had seen in Sec.5.5. For fermions this amplitude is

$$C(k) \propto \frac{2}{V} \sum_p f(p)[1 - f(p+k)] \tag{7.6.3}$$

where $f(p) = [e^{(\epsilon(p)-\mu)\beta}+1]^{-1}$ (the factor 2 is the spin degeneracy). It describes a process where a particle comes with momentum p exchanges momentum k with the scatterer and goes off with momentum $p+k$. The factor $[1 - f(p+k)]$ ensures that there are no particles with momentum $p+k$ in the final state. The correlation function in coordinate space is found by inverting Eq.(7.6.2) as

$$n(r_{12}) = \int \frac{d^3 k}{(2\pi)^3} C(k) = n\delta(r_{12}) - 2|f(r_{12})|^2 \tag{7.6.4}$$

where

$$f(r) = \frac{1}{V} \sum_p f(p)e^{ip.r} = \frac{I(r)}{2\pi^2 r} \tag{7.6.5}$$

with

$$I(r) = \int\limits_{0}^{\infty} dp \frac{p.\sin(pr)}{e^{(\epsilon - \mu)\beta} + 1}.$$

The integral $I(r)$ can be found in Vol.5 of Theoretical Physics by Landau and Lifshitz and is given by

$$I(r) = \frac{\partial}{\partial r} \left[\frac{\sin(p_F r)}{2\xi \, \sinh \, (r/2\xi)} \right] \tag{7.6.6}$$

where

$$2\xi = p_F/mk_B T \quad \text{and} \quad \mu = p_F^2/2m.$$

As $T \to 0$, the length scale ξ becomes extremely large and we can always take $r \ll \xi$. In this limit,

$$I(r) \cong \frac{\partial}{\partial r} \left(\frac{1}{r} \sin \, p_F r \right) \tag{7.6.7}$$

and

$$C(r) \cong n\delta(r) - \frac{p_F^4}{2\pi^4 r^2} \left[\frac{\cos \, p_F r}{p_F r} - \frac{\sin \, p_F r}{(p_F r)^2} \right]^2. \tag{7.6.8}$$

We note that for large r, there is no correlation length, the correlation falls of as r^{-4}. Also to be noted is the appearance of the periodic factors $\cos^2(p_F r)$ etc., which implies a spatial periodicity with period $2\pi/2p_F$. This is an effect of the existence of a sharp Fermi surface. The oscillations in the correlation function are related to the Friedel oscillations in the electron gas. For $T > 0$, there is a finite correlation length ξ. For $\xi \gg p_F^{-1}$ and $r/\xi \gg 1$, we find

$$C(r) \sim \frac{p_F^2 e^{-r/\xi}}{2\pi^5 \xi^2} \sin^2 p_F r. \tag{7.6.9}$$

which represents an exponential decay modulated by a spatial periodicity of π/p_F.

7.7 Electrons in Graphene

In this Section we present a very elementary introduction to the unusual properties of electrons in a two-dimensional allotrope of carbon called graphene. Since this field has attracted a lot of attention in the last few years, it does warrant a discussion. One of the reasons for this excitement (the one we will focus on) is the extreme relativistic dispersion (appropriate to massless fermions) exhibited by low energy electrons of graphene in the long wavelength limit. This surprising emergence of Dirac equation in a completely condensed matter setting has

allowed for experimental investigations of certain niceties of the Dirac equation
(e.g., Klein paradox, Zitterbewegung etc.) which have not been possible in the
high energy context. We break up the Section into two subsections: in subsection A we describe the tight-binding approximation and in subsection B we use
it to obtain the energy band of graphene.

7.7.1   The Tight-binding Approximation

In almost all our discussions of electrons so far, the crystal potential in a solid
has played no role except in a weak limit for the Peierls' instability. We end this
Chapter with the discussion of case where the crystal potential has a very important role to play. For that we need to discuss the state of an electron when it
is very tightly bound in a metal. The crystal potential (i.e., the ionic potential)
is now strong and during its motion through the lattice the electron spends a
long time in the vicinity of a given ion before tunneling to the neighbouring
one. A one-dimensional depiction is given in Figure(7.5).

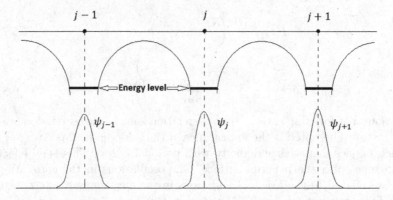

Figure 7.5: One dimensional potential and the localized states around each site.

When in the vicinity of a given ion 'j', the wavefunction of the electron
is approximately that of an atomic orbital $\psi_j(x)$. We will consider the electron
as primarily spending its time near a particular ion (say j) and because of this
description that follows is called the tight-binding model (good for low lying
narrow bands with shell radius much smaller than lattice spacing). The atomic
orbital that we consider for the j-th atom has energy E_α and wavefunction
$\phi_\alpha(x)$. We will write down a reasonable Bloch function

$$\psi_k(x) = \frac{1}{\sqrt{N}} e^{ikx} \sum_{j=1}^{N} e^{-ik(x-x_j)} \phi_\alpha(x - x_j). \qquad (7.7.1)$$

The factor within the sum is periodic under the assumption that there
is very little overlap between sites j and $j \pm 1$ and the function $\phi_\alpha(x - x_j)$
is localized around $x = x_j$ (see Figure(7.5)). Near $x = x_j$, we have $\psi_k(x) \simeq$

$e^{ikx_j}\phi_\alpha(x - x_j)$, i.e., proportional to the atomic orbital, and thus $\psi_k(x)$ is a good approximation for a crystal wavefunction.

The energy $E(k)$ of the electron in the state $\psi_k(x)$ is found to be

$$
\begin{aligned}
E_k &= \langle \psi_k | H | \psi_k \rangle \\
&= \frac{1}{N} \sum_{j,l} e^{ik(x_j - x_l)} \langle \phi_\alpha(x - x_l) | H | \phi_\alpha(x - x_j) \rangle.
\end{aligned} \tag{7.7.2}
$$

The double sum over j and l is over all N sites. Each term in the above sum will be a function of $(x_l - x_j)$ and not of x_j and x_l individually and hence we can fix l at some given value and do a sum over j. There would be N such terms which would be identical and hence

$$
E_k = \sum_{-N/2}^{(N-1)/2} e^{ikx_j} \langle \phi_\alpha(x) | H | \phi_\alpha(x - x_j) \rangle \tag{7.7.3}
$$

In the sum above we need to separate $x_j = 0$ from the others. For $x_j = 0$ we have a situation where the energy comes from a localization around $x_j = 0$ alone (this is the single site energy E_s), while all others involve tunneling and contribute an amount E_t. We accordingly write

$$
\begin{aligned}
E(k) &= \langle \phi_\alpha(x) | H | \phi_\alpha(x) \rangle + \sum_{j}' e^{ikx_j} \langle \phi_\alpha(x) | H | \phi_\alpha(x - x_j) \rangle \\
&= E_s + E_t.
\end{aligned} \tag{7.7.4}
$$

In the second term above a tight binding would make $x_j = \pm a$ (where a is the lattice spacing) the most important terms. As for H we have $H = -\frac{\hbar^2}{2m}\frac{d^2}{dx^2} + V(x)$, where $V(x)$ is the crystal potential and as seen from Figure(7.5), is localized around x_j so that $V(x) = \sum_j v(x - x_j)$.

To begin evaluating $E(k)$, we start with the first term of Eq.(7.7.4). The potential $V(x)$ around $x_j = 0$ is split as $v(x)$ and $V'(x)$, the latter coming from all other x_j, i.e., $V(x) = v(x) + V'(x)$ and we have

$$
\begin{aligned}
E_s &= \langle \phi_\alpha(x) | -\frac{\hbar^2}{2m}\frac{d^2}{dx^2} + V(x) | \phi_\alpha(x) \rangle + \langle \phi_\alpha(x) | V'(x) | \phi_\alpha(x) \rangle \\
&= E_\alpha - \beta.
\end{aligned} \tag{7.7.5}
$$

Here E_α is just the atomic energy for a single particle problem and β stands for the effect of the other ions and clearly $\beta > 0$ as $V'(x)$ os negative.

In the tight-binding situation we need to consider only $j = \pm 1$, i.e., $x_j = \pm a$, i.e., the contribution to the tunneling energy would be

$$
E_t = \langle \phi_\alpha(x) | H | \phi_\alpha(x - a) \rangle + \langle \phi_\alpha(x) | H | \phi_\alpha(x + a) \rangle. \tag{7.7.6}
$$

With the potential still split as $V = v + V'$ we see that $\langle \phi_\alpha(x) | -\frac{\hbar^2}{2m}\frac{d^2}{dx^2} + v(x) | \phi_\alpha(x \pm a) \rangle = E_\alpha \langle \phi_\alpha(x) | \phi_\alpha(x \pm a) \rangle \simeq 0$, since the wavefunctions centred at

$x = 0$ and $x = \pm a$ barely overlap. The dominant contribution is $\int \phi_\alpha^*(x) V'(x - a)\phi_\alpha(x - a)$ since $V'(x - a)$ is appreciable near $x = 0$. This integral is the overlap integral and denoted by $-\gamma$ so that

$$
\begin{aligned}
E(k) &= E_\alpha - \beta - \gamma(e^{ika} + e^{-ika}) \\
&= E_\alpha - \beta - 2\gamma\cos(ika)
\end{aligned}
\tag{7.7.7}
$$

This gives the band energy in terms of well defined parameters which can be obtained from the knowledge of atomic energy and atomic orbitals.

It is convenient to write the above as

$$
E(k) = E_0 - 4\gamma\sin^2\frac{ka}{2}
\tag{7.7.8}
$$

$$
E_0 = E_\alpha - \beta - 2\gamma.
$$

The original atomic level E_α has broadened to an energy band (Figure(7.6)).

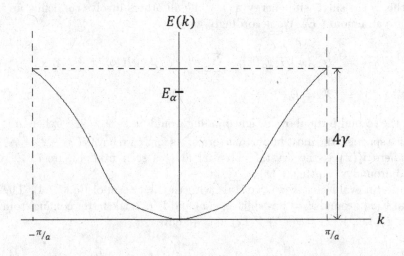

Figure 7.6: Energy band in one-dimensional tight-binding approximation.

The bottom of the band is $E_0 < E_\alpha$ as expected, since one of the effects of the other atoms is to produce a negative potential. Extra kinetic energy comes from the electron's ability to move through the crystal and this shows up as the second term in Eq.(7.7.8). The bandwidth is 4γ and is proportional to the overlap integral. The above result is easily extended to three dimensions and for the simple cubic lattice the result would be

$$
E(k) = E_0 - 4\gamma\left[\sin^2\frac{k_x a}{2} + \sin^2\frac{k_y a}{2} + \sin^2\frac{k_z a}{2}\right].
\tag{7.7.9}
$$

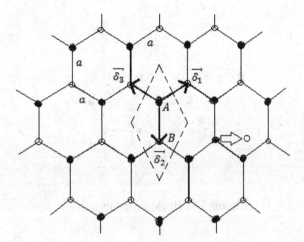

Figure 7.7: Honeycomb lattice: hexagonal unit cell with two atoms A and B. $\vec{\delta}_{1,2,3}$ are nearest neighbour distance vectors.

7.7.2 Application to Graphene

Over the last few years an extremely interesting application of the above picture has become tremendously important. This relates to virtually single atomic layer material called graphene (two-dimensional graphite) which is made of carbon atoms arranged in a honeycomb lattice (see Figure(7.7)) that can be described as a regular hexagonal pattern of sides a with two atoms A and B per unit cell.

The Bravais lattice is as shown in Figure(7.7) by dashed lines and its primitive lattice vectors are $\vec{a}_1$ and $\vec{a}_2$ as shown in Figure(7.8). Clearly $\vec{a}_1 = a(\frac{\sqrt{3}}{2}, \frac{3}{2})$ and $\vec{a}_2 = a(-\frac{\sqrt{3}}{2}, \frac{3}{2})$. The reciprocal lattice vectors are $\vec{b}_1$ and $\vec{b}_2$ defined by $\vec{a}_i.\vec{b}_j = 2\pi\delta_{ij}$, so that $\vec{b}_1 = \frac{2\pi}{3a}(\sqrt{3}, 1)$ and $\vec{b}_2 = \frac{2\pi}{3a}(\sqrt{3}, -1)$. The first Brillouin zone is defined by drawing the perpendicular bisectors of the vectors to the nearest lattice point and thus has the form shown in Figure(7.9). The corner points K_1 and K_2 have the coordinates $\frac{2\pi}{3a}(\frac{1}{\sqrt{3}}, 1)$ and $\frac{2\pi}{3a}(-\frac{1}{\sqrt{3}}, 1)$ and they are typical members of the two equivalent groups of three in the six cornered Brillouin zone.

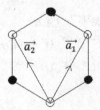

Figure 7.8: The primitive basis $\vec{a}_{1,2}$.

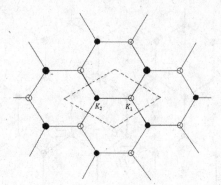

Figure 7.9: Brillouin zone.

We note that in real space the atom A has three nearest neighbours with the position vectors $\vec{\delta}_1 = \frac{a}{2}(\sqrt{3},1)$, $\vec{\delta}_2 = a(0,-1)$ and $\vec{\delta}_3 = \frac{a}{2}(-\sqrt{3},1)$. Now the tight-binding approximation can be used. Given that there are two kinds of atoms A and B, we need to write the Bloch wavefunction as

$$\psi = e^{i\vec{k}.\vec{r}}\left[\sum_{\vec{r}_A} e^{i\vec{k}.(\vec{r}-\vec{r}_A)}\chi(\vec{r}-\vec{r}_A) + \lambda\sum_{\vec{r}_B} e^{i\vec{k}.(\vec{r}-\vec{r}_B)}\chi(\vec{r}-\vec{r}_B)\right] \quad (7.7.10)$$

where λ is a parameter that we do not know and the sums are over all positions of the atoms A and B.

We will use the method of the previous subsection and write $H\psi = E\psi$ with $H = \frac{p^2}{2m} + v(x) + V'(x)$, where $v(x)$ is the potential due to a specific ion and $V'(x)$ is the potential created by the other ions. We write

$$\begin{aligned} S &= \int d^2r \chi^*(\vec{r}-\vec{r}_A)\chi(\vec{r}-\vec{r}_A) \\ &= \int d^2r \chi^*(\vec{r}-\vec{r}_B)\chi(\vec{r}-\vec{r}_B) \end{aligned} \quad (7.7.11)$$

and the matrix elements of H as

$$\begin{aligned} H_{11} &= e^{i\vec{k}.(\vec{r}-\vec{r}_{A,B})}\chi^*(\vec{r}-\vec{r}_{A,B}) \\ &\quad \times\ He^{-i\vec{k}.(\vec{r}-\vec{r}_{A,B})}\chi(\vec{r}-\vec{r}_{A,B}) \end{aligned} \quad (7.7.12)$$

$$H_{12} = e^{i\vec{k}.(\vec{r}-\vec{r}_A)}\chi^*(\vec{r}-\vec{r}_A)He^{-i\vec{k}.(\vec{r}-\vec{r}_B)}\chi(\vec{r}-\vec{r}_B) \quad (7.7.13)$$

and note that

$$\int d^2r \chi^*(\vec{r}-\vec{r}_A)\chi(\vec{r}-\vec{r}_A) = 0. \quad (7.7.14)$$

Multiplying $H\psi = E\psi$ by $e^{i\vec{k}\cdot(\vec{r}-\vec{r}_A)}\chi(\vec{r}-\vec{r}_A)$ and integrating and then following the same procedure with $e^{i\vec{k}\cdot(\vec{r}-\vec{r}_B)}\chi(\vec{r}-\vec{r}_B)$ we arrive at

$$ES = H_{11} + \lambda H_{12} \qquad (7.7.15)$$
$$\lambda ES = H_{21} + \lambda H_{11} \qquad (7.7.16)$$

giving

$$E = H_{11} + \sqrt{|H_{12}|^2} \qquad (7.7.17)$$

Looking at H_{11} we see that exactly as in the previous subsection, it will contribute $E_\alpha - \beta - \gamma \sum e^{-i\vec{k}\cdot\vec{r}}$, where the sum has been taken over nearest neighbours. E_α is the single atom, β is produced by the potential of the other atoms and γ is the overlap integral restricted to nearest neighbours. The nearest neighbour vectors (AA and BB) are $(\frac{\sqrt{3}a}{2}, \frac{3a}{2}), (-\frac{\sqrt{3}a}{2}, \frac{3a}{2}); (\frac{\sqrt{3}a}{2}, -\frac{3a}{2}), (-\frac{\sqrt{3}a}{2}, -\frac{3a}{2});$ $(\sqrt{3}a, 0), (-\sqrt{3}a, 0)$. This makes

$$
\begin{aligned}
H_{11} &= E_\alpha - \beta \\
&\quad -\gamma \left[2e^{3iak_y/2} \cos \frac{\sqrt{3}ak_x}{2} + 2e^{-3iak_y/2} \cos \frac{\sqrt{3}ak_x}{2} + 2\cos\sqrt{3}ak_x \right] \\
&= E_\alpha - \beta - \gamma \left[4\cos\frac{\sqrt{3}ak_x}{2} \cos\frac{3ak_y}{2} + 2\cos\sqrt{3}ak_x \right]. \qquad (7.7.18)
\end{aligned}
$$

On the other hand H_{12} has the overlap part only, and is determined by the nearest neighbours of the AB variety, which are given by our δ_1, δ_2 and δ_3. If the overlap integral is denoted by t then

$$H_{12} = -t\left[e^{iak_y} + 2e^{-\frac{iak_y}{2}} \cos\frac{\sqrt{3}ak_x}{2} \right]. \qquad (7.7.19)$$

For the magnitude of H_{12} we get

$$|H_{12}|^2 = t^2 \left[1 + 4\cos^2\frac{\sqrt{3}ak_x}{2} + 4\cos\frac{\sqrt{3}ak_x}{2}\frac{3ak_y}{2} \right]. \qquad (7.7.20)$$

We notice from above that H_{12} has zeros at $k_x^c = \frac{2\pi}{3\sqrt{3}}$ and $k_y^c = \pm\frac{2\pi}{3}$, i.e., at the points K_1 and K_2 of the Brillouin zone. If we expand about the corner points of the Brillouin zone, i.e., $\delta k_x = k_x - k_x^c$ and $\delta k_y = k_y - k_y^c$ then

$$
\begin{aligned}
|H_{12}|^2 &= \frac{9}{4a^2}t^2[(\delta k_x)^2 + (\delta k_y)^2] + \text{higher order terms} \\
&\simeq \frac{9t^2}{4a^2}. \qquad (7.7.21)
\end{aligned}
$$

Confining ourselves simply to the nearest neighbours, the AB neighbours alone will dominate and in that approximation we ignore H_{11} in comparison to $|H_{12}|$ and

$$E = \pm\frac{3}{2a}t|\delta k|. \qquad (7.7.22)$$

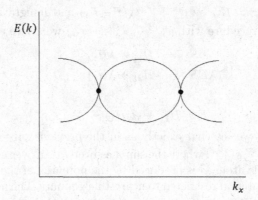

Figure 7.10: Cross-sections of Dirac cones with constant k_y.

This is a remarkable result since the energy spectrum is like that of a massless Dirac particle. To look at this more closely, we can expand the matrix elements of the Hamiltonian more carefully and find that to the lowest order

$$H_{12} = -\frac{3at}{2}e^{2\pi i/3}(\delta k_x + i\delta k_y) \qquad (7.7.23)$$

$$= -\frac{3at}{2}e^{-2\pi i/3}(\delta k_x - i\delta k_y) \qquad (7.7.24)$$

Hence around the point K_1 in the Brillouin zone we can write the Hamiltonian as

$$H = -\frac{3ta}{2}\begin{pmatrix} 0 & e^{2\pi i/3}(\delta k_x + i\delta k_y) \\ e^{-2\pi i/3}(\delta k_x - i\delta k_y) & 0 \end{pmatrix}. \qquad (7.7.25)$$

Writing $p_{x,y} = \hbar k_{x,y}$ and changing the phases of the basis wavefunctions to absorb $e^{\pm 2\pi i/3}$, we have

$$H = \hbar v_F \begin{pmatrix} 0 & p_x + ip_y \\ p_x - ip_y & 0 \end{pmatrix}$$

$$= \hbar v_F (p_x \sigma_\mathbf{x} + p_y \sigma_\mathbf{y}) \qquad (7.7.26)$$

where $v_F = \frac{3at}{2a\hbar}$ and $\sigma_\mathbf{x}$ and $\sigma_\mathbf{y}$ are the Pauli spin matrices. This is the Hamiltonian for a massless Dirac particle in two dimensions. The eigenvalues and eigenfunctions are

$$E = \pm \hbar v_F |\vec{p}|$$

$$\psi_\pm = \frac{1}{\sqrt{2}}(e^{i\theta_p/2}, \pm e^{-i\theta_p/2}) \qquad (7.7.27)$$

where θ_p is the angle of the planar vector $\vec{p}$ with the x-axis.

The phase of the wavefunction changes by π as $\vec{p}$ winds around the origin once and completes a full winding if one winds twice around the origin. Thus the phase can be viewed as a vortex and the corresponding winding number is an integer which is an example of a topological property – a property insensitive to minor deviations (here the path of the winding).

It should be realized that something rather remarkable has happened. Starting from a problem of nonrelativistic electrons, we have arrived at a long wavelength low energy description that corresponds to relativistic fermions. This allows for testing of some particularly relativistic effects like the Klein paradox – the transmission of a relativistic electron satisfying the Dirac equation being fully transmitted after hitting a strong potential barrier.

If we draw the band structure similar to that shown in Figure(7.6), then the section of constant k_y (setting $ak_y = \pm 2\pi/3$) has the form shown in Figure(7.10). Focussing on the region near K_1 (or K_2), we see that the structure consists of two V-shapes (one inverted) meeting at K_1 or K_2. Because of the V-shaped sections these regions are known as the Dirac cones. The condition of meeting of the energy bands ($E = 0$) is valid only at the two points K_1 and K_2 and not over a complete surface. Hence at exactly half filling of the band, the density of states at the Fermi level is exactly zero. In the absence of doping graphene has exactly two electrons per unit cell and hence taking spin into account, the band is indeed half-filled. Thus undoped graphene is a perfect semimetal.

7.8 Problems for Chapter 7

1. For a two-dimensional gas of non-interacting fermions find the chemical potential μ at any temperature T in terms of the number density n. Check to see that as $T \to 0$, the chemical potential correctly reduces to the two-dimensional free energy and that as $T \to \infty$, the appropriate classical limit is obtained.

2. Show that for a three-dimensional electron gas, the ratio of constant pressure specific heat C_p to the constant volume specific heat C_v is given by

$$\frac{C_p}{C_v} - 1 = \frac{\pi^2}{3}\left(\frac{k_B T}{\epsilon_F}\right)^2 + \cdots$$

   for $k_B T < \epsilon_F$

3. Show that for $k_B T \ll \epsilon_F$, if the number of electrons changes by one at fixed temperature, the chemical potential changes by $\Delta\mu = [Vg(\epsilon_F)]^{-1}$, where $g(\epsilon_F)$ is the density of states. Consequently, show hat the maximum change in the occupation probability of a level is $\Delta f = \frac{\epsilon}{6k_B T}\cdot\frac{1}{N}$. This shows that the distribution function is insensitive to a small change in the number of electrons.

4. Find the equation of state of a relativistic completely degenerate electron gas (energy momentum relation is $E^2 = p^2c^2 + m^2c^4$).

5. A sufficiently dense neutrino (rest mass zero) star may be treated as a degenerate gas of relativistic fermions. Find the relation between mass and radius for the equilibrium of such a star.

6. Find the specific heat of a degenerate extreme relativistic electron gas.

7. For photoelectric effect the escape condition for electrons is $\left(\frac{p_z^2}{2m} + h\nu \right) > w$. Show that for $h(\nu - \nu_0) \gg k_B T$, where $h\nu_0 = w - \mu \approx w - \epsilon_F = \phi$, the current density is $J \approx \frac{2\pi m e}{h}(\nu - \nu_0)^2$.

8

Electrons in a Magnetic Field

8.1 Introduction

In this chapter we will deal with various situations that can arise in an electron gas in the presence of a magnetic field. To do so we must first deal with the energy levels of a system of otherwise free electrons put in a uniform magnetic field $\mathbf{B}$ taken to be in the z-direction. The Hamiltonian can be obtained by replacing the momentum $\mathbf{p}$ by $\mathbf{p} + e\mathbf{A}/c$, where $-e$ $(e > 0)$ is the charge of the electron and $\mathbf{A}$ is the vector potential, and hence the Schrodinger equation that needs to be solved is

$$\frac{1}{2m}\left(\mathbf{p} + \frac{e\mathbf{A}}{c}\right)^2 \psi = E\psi. \tag{8.1.1}$$

We now have to fix a gauge. It should be noted that the Schrodinger equation above is gauge invariant in that the transformations $\mathbf{A}(r) \to \mathbf{A}(r) - \nabla\alpha(r)$ and $\psi(r) \to e^{-ie\alpha(r)/\hbar c}\psi(r)$, leaves Eq.(8.1.1) unchanged. We fix the gauge by choosing $A_x = -By$ and $A_y = A_z = 0$ (Landau gauge).

In the Landau gauge, Eq.(8.1.1) becomes

$$\left[\frac{1}{2m}\left(p_x - \frac{eBy}{c}\right)^2 + \frac{p_y^2}{2m} + \frac{p_z^2}{2m}\right]\psi = E\psi. \tag{8.1.2}$$

In the z-direction the motion is that of a free particle with energies given by $E' = p_z^2/2m$, appropriately quantized in a large one dimensional box in that direction. In the $x - y$ plane, the energy eigenvalue equation is

$$\left[\frac{1}{2m}\left(p_x - \frac{eBy}{c}\right)^2 + \frac{p_y^2}{2m}\right]F(x,y) = \epsilon F(x,y). \tag{8.1.3}$$

We note that p_x is clearly a constant of motion and hence for a given value of p_x, ($p_x = 2\pi\hbar n_x/L_x$ is the appropriate quantization), we can write

Eq.(8.1.3) as

$$\left[\frac{p_y^2}{2m} + \frac{1}{2}m\left(\frac{eB}{mc}\right)^2 (y - y_0)^2\right] 2f(y) = \epsilon f(y) \tag{8.1.4}$$

where $y_0 = cp_x/eB$. The energy eigenvalues are

$$\epsilon = \left(n + \frac{1}{2}\right)\frac{eB\hbar}{mc} \tag{8.1.5}$$

and the two-dimensional wavefunction is

$$F(x,y) = \frac{1}{\sqrt{L_x}}e^{ik_x x}N_n \exp\left[-\frac{eB(y-y_0)^2}{\hbar c}\right]H_n\left(\frac{eB(y-y_0)}{\hbar c}\right)$$

where $H_n(\xi)$ (ξ being $\frac{eB(y-y_0)}{\hbar c}$) is the Hermite polynomial of order n and N_n the appropriate normalization factor and $k_x = p_x/\hbar$.

The above energy levels are known as Landau levels and since the operator p_x is a constant of motion they must be degenerate. To find the degeneracy, we note that the energy eigenvalue is independent of k_x and so k_x can take all the possible values that ensure $y_0 = (\hbar c/eB).k_x$ does not exceed the size L_y of the system in the y-direction. With $k_x = 2\pi n_x/L_x$, $y_0 = (\hbar c/BL_x).n_x$ and $\Delta y_0 = (\hbar c/eBL_x)\Delta n_x$, the degeneracy g, which is clearly the maximum possible value of Δn_x, is given by

$$\begin{aligned} g &= (\Delta n_x)_{\text{max}} = \frac{eBL_x}{\hbar c}(\Delta y)_{\text{max}} \\ &= eBL_xL_y/\hbar c = eAB/\hbar c. \end{aligned} \tag{8.1.6}$$

Thus the degeneracy of a Landau level is proportional to the area of the $x - y$ plane (this corresponds to the fact that the trajectory can be centred around any point in the $x - y$ plane) and the applied magnetic field $\mathbf{B}$.

The Landau levels with their degeneracy (independent of the level) are the only things required for a study of the magnetic properties but to conclude this section it is worth pointing out certain features of our result. Imagine a plane with a circular hole in the centre (Fig.8.1).

There is a certain amount of magnetic flux Φ passing through the hole but everywhere in the plane we take $\mathbf{B} = 0$. If $\mathbf{B} = 0$, then the vector potential $\mathbf{A}$ must be pure gauge, i.e. we can write

$$\mathbf{A} = \nabla\alpha. \tag{8.1.7}$$

It must be noted that it is not possible to carry out a gauge transformation that reduces $\mathbf{A}$ to zero everywhere since we must have for any closed path C about the centre in the plane

$$\int_C \mathbf{A}.d\mathbf{l} = \int_S (\nabla \times \mathbf{A}).d\mathbf{S} = \int_S \mathbf{B}.d\mathbf{S} = \Phi \tag{8.1.8}$$

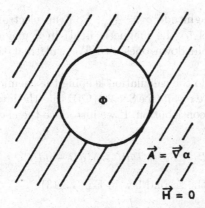

Figure 8.1: $\mathbf{B} = 0$ everywhere in the plane except in a circular region where there is a magnetic flux Φ.

where S is an open surface whose boundary curve is C and hence the surface has to intersect the flux Φ. The solution for α is

$$\alpha = \frac{\Phi\theta}{2\pi}. \qquad (8.1.9)$$

We now consider the solution of the Schrodinger equation for an electron in the plane with the boundary condition that the wave-function has to vanish in the hole. Because of the gauge invariance of the Schrodinger equation, we can try to remove $\mathbf{A}$ from the equation by the transformation $\mathbf{A} \to \mathbf{A} - \nabla\alpha$ but then the wavefunction undergoes the transformation $\psi \to \psi e^{-ie\alpha/\hbar c} = \psi e^{-i\theta e\Phi/\hbar c}$, which makes it discontinuous in space in general, i.e. when $\theta \to \theta + 2\pi$ we do not regain the same wavefunction. There are two ways out of this difficulty,

1. If the wavefunction ψ is localized and hence can vanish where the discontinuity is to occur (i.e. the wavefunction does not spread over a range of $\theta = 2\pi$) then the difficulty is trivially removed. This can be true for electrons trapped by impurities but not for free electrons.

2. If the flux Φ is such that

$$\frac{e\Phi}{hc} = n \cdot \qquad (8.1.10)$$

where n is an integer. In this case changing θ by 2π does not affect anything since $e^{2n\pi i} = 1$. The vector potential can now indeed be gauged away and the electron does not know about the flux. This is the flux quantization in superconductivity.

8.2 Magnetic Properties at $T = 0$

The magnetism of the electron gas comes from two sources – the orbital motion and the intrinsic magnetic moment of the electron. The electron has an intrinsic magnetic moment of magnitude $\sigma = \frac{|e|\hbar}{2mc}$. Spin up electrons (i.e. magnetic moment, aligned parallel to the magnetic field) in a magnetic field B have energy $-\sigma B$ and the spin down ones σB. We will treat the two sources of magnetism separately.

It is convenient to do the calculation keeping the chemical potential fixed. For a magnetic system, we recall (see Sec.2.1(C))) that the internal energy needs to include the magnetic contribution. If we introduce the chemical potential in Eq.(2.1.25), then we have

$$TdS = dE + PdV + MdB - \mu dN. \tag{8.2.1}$$

Using this form of the second law in Eq.(3.3.13),

$$
\begin{aligned}
d\tilde{F} &= dE - TdS - SdT - \mu dN - Nd\mu \\
&= -Nd\mu - MdB - SdT - PdV.
\end{aligned} \tag{8.2.2}
$$

Hence,

$$M = -\left(\frac{\partial \tilde{F}}{\partial B}\right)_{\mu,T,V}. \tag{8.2.3}$$

As shown in Eq.(3.3.13), $\tilde{F} = -PV$ and we arrive at the final formula,

$$M = \left[\frac{\partial(PV)}{\partial B}\right]_{\mu,T,V} \tag{8.2.4}$$

which we are going to use for the orbital contribution.

8.2.1   Intrinsic Magnetic Moment

If the total number of electrons is N, we denote the number of up spins by N_1 and down spins by N_2, so that $N_1 + N_2 = N$. The fermi energy ϵ_F will be as written down in Eq.(7.1.3) and is determined by N and V. Since the energy of an up-spin electron is $\frac{p^2}{2m} - \sigma B$ the range of kinetic energy available is from 0 to $\mu + \sigma B$ where μ is the chemical potential. The number N_1 of up- spins is consequently

$$N_1 = 2\pi V \left(\frac{2m}{h^2}\right)^{3/2} \int_0^{\mu+\sigma B} \epsilon d\epsilon = \frac{4\pi V}{3}\left(\frac{2m}{h^2}\right)^{3/2}(\mu+\sigma B)^{3/2}. \tag{8.2.5}$$

Similarly,

$$N_2 = \frac{4\pi V}{3}\left(\frac{2m}{h^2}\right)^{3/2}(\mu-\sigma B)^{3/2}$$

and hence

$$N = \frac{4\pi V}{3}\left(\frac{2m}{h^2}\right)^{3/2}\left[(\mu+\sigma B)^{3/2} + (\mu-\sigma B)^{3/2}\right]. \tag{8.2.6}$$

Our interest is in small values of B, such that $\sigma B \ll \mu$ since we want to calculate zero field susceptibilities. Accordingly, we can carry out a low field expansion of Eq.(8.2.2) as

$$N = \frac{8\pi V}{3}\left(\frac{2m}{h^2}\right)^{3/2}\mu^{3/2}\left[1 + \frac{3}{8}\left(\frac{\sigma B}{\mu}\right)^2 + \cdots\right]. \qquad (8.2.7)$$

Using the expression for fermi energy from Eq.(7.1.2), we have

$$\mu = \epsilon_F\left[1 - \frac{1}{4}\left(\frac{\sigma B}{\epsilon_F}\right)^2 + \cdots\right]. \qquad (8.2.8)$$

The difference between the number of up- spin and down-spin electrons is given by

$$N_1 - N_2 \approx \frac{3}{2}N\frac{\sigma B}{\epsilon_F}. \qquad (8.2.9)$$

It should be noted that for static spins (as we are considering now), the magnetization $M = \sigma(N_1 - N_2)$ and hence the above expression gives

$$M = \frac{3}{2}N\frac{\sigma^2 B}{\epsilon_F} \qquad (8.2.10)$$

and a spin susceptibility of

$$\chi_s = \frac{3}{2}\frac{N\sigma^2}{\epsilon_F}. \qquad (8.2.11)$$

To obtain the same result differently, we need the energy E. Since up-spin electrons have an energy $\epsilon - \sigma B$ (here ϵ is the kinetic energy) and spin down electrons have an energy $\epsilon + \sigma B$, the total energy of the electron gas is

$$\begin{aligned}
E &= 2\pi V\left(\frac{2m}{h^2}\right)^{3/2}\left[\int_0^{\mu+\sigma B}(\epsilon^{3/2} - \epsilon^{1/2}\sigma B)d\epsilon + \int_0^{\mu-\sigma B}(\epsilon^{3/2} + \epsilon^{1/2}\sigma B)d\epsilon\right] \\
&= \frac{4\pi V}{5}\left(\frac{2m}{h^2}\right)^{3/2}\left[(\mu + \sigma B)^{5/2} + (\mu - \sigma B)^{5/2}\right] - \frac{3}{2}N\epsilon_F\left(\frac{\sigma B}{\epsilon_F}\right)^2 \\
&= \frac{8\pi V}{5}\left(\frac{2m}{h^2}\right)^{3/2}\epsilon_F^{5/2}\left[1 + \frac{5}{4}\left(\frac{\sigma B}{\epsilon_F}\right)^2\right] - \frac{3}{2}N\epsilon_F\left(\frac{\sigma B}{\epsilon_F}\right)^2 \qquad (8.2.12)
\end{aligned}$$

where use has been made of Eq.(8.2.8).

Magnetic susceptibility due to the spin is $\chi_s = -\frac{\partial^2 E}{\partial B^2}$ and is found to be

$$\chi_s = \frac{3}{2}N\frac{\sigma^2}{\epsilon_F} \qquad (8.2.13)$$

entirely in agreement with Eq.(8.2.11). This is the Pauli paramagnetic susceptibility.

8.2.2   Orbital Magnetic Moment

The motion of the electron in the magnetic field B, as seen from Sec.8.1, leads to the allowed energies as

$$E = \left(n + \frac{1}{2}\right)\frac{e\hbar}{mc}.B + \frac{p_z^2}{2m} = \left(n + \frac{1}{2}\right)2\sigma B + \frac{p_z^2}{2m} \tag{8.2.14}$$

We see from Eq.(6.3.1) [the sum over "i" modified to accommodate the correct degeneracy factor] that for the electron in a magnetic field

$$PV = k_B T \sum_i \ln Q_F^{(i)} = 2k_B T.\frac{e\hbar}{2mc}.AB\frac{2\pi}{h}.\frac{2m}{h^2}L$$

$$\times \int_{-\infty}^{\infty} dp_z \sum_{n=0}^{\infty} \ln\left[1 + \exp\left\{\beta\left(\mu - \frac{p_z^2}{2m} - (n + \frac{1}{2}).2\sigma B\right)\right\}\right].$$

$$\tag{8.2.15}$$

To manipulate the right hand side of the above equation, we need to use the Euler Maclaurin formula to convert sums to integrals. For a function $f(x)$, this reads

$$\sum_{n=0}^{\infty} f(a + n) = \int_a^{\infty} f(x)dx + \frac{1}{2}f(a) - \frac{1}{12}f'(a) + \dots \tag{8.2.16}$$

when we have $f(\infty) = f'(\infty) = 0$ as we have in Eq.(8.2.15). We need $a = 1/2$ for Eq.(8.2.15), so that

$$\sum_{n=0}^{\infty} f\left(n + \frac{1}{2}\right) = \int_{1/2}^{\infty} f(x)dx + \frac{1}{2}f(1/2) - \frac{1}{12}f'(1/2)$$

$$= \int_0^{\infty} f(x)dx - \int_0^{1/2} f(x)dx$$

$$+ \frac{1}{2}f(1/2) - \frac{1}{12}f'(1/2). \tag{8.2.17}$$

In the range $0 \leq x \leq 1/2$ we expand

$$f(x) = f(0) + xf'(0) + \dots$$

to write

$$\int_0^{1/2} f(x)dx = \frac{1}{2}f(0) + \frac{1}{8}f'(0) + \dots.$$

Similarly,

$$f(1/2) = f(0) + \frac{1}{2}f'(0)$$

and

$$f'(1/2) = f'(0).$$

Working to first order in the derivative,

$$
\begin{aligned}
\sum_{n=0}^{\infty} f\left(n + \frac{1}{2}\right) &= \int_0^{\infty} f(x)dx - \frac{1}{2}f(0) + \frac{1}{8}f'(0) \\
&+ \frac{1}{2}f(0) + \frac{1}{4}f'(0) - \frac{1}{12}f'(0) \\
&= \int_0^{\infty} f(x)dx + \frac{1}{24}f'(0) + O(f''(0)). \quad (8.2.18)
\end{aligned}
$$

Returning to Eq.(8.2.15) and using Eq.(8.2.18), (note that $AL = V$, the volume) we have

$$
\begin{aligned}
PV &= \frac{4\pi V}{h} \cdot \frac{2m}{h^2} \sigma B k_B T \int_{-\infty}^{\infty} dp_z \int_0^{\infty} dx \ln\left[1 + e^{\beta(\mu - \frac{p_z^2}{2m} - 2\sigma Bx)}\right] \\
&- \frac{4\pi V}{h} \cdot \frac{2m}{h^2} \cdot \frac{(\sigma B)^2}{12} \int_{-\infty}^{\infty} \frac{dp_z e^{\beta(\mu - \frac{p_z^2}{2m})}}{1 + e^{\beta(\mu - \frac{p_z^2}{2m})}} + O(B^3). \quad (8.2.19)
\end{aligned}
$$

The first term on the right hand side above is a function of μ alone (write $y = \sigma Bx$ to see this). For $T \to 0$ the second term is non zero (equal to unity) only if $|p_z| \leq \sqrt{2m\mu}$. Hence Eq.(8.2.19) becomes

$$PV = G(\mu) - \frac{8\pi V}{12}(\sigma B)^2 \left(\frac{2m}{h^2}\right)^{3/2} \epsilon_F^{1/2} + O(B^3) \quad (8.2.20)$$

The susceptibility χ_{orb} is given by $\left.\frac{\partial^2}{\partial B^2}(PV)\right|_{\mu,T,V}$ from Eq.(8.2.4). We find

$$
\begin{aligned}
\chi_{orb} &= -\frac{8\pi V}{6}\sigma^2 \left(\frac{2m}{h^2}\right)^{3/2} \epsilon_F^{1/2} \\
&= -\frac{N\sigma^2}{2\epsilon_F} = -\frac{1}{3}\chi_s. \quad (8.2.21)
\end{aligned}
$$

This is the Landau diamagnetism coming from the orbital motion. The total magnetic response of the electron gas is $\chi_{orb} + \chi_s = \frac{2}{3}\chi_s$, which is a paramagnetic response.

8.3   Magnetic Properties at $T \gg T_F$

At temperatures high enough that $T \gg T_F$, the average number of fermions is $\langle n_i \rangle = e^{-\beta(\epsilon_i - \mu)}$. The chemical potential is large and negative in this limit and that triggers this approximation. The number of spin up electrons ($\epsilon_i = \frac{p^2}{2m} - \sigma B$) is

$$
\begin{aligned}
N_1 &= \frac{V}{h^3} (2m)^{3/2} e^{(\mu + \sigma B)/k_B T} \int_0^{\infty} \frac{dp}{\sqrt{2m}} \cdot \frac{4\pi p^2}{2m} \cdot \exp\left(-\frac{p^2}{2mk_B T}\right) \\
&= V \left(\frac{2\pi mk_B T}{h^2}\right)^{3/2} e^{(\mu + \sigma B)/k_B T}.
\end{aligned}
\tag{8.3.1}
$$

Similarly,

$$
N_2 = V \left(\frac{2\pi mk_B T}{h^2}\right)^{3/2} e^{(\mu - \sigma B)/k_B T}.
\tag{8.3.2}
$$

For $\sigma B \ll k_B T$, the total number of spins is

$$
N = N_1 + N_2 = 2V \left(\frac{2\pi mk_B T}{h^2}\right)^{3/2} e^{\mu/k_B T} + O(B^2).
\tag{8.3.3}
$$

The spin imbalance is

$$
\begin{aligned}
N_1 - N_2 &= 2V \left(\frac{2\pi mk_B T}{h^2}\right)^{3/2} e^{\mu/k_B T} \cdot \frac{\sigma B}{k_B T} \\
&= N \frac{\sigma B}{k_B T}.
\end{aligned}
\tag{8.3.4}
$$

The corresponding magnetization is $\sigma^2 N B / kT$ and the susceptibility is

$$
\chi_s(T \gg T_F) = \frac{\sigma^2 N}{k_B T}.
\tag{8.3.5}
$$

The orbital magnetism is obtained from Eq.(8.2.19) . The second term on the right hand side now needs an approximation appropriate to large and negative μ i.e., $e^{\mu/k_B T} \ll 1$. Consequently

$$
\begin{aligned}
PV &= G(\mu, T \gg T_F) - \frac{4\pi V e^{\beta\mu}}{h} \cdot \frac{2m}{h^2} \frac{(\sigma B)^2}{12} \int_{-\infty}^{\infty} \exp\left(-\frac{p_z^2}{2mkT}\right) dp_z \\
&= G(\mu, T \gg T_F) - \frac{4\pi V e^{\beta\mu}}{h} \cdot \frac{2m}{h^2} \frac{(\sigma B)^2}{12} \cdot \sqrt{2\pi mkT} \\
&= G(\mu, T \gg T_F) - 4V \left(\frac{2\pi mkT}{h^2}\right)^{3/2} \frac{e^{\beta\mu}}{kT} \cdot \frac{(\sigma B)^2}{12} \\
&= G(\mu, T \gg T_F) - \frac{N}{kT} \frac{(\sigma B)^2}{6}.
\end{aligned}
\tag{8.3.6}
$$

The orbital susceptibility follows as

$$\chi_{orb}(T \gg T_F) = -\frac{N\sigma^2}{3kT} = -\frac{\chi_s}{3}(T \gg T_F). \qquad (8.3.7)$$

Once again the total susceptibility is $\frac{2}{3}\chi_s$. It can he shown that at all temperatures

$$\chi_{orb} = -\frac{\chi_s}{3}.$$

8.4   De Haas - Van Alphen Effect: $T = 0$

We now consider the effect of a finite field at $T = 0$. Clearly in this limit the sum formula of Eq.(8.2.17) is no longer valid as the $f(x)$ of Eq.(8.2.15) need not change by a small amount when n goes to $n+1$. This will happen when the argument approaches zero and it is for this condition that it is essential that $\sigma B \ll kT$ (the criterion that ensures a small B) be valid. Since this cannot be ensured for $T = 0$ we do not use Eq.(8.2.17) in this section. We calculate the total energy E of N electrons confined to a plane with single particle energies being given by Eq.(8.1.5) as

$$\epsilon_j = \left(j + \frac{1}{2}\right)2\sigma B. \qquad (8.4.1)$$

The degeneracy factor (including the spin degeneracy)

$$g = 2eAH/hc = N.B/B_0 \qquad (8.4.2)$$

where N is the number of particles and $B_0 = nhc/2e$, with $n = N/A$ being the number density of the electrons.

At $T = 0$, all the particles tend to settle in the lowest energy states and clearly if $B > B_0$, $g > N$ and hence all the electrons can be accommodated in the lowest Landau level. Hence if $B > B_0$,

$$E_0 = N\sigma B \qquad (8.4.3)$$

while for $B < B_0$, not all particles can be accommodated in the ground Landau level. Suppose that B is such that j-th Landau level is completely filled while some spill over to the $(j + 1)$-th level. Clearly then

$$(j + 1)g < N < (j + 2)g$$

or

$$\frac{1}{j+2} < \frac{B}{B_0} < \frac{1}{j+1}. \qquad (8.4.4)$$

For B in this interval, the energy is

$$
\begin{aligned}
E &= g\sum_{i=0}^{j}(2j+1)\sigma B + [N - g(j+1)](2j+3)\sigma B \\
&= \sigma B[gj(j+1) + (j+1)g - g(j+1)(2j+3) + N(2j+3)] \\
&= \sigma B[N(2j+3) - g(j+1)(2j+2)] \\
&= \sigma B[N(2j+3) - N(B/B_0)(j+1)(j+2)] \\
&= N\sigma B_0\left[(2j+3)\frac{B}{B_0} - (j+1)(j+2)\left(\frac{B}{B_0}\right)^2\right].
\end{aligned}
\tag{8.4.5}
$$

Writing $x = B/B_0$, we summarize the above results as

$$
E = \begin{cases} N\mu_0 B_0 x & \text{if } x > 1 \\ N\mu_0 B_0[(2j+3)x - (j+1)(j+1)(j+2)x^2] & \text{if } \frac{1}{j+2} < x < \frac{1}{j+1} \end{cases}.
\tag{8.4.6}
$$

The magnetization can be calculated from the relation $M = -\partial(E/B_0)/\partial x$ and susceptibility from $\chi = \partial(M/B_0)/\partial x$. Hence

$$
M = \begin{cases} -N\mu_0 & \text{if } x > 1 \\ -N\mu_0[2j+3 - 2(j+1)x] & \text{if } \frac{1}{j+2} < x < \frac{1}{j+1} \end{cases}
\tag{8.4.7}
$$

and

$$
\chi = \begin{cases} 0 & \text{if } x > 1 \\ 2(j+1)(j+2)N\mu_0/B_0 & \text{if } \frac{1}{j+2} < x < \frac{1}{j+1} \end{cases}.
\tag{8.4.8}
$$

As shown in Fig.8.2, the magnetization M in units of $\mu_0 N$ oscillates between $+1$ and -1 as B is varied in the range $0 < B < B_0$ and the grows linearly.

The susceptibility exhibits steps as a function of B decreasing to a zero value for $B \geq B_0$ (Fig.8.3). The oscillations in magnetization were discovered experimentally in a metal by De Haas and Van Alphen in 1931 and predicted for an electron gas by Landau in 1930.

8.5 Quantum Hall Effect

8.5.1 Integer Hall Effect

Hall effect deals with the situation of crossed electric and magnetic fields. We imagine a situation with the electric field $\mathbf{E}$ in the x-direction and the magnetic field $\mathbf{B}$ in the z-direction. The electrons which acquire a velocity in the x-direction due to the electric field are deflected in the y-direction due to the Lorentz force arising from the x-velocity and the magnetic field along the z-axis. This sets up a current in the y-direction. We will immediately specialize to a two-dimensional situation embedded in a three-dimensional space with the magnetic field chosen in the z-direction and the velocity of the charge carriers

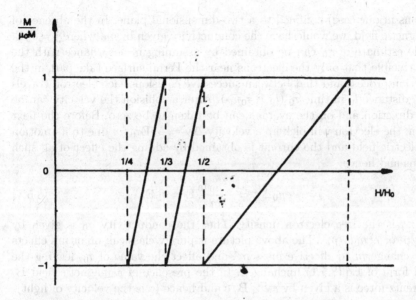

Figure 8.2: Oscillations in magnetization.

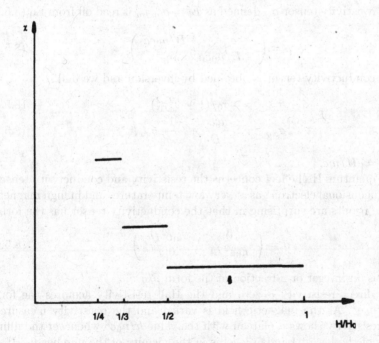

Figure 8.3: The susceptibility at $T = 0$ as a function of B/B_0.

(electrons in our case) confined to a two-dimensional plane. In the absence of the magnetic field, we would have the conductivity given by σ_0, where $\mathbf{j} = \sigma_0 \mathbf{E}$. A crude estimate of σ_0 can be obtained by assuming in accordance with the Pauli principle that only the electrons near the Fermi surface take part in the conduction process and that between successive collisions each electron travels a mean distance l_0 in time τ_0 ($l_0 \cong \tau_0 v_F$). After a collision the velocity can be in any direction and on the average can be taken to be zero. Before the next collision, the electron will pick up a velocity $\Delta \mathbf{v} = -e\mathbf{E}\tau_0/m$ due to its motion in an electric field and the current is obtained by adding the effect of all such electrons and hence

$$\mathbf{j} = -en_0 \Delta \mathbf{v} = \frac{e^2 n_0 \tau_0}{m} \mathbf{E} = \sigma_0 \mathbf{E}. \qquad (8.5.1)$$

where n_0 is the free electron density. Thus the conductivity σ_0 is given by $n_0 e^2 \tau_0/m$ or $e^2 n_0 l_0/p_F$. The above picture is purely classical: quantum effects usually change m to effective mass m^* and affect the value of τ_0, leaving the general form of Eq.(8.5.1) unchanged. In the presence of a magnetic field $\mathbf{B}$, the Lorentz force is a given by $e\mathbf{v} \times \mathbf{B}/c$ and hence (c is the velocity of light)

$$\mathbf{j} = \sigma_0 [\mathbf{E} - \mathbf{j} \times \mathbf{B}/n_0 ec]. \qquad (8.5.2)$$

The resistivity tensor ρ_{ij} defined as $E_l = \rho_{lm} j_m$ is read off from Eq.(8.5.2) as

$$\rho = \begin{pmatrix} \sigma_0^{-1} & B/n_0 ec \\ -B/n_0 ec & \sigma_0^{-1} \end{pmatrix} \qquad (8.5.3)$$

The conductivity tensor is obtained by inversion and we find

$$\sigma_{xx} = \sigma_0/(1 + \omega_c^2 \tau_0^2) \qquad (8.5.4)$$

$$\sigma_{xy} = \frac{n_0 ec}{B} + \frac{1}{\omega_c \tau_0} \sigma_{xx}. \qquad (8.5.5)$$

where $\omega_c = eH/mc$.

The quantum Hall effect concerns the resistivity and conductivity tensors of a two dimensional electron gas at very low temperatures and in high magnetic fields. The results are surprising in that the conductivity tensor has the form

$$\sigma = \begin{pmatrix} 0 & -n_0 e^2/h \\ n_0 e^2/h & 0 \end{pmatrix} \qquad (8.5.6)$$

where n_0 is an integer or a fraction of the form p/q.

The direct resistivity is zero and the Hall resistivity acquires the form $\rho_{xy} = h/n_0 e^2$. As the magnetic field is varied and the resistivity measured, the Hall resistivity shows a plateau with the value $h/n_0 e^2$ whenever the filling fraction ν of the lowest Landau level is in the vicinity of the number n_0. If n_0 is the number density of the charge carriers, then from the degeneracy of the Landau levels, we have

$$\nu = \frac{hcn_0}{eH} \qquad (8.5.7)$$

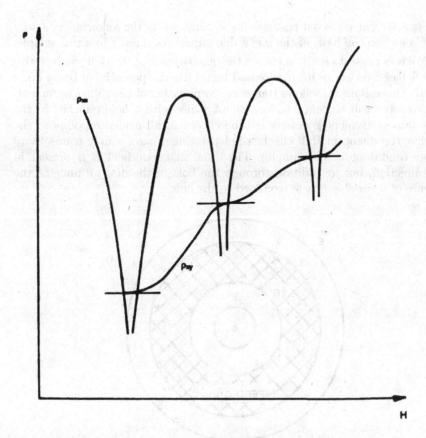

Figure 8.4: Resistivities as functions of magnetic fields.

and whenever ν acquires a value near n_0 with changing B, one observes a plateau in ρ_{xy} with ρ_{xx} simultaneously dropping down to zero. This situation is shown in Fig.8.4. The vanishing of σ_{xx} makes the system dissipationless and hence related to superfluidity and superconductivity. The vanishing of σ_{xx} is essential in understanding the quantization of σ_{xy}.

To set up a naive picture for the integer quantum Hall effect, so called when n_0 is an integer, we imagine varying B and arriving at a value for which $\nu = 1$, i.e. the lowest Landau level is completely filled. Then the Fermi level must lie exactly halfway between the first and second Landau levels and hence one would require a finite amount of energy to make new states available to the electrons. Scatterings would become difficult, τ_0 would tend to infinity and consequently, we would have $\rho_{xx} = \rho_{yy} = 0$ and ρ_{xy} quantized as seen. This would occur for all integral filling fractions and we could have an explanation for the integral effect. However, this can only explain the quantized values of σ_{xy} and dissipationless situation at certain discrete values of B. Experimentally what one observes are plateaus in the conductivity (or resistivity) vs. B plot (see Fig.8.4). The observed plateaus, as we shall see in the following, can only be explained on the basis of the inevitable impurities present in a real sample.

We now present an argument for the quantization of the Hall conductivity which is based on gauge invariance and hence has the potential of being quite general. This discussion follows Halperin's formulation of Laughlin's argument. The geometry will be taken to be that of a disc with a hole punched in the centre (any convenient geometry can be chosen as all evidence points to the fact that the quantum Hall effect is independent of size, shape, connectivity or edge conditions in the sample). The usual magnetic field **B** is present in the z-direction, but in addition through the hole in the disc we imagine the existence of a variable flux Φ (confined to the hole).

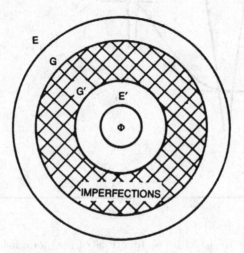

Figure 8.5: The geometry for the discussion of resistivity quantization.

As shown in Fig.8.5, the sample is a disc of realistic region with impurities, the boundaries being designated by GG'. The regions EG and $G'E'$ are completely devoid of impurities and act as guard rings for the impurities. We imagine that the total number of electrons is such that exactly n_0 Landau levels are filled in the guard rings. Since there are no impurities in the guard rings, the Fermi level must lie between two Landau levels in the guard rings. The guard rings serve to relate the conducting properties of the realistic region to that of the ideal region.

We now consider changing the flux Φ adiabatically. There will exist a circumferential electric field **E** everywhere in the disc now, such that if we take any closed contour surrounding the hole, we must have

$$\frac{1}{c}\frac{d\Phi}{dt} = \frac{1}{c}\int \frac{\partial \mathbf{B}}{\partial t}.d\mathbf{S} = -\frac{1}{c}\oint \mathbf{E}.d\mathbf{l} = -\frac{1}{c}\oint E_\theta R d\theta = -\frac{1}{c}\oint \rho_{\theta r} j_r R d\theta$$

$$(8.5.8)$$

where E_θ is the circumferential electric field, $\rho_{\theta r}$ is the off-diagonal element of the resistivity tensor, and j_r is the radial current.

Let the loop C be in one of the guard rings (say EG). The resistivity is now given by $\rho_{\theta\theta} = \rho_{rr} = 0$ and $\rho_{\theta r} = h/ne^2$ and if the flux change is $\Delta\Phi = \Phi_0 = hc/e$, then we have (a circumferential E field in this case means a Hall current in the radial direction)

$$\frac{h}{e} = \frac{1}{c} \cdot \Delta\Phi = \int dt \oint_C j_r \frac{h}{n_0 e^2} dl$$

or

$$n_0 e = \int dt \oint_C j_r dl$$

$$= \text{Charge transferred across the loop } C. \qquad (8.5.9)$$

The electrons transported across EG must have come from the edge E (say) and will cross into the sample at G. At a physical edge such as E, there is a quasi continuum of edge states for each Landau level as illustrated in Fig.8.6.

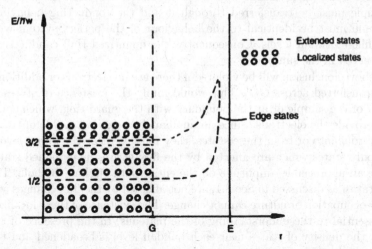

Figure 8.6: Energy levels of the ring system vs. radius.

These edge states will be filled to the Fermi level. We now explain how the current flows across the guard ring. The electron density across the region EG with a completely filled Landau level will appear as shown in Fig.8.7. As we start changing the flux, the electrons have to respond to the change, until the change in flux becomes $\Delta\Phi = \Phi_0$. Gauge invariance implies that if $\Delta\Phi = \Phi_0$, the effect of the vector potential can be removed by a gauge transformation and thus the electron distribution has to look exactly as it did before the flux change was turned on.

So the distribution of Fig.8.7 has to be regained for $\Delta\Phi = \Phi_0$, which can occur if an integral number of electrons have moved across the region EG. This movement, as expected, results in an infinitesimal change of energy.

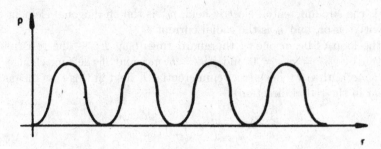

Figure 8.7: Level density as a function of radius in the impurity-free region.

Now assume that there is no dissipation in the sample GG'. This will happen if the Fermi level lies in the region of the localized states in Fig.8.6. This assumption is equivalent to saying that the current is normal to the electric field or else the diagonal component of the conductivity tensor vanishes. Now if there is no build-up of charge in GG' or its edges, the electrons delivered to the sample must be transferred through it and the conducting behaviour of the sample must be identical to the behaviour of the perfect system with integrally filled Landau levels. Consequently, the quantized Hall conductivity will be observed in the sample.

The above conclusion will be violated if there are no electrons or additional electrons transferred across GG'. This would imply the existence of states in the interior of the sample or at the boundary with the guard ring, which could accept or provide the electrons. Since an adiabatic process is capable of transferring electrons into or from these states, they must be near the Fermi level.

The only states which are affected by the flux Φ are the extended states which have an appreciable amplitude on the entire path circling the hole. The localized states, as discussed in Sec.8.1, are not affected by the flux changes and hence their occupation or nature cannot change. The existence of these localized states is essential to the understanding of the plateaus. In the presence of the impurities, the density of states near each Landau level is broadened and the situation is as shown in Fig.8.8.

Now as the magnetic field is decreased (i.e. the density is increased) the localized states gradually fill up without changing the occupation of the extended states and hence with no change in the Hall conductance. Thus, over a range of magnetic field values, while the Fermi level E_F lies between the core of the extended states in two successive Landau levels, the Hall conductivity is a constant and it is only when E_F crosses the core of the extended states on the next Landau level that the conductivity jumps to the next plateau.

Thus if the Fermi level lies in a mobility gap, so that the diagonal conductivity vanishes, the Hall conductivity gets quantized. The trivial case, $n_0 = 0$, corresponds to an insulator. If the impurity is weak enough and the impurity region connects smoothly to an impurity-free region, the non-trivial cases $n_0 = 1, 2, 3, \ldots$ etc. are realized and we have the integral quantum Hall effect.

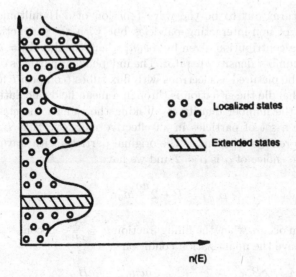

Figure 8.8: Density of states in the presence of impurities.

8.5.2 Fractional Hall effect

The fractional quantum Hall effect occurs when the n_0 of the previous subsection, instead of being an integer, is a fraction of the form p/q, where p and q are integers. The fractional quantum Hall effect requires the inclusion of the electron-electron interaction for its explanation and cannot be couched in the simple terms given above. The nature of the electronic states in the guard ring would depend on the Coulomb repulsion of electrons and to explain the fractional Hall effect in the above terms one would have to show that for the interacting case, a transfer of fractional number of electrons occurs if the flux changes by an integer number of flux quanta. This particular formulation of the fractional effect is not yet available, although Laughlin's picture of the incompressible quantum fluid when the filling fraction ν takes on certain values like 1/3, 2/3, 2/5, etc. gives a satisfactory picture of the phenomenon.

The fractional fillings can be handled if one takes into account the Coulomb interaction between the electrons. For a N electron system, the hamiltonian is

$$B = \sum_i \frac{1}{2m}\left(\mathbf{p}_i - e\frac{\mathbf{A}_i}{c}\right)^2 + \sum_{i\neq j} \frac{e^2}{|\mathbf{r}_i - \mathbf{r}_j|}. \tag{8.5.10}$$

The free particle part of the hamiltonian being highly degenerate, perturbation theory in not practical. If the wave function of the above hamiltonian is $\Psi(\{\mathbf{r}_i\})$, one makes a transformation to a function $\chi(\{\mathbf{r}_i\})$, such that

$$\chi(\{\mathbf{r}_i\}) = \Psi(\{\mathbf{r}_i\})\,\Pi_{i\neq j}\exp\left(i\alpha.\arg(\mathbf{r}_i - \mathbf{r}_j)\right). \tag{8.5.11}$$

In the above, α is a parameter which determines the statistics of χ and $\arg(\mathbf{r}_i - \mathbf{r}_j)$ is the angle θ which the vector $\mathbf{r}_i - \mathbf{r}_j$ makes with the x-axis.

The function χ turns out to be the wave function of a Hamiltonian which is a Hamiltonian of non-interacting particles but with the magnetic field $\mathbf{B}$ acquiring an extra contribution given by $\alpha \Phi_0 \hat{n}$, where the flux $\Phi_0 = hc/e$, and $\hat{n}$ is the electron number density operator. The independent particles of the new Hamiltonian can be pictured as electrons with flux tubes (composite fermions).

The way to handle the situation is through a mean field calculation where $\hat{n}$ is replaced by the number density n_0. Taking the sign of the charges into account, we have a set of particles in an effective magnetic field B_{eff}. If the χ-particles are to remain fermions as the original particles with wave function Ψ, then a possible choice of α is $\alpha = 2$ and we have

$$B_{eff} = B - 2\frac{hc}{e}.n_0. \tag{8.5.12}$$

Quantization occurs when the filling fraction $\nu = \frac{hcn_0}{eB_{\mathrm{eff}}}$ becomes an integer p and hence we have the quantization condition

$$\frac{1}{p} = \frac{eB_{\mathrm{eff}}}{n_0 hc} = \frac{e}{n_0 hc}(B - 2\frac{hcn_0}{e}) = \frac{eB}{n_0 hc} - 2. \tag{8.5.13}$$

The critical values of $eB/n_0 hc$ for quantization are

$$\frac{eB}{n_0 hc} = \frac{p}{1 + 2p}, \qquad \text{where } p \text{ is an integer.} \tag{8.5.14}$$

This explains the occurrence of such filling fractions as $1/3$, $2/5$... etc.

The above picture developed by Jain and collaborators has gradually been extended to include a much larger number of filling fractions.

9

Bose-Einstein Distribution: Examples

9.1 Bose-Einstein Condensation

9.1.1 The Condensation Point

In Chapter 6, we dealt with the dilute Bose gas where $\lambda^3 n \ll 1$. In this section, we turn to the other limit where the quantum effects are all important. We shall deal with spin 0 particles for simplicity and $2s + 1 = 1$ in the density of states of Eq.(6.2.10). The fundamental equation which we want to explore in detail is

$$N = V.2\pi \left(\frac{2m}{h^2}\right)^{3/2} \int\limits_0^\infty \frac{\epsilon^{1/2} d\epsilon}{e^{(\epsilon-\mu)\beta} - 1}. \tag{9.1.1}$$

For the dilute Bose gas, the appropriate analysis begins with the assumption $e^{-\mu\beta} \gg 1$. In the case we want to discuss now, $e^{-\mu\beta}$ will be near its minimum possible value, namely unity. At high temperatures $e^{-\mu\beta} \gg 1$ and its value decreases as we lower the temperature [see Eq.(6.3.8)]. If $e^{-\mu\beta}$ should become equal to unity at some temperature T_c, Eq.(9.1.1) does not have a solution for temperatures $T < T_c$, since $e^{-\mu\beta}$ cannot be smaller than unity. Separating the $\epsilon = 0$ term, we write Eq.(9.1.1) as

$$N = N_0(\mu) + 2\pi V \left(\frac{2m}{h^2}\right)^{3/2} \int\limits_0^\infty \frac{\epsilon^{1/2} d\epsilon}{e^{(\epsilon-\mu)\beta} - 1} \tag{9.1.2}$$

where $N_0(\mu)$ is the ground state occupancy given by $N_0(\mu) = e^{\beta\mu}/(1 - e^{\beta\mu})$. Writing $\beta\epsilon = x$, Eq.(9.1.2) becomes

$$N = N_0(\mu) + 2\pi V \left(\frac{2mk_B T}{h^2}\right)^{3/2} \int\limits_0^\infty \frac{x^{1/2} dx}{e^{-\mu\beta}.e^x - 1}. \tag{9.1.3}$$

The difficulty mentioned above becomes transparent from Eq.(9.1.3). If at the temperature T_c, $e^{\mu\beta} = 1$ then Eq. (9.1.1) becomes

$$N = 2\pi V \left(\frac{2mk_BT_c}{h^2}\right)^{3/2} \int_0^\infty \frac{x^{1/2}dx}{e^x - 1} = 2\pi V \left(\frac{2mk_BT_c}{h^2}\right)^{3/2} \zeta_{3/2}(1). \quad (9.1.4)$$

where the function

$$\xi_{3/2}(z) = \sum_{n=0}^\infty \frac{z^{n+1}}{(n+1)^{3/2}} \quad (9.1.5)$$

is clearly a monotonically increasing function of z for $z > 0$ and rises from 0 for $z = 0$ to $\xi_{3/2}(1) = 2.612$ at $z = 1$ (the derivative is discontinuous at $z = 1$).

For $T < T_c$ the prefactor of the integral in Eq.(9.1.3) has to decrease and hence the integral itself has to increase to keep the total number of particles fixed at N. The integral can increase only if $e^{-\mu\beta} < 1$ for $T < T_c$ but such values of $e^{-\mu\beta}$ are not allowed in Bose-Einstein statistics. In the presence of $N_0(\mu)$, we arrive at a solution to this problem by maintaining $\mu = 0$ for $T < T_c$ and letting $N_0(\mu)$ vary to keep Eq.(9.1.3) satisfied with $e^{-\mu\beta} = 1$. Writing $e^{\mu\beta} = 1$, we can write Eq.(9.1.3) as

$$
\begin{aligned}
N &= N_0 + 2\pi V \left(\frac{2mk_BT}{h^2}\right) \int_0^\infty \frac{x^{1/2}e^{-x}}{1 - e^{-x}} dx \\
&= N_0 + 2\pi V \left(\frac{2mk_BT}{h^2}\right) \int_0^\infty \sum_{n=0}^\infty x^{1/2}e^{-(n+1)x} dx \\
&= N_0 + V \left(\frac{2\pi mk_BT}{h^2}\right)^{3/2} \sum_{n=0}^\infty \frac{1}{(n+1)^{3/2}} \\
&= N_0 + V \left(\frac{2\pi mk_BT}{h^2}\right)^{3/2} \xi_{3/2}(1). \quad (9.1.6)
\end{aligned}
$$

We now write Eq.(9.1.6) as

$$N_0 = N - V \left(\frac{2\pi mk_BT}{h^2}\right)^{3/2} \xi_{3/2}(1) \quad (9.1.7)$$

and $N_0 > 0$ for $T < T_c$, where

$$\left(\frac{2\pi mk_BT_c}{h^2}\right)^{3/2} \xi_{3/2}(1) = \frac{N}{V} = n. \quad (9.1.8)$$

For $T < T_c$, the particle density in the ground state is given by

$$n_0(T) = \frac{N_0}{V} = \frac{N}{V} - \left(\frac{2\pi mk_BT}{h^2}\right)^{3/2} \xi^{3/2}(1) = n\left[1 - \left(\frac{T}{T_c}\right)^{3/2}\right]. \quad (9.1.9)$$

The ground state, i.e., $p = 0$ $(\epsilon_p = 0)$ now has a finite fraction of particles (in the thermodynamic limit). We can imagine that this Bose gas has two possible phases – one in which the particles are spread thinly over all the different energy levels and another phase in which a finite density of particles is obtained in the ground state which is now macroscopically occupied, while over the rest of the states the particles are thinly spread. The transformation from one phase to another occurs at $T = T_c$ and results in a ground state condensate for $T < T_c$, with the condensate accounting for all the particles in the system at $T = 0$ This macroscopic occupation of a single quantum state is called the Bose-Einstein condensation.

9.1.2 Order of the Transition

To get a better understanding of the transition, we calculate the thermodynamic quantities for the two phases.

i) $T < T_c$:   In this range, the total energy is obtained by noting that $\mu = 0$ and only the particles outside the condensate contribute, i.e. in Eq.(9.1.3), we do not have any contribution from N_0 to the energy. Consequently,

$$
\begin{aligned}
E &= 2\pi V \left(\frac{2mk_BT}{h^2}\right)^{3/2} k_BT \int_0^\infty \frac{x^{3/2}}{e^x - 1} dx \\
&= 2\pi V \left(\frac{2mk_BT}{h^2}\right)^{3/2} k_BT \frac{3}{2}\pi^{1/2}\xi_{3/2} \\
&= \frac{3}{2}V \left(\frac{2\pi mk_BT}{h^2}\right)^{3/2} k_BT\xi_{5/2} \\
&= \frac{3}{2}NkT \left(\frac{T}{T_c}\right)^{3/2} \frac{\xi_{5/2}}{\xi_{3/2}} \\
&= \frac{3}{2}\cdot\frac{1.341}{2.312}NkT \left(\frac{T}{T_c}\right)^{3/2} \\
&= 0.770NkT \left(\frac{T}{T_c}\right)^{3/2}
\end{aligned}
\tag{9.1.10}
$$

where, the second line is obtained from the first one by following the same sequence of steps that was used for the second term on the right hand side of Eq.(9.1.5). The equation of state follows from $PV = \frac{2}{3}E$ as

$$
P = \left(\frac{2\pi mk}{h^2}\right)^{3/2} k_BT^{5/2}\xi_{5/2}.
\tag{9.1.11}
$$

Differentiating the energy E yields the specific heat at constant volume as

$$
C_V = 1.925Nk \left(\frac{T}{T_c}\right)^{3/2}.
\tag{9.1.12}
$$

The entropy is obtained by integrating C_V and hence

$$S = 1.282 Nk \left(\frac{T}{T_c} \right)^{3/2}. \tag{9.1.13}$$

ii) $T > T_c$:   Now $N_0 = 0$ and z has to be fixed from Eq.(9.1.3), which yields

$$N = V \left(\frac{2\pi m k_B T}{h^2} \right)^{3/2} \xi_{3/2}(z). \tag{9.1.14}$$

The energy follows from Eq.(9.1.3) as

$$
\begin{aligned}
E &= 2\pi V \left(\frac{2 m k_B T}{h^2} \right)^{3/2} \int_0^\infty \frac{\chi^{3/2} d\chi}{e^\chi - 1} \\
&= \frac{3}{2} V \left(\frac{2\pi m k_B T}{h^2} \right)^{3/2} \xi_{5/2}(z) k_B T.
\end{aligned}
\tag{9.1.15}
$$

To find the entropy we note that in the grand canonical ensemble

$$E - TS - \mu N = -PV = -\frac{2}{3} E$$

or $TS = \frac{5}{3} E - \mu N$, which in turn implies

$$S = \frac{5}{2} k_B V \left(\frac{2\pi m k_B T}{h^2} \right)^{3/2} \xi_{5/2}(z) - N k_B \ln z. \tag{9.1.16}$$

The specific heat can be obtained by differentiating the energy at constant volume

$$C_V = \frac{15}{4} kV \left(\frac{2\pi m k_B T}{h^2} \right)^{3/2} \xi_{5/2}(z) + \frac{3}{2} VkT \left(\frac{2\pi m k_B T}{h^2} \right)^{3/2} \cdot \left(\frac{d}{dz} \xi_{5/2}(z) \right) \cdot \frac{dz}{dT}. \tag{9.1.17}$$

To find dz/dT at constant V and N, we differentiate Eq.(9.1.14) to write

$$0 = \frac{d}{dT} \left(\frac{N}{V} \right) = \frac{d}{dT} \left(\frac{2\pi m k_B T}{h^2} \right)^{3/2} \xi_{3/2}(z)$$

leading to

$$\frac{3}{2} T^{1/2} \xi_{3/2}(z) + T^{3/2} \left(\frac{d}{dz} \xi_{3/2}(z) \right) \frac{dz}{dT} = 0$$

or

$$\frac{dz}{dT} = -\frac{3}{2T} \left[\frac{d}{dz} \xi_{3/2}(z) \right]^{-1} \cdot \xi^{3/2}(z). \tag{9.1.18}$$

We now find from the definition of Eq.(9.1.6),

$$z \cdot \frac{d}{dz} \xi_{5/2}(z) = \xi_{3/2}(z) \tag{9.1.19}$$

and

$$z \cdot \frac{d}{dz}\xi_{3/2}(z) = \xi_{1/2}(z). \tag{9.1.20}$$

Consequently,

$$\frac{1}{2}\frac{dz}{dT} = -\frac{3}{2T}\frac{\xi_{3/2}(z)}{\xi_{1/2}(z)} \tag{9.1.21}$$

and Eq.(9.1.17) becomes

$$C_V = \frac{15}{4}kV\left(\frac{2\pi m k_B T}{h^2}\right)^{3/2}\xi_{5/2}(z) - \frac{9}{4}Nk\frac{\xi_{3/2}(z)}{\xi_{1/2}(z)}. \tag{9.1.22}$$

If we compare the thermodynamic quantities E, S and C_V above and below the transition, then they are continuous at $T = T_c$.

However, we might note the fact that in the low temperature phase, i.e., for $T < T_c$, the pressure is independent of volume and hence studying isotherms could be interesting. If we hold T fixed and vary V starting from large values to smaller ones (going from dilute gas to the degenerate gas) then at a critical V ($= V_c$) which satisfies

$$\left(\frac{2\pi m k_B T_c}{h^2}\right)^{3/2}\xi_{3/2}(1) = \frac{N}{V_c}. \tag{9.1.23}$$

the condensation phenomenon will set in and for $V < V_c$, there will be a finite condensate N, determined by

$$N_0 = N\left(1 - \frac{V}{V_c}\right) \tag{9.1.24}$$

In the $P - V$ diagram at a fixed T (isotherms shown in Fig.9.1), the pressure is unchanging over the range $0 < V < V_c$ and then decreases with increasing V.

For higher temperatures, the critical V_c occurs at smaller and smaller values. The flat part of each curve can be thought of as the coexistence region for the two phases, which is a standard feature of the first order transition. Further the pressure is determined solely by the temperature in this region, which is characteristic of the vapour pressure and from Eq.(9.1.11), we find the slope of the vapour pressure curve to be

$$\frac{dP}{dT} = \frac{5}{2}\xi_{5/2}(1)\left(\frac{2\pi m k_B T}{h^2}\right)^{3/2} \cdot k \tag{9.1.25}$$

Now for an isotherm at the temperature T, the specific volumes are zero and V_c/N for the condensate and the gas phase respectively. Using the Clausius-Clapeyron relation, we can now get the latent heat of the transition per particle

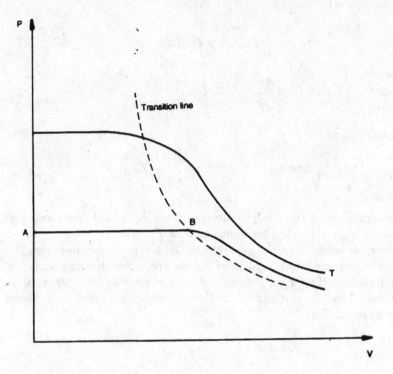

Figure 9.1: Isotherms for the Bose gas. The dashed line corresponds to the condensation.

as

$$L = T\left(\frac{V_c}{N} - 0\right) \cdot \frac{dP}{dT} = T\frac{V_c}{N} \cdot \frac{dP}{dT}$$

$$= \frac{5}{2} \cdot \frac{\xi_{5/2}(1)}{\xi_{3/2}(z)} \cdot k_B T. \qquad (9.1.26)$$

Thus the condensation has certain features of a first-order transition.

9.1.3   Near the Transition Point

Without trying to classify this transition as first order or second order, we will now go to T slightly above T_c. and imagine approaching T_c from above. Clearly the chemical potential will be very close to zero for $T \cong T_c$ and we want to determine the exact manner in which μ approaches zero as $\delta T = (T - T_c)$ tends to zero. To do this we need to analyze Eq.(9.1.21) near $z = 1$ and $T = T_c$. However, $\xi_{1/2}(z)$ is divergent as $z \to 1$ and hence we need to go back to Eq.(9.1.14) which says that for fixed N and V,

$$T^{3/2}\xi_{3/2}(z) = \text{constant}. \qquad (9.1.27)$$

Taking differentials near $z = 1$ and $T = T_c$

$$\frac{3}{2}\xi_{3/2}(1)\frac{\delta T}{T_c} = -\left.\frac{d\xi_{3/2}}{dz}\right|_{z=1}\delta z. \tag{9.1.28}$$

Since $\left.\frac{d\xi_{3/2}(z)}{dz}\right|_{z=1}$ diverges at $z = 1$, it is proper to start with the definition of derivatives and write

$$\left.\frac{d}{dz}\xi_{3/2}(z)\right|_{z=1}\delta z = \xi_{3/2}(1 + \delta z) - \xi_{3/2}(1). \tag{9.1.29}$$

An integral representation of Eq.(9.1.29) is clearly

$$\begin{aligned}
\left.\frac{d}{dz}\xi_{3/2}(z)\right|_{z=1}\delta z &= \frac{1}{\Gamma(3/2)}\int_0^\infty dx.x^{1/2}\left[\frac{1}{\frac{e^x}{1+\delta z} - 1} - \frac{1}{e^x - 1}\right] \\
&= \frac{1}{\Gamma(3/2)}\int_0^\infty \frac{dx.x^{1/2}e^x}{(e^x - 1)\left(\frac{e^x}{1+\delta z} - 1\right)}\delta z.
\end{aligned} \tag{9.1.30}$$

We note two features of Eq.(9.1.30).

1. As $\delta z \to 0$, the integral is completely dominated by the small-x regime and hence to get the leading term in δz, it suffices to linearize e^x.

2. The integral is well defined only for negative δz, which is as it should be since the maximum value of z (i.e., $z = 1$) is achieved at $T = T_c$ and z has to decrease as T increases above T_c,

In accordance with (ii) we set $\delta z = -\epsilon(-\epsilon < 0)$ and find by linearizing e^x,

$$\left.\epsilon\frac{d}{dz}\xi_{3/2}(z)\right|_{z=1} \cong \frac{\epsilon}{\Gamma(3/2)}\int_0^\infty \frac{dx}{x^{1/2}}\cdot\frac{1}{x+\epsilon} = \frac{\pi\epsilon^{1/2}}{\Gamma(3/2)}. \tag{9.1.31}$$

Returning to Eq.(9.1.28), we find

$$\epsilon^{1/2} = \frac{3}{2}\frac{\xi_{3/2}(1)\Gamma(3/2)}{\pi}\frac{\delta T}{T_c}. \tag{9.1.32}$$

Now, $z = e^{\mu/k_B T}$ leads to $\delta z = z[\frac{\delta\mu}{k_B T} - \frac{\mu}{k_B T}\cdot\frac{\delta T}{T}]$. Near $T = T_c, \mu = 0$, $z = 1$ and we have $\delta z \sim \delta\mu/k_B T_c$. Combined with Eq.(9.1.32), this leads to (keeping in mind $|\delta z| = \epsilon$)

$$\delta\mu = \frac{9}{4}k_B T_c \cdot \left(\frac{\xi_{3/2}(1)(3/2)^2}{\pi}\right)^2\left(\frac{\delta T}{T_c}\right)^2. \tag{9.1.33}$$

The slope of the chemical potential is consequently discontinuous at $T = T_c$. This leads to interesting consequences for the slope of the specific heat.

We turn to the energy expression, viewing it as a function of the independent variables z and T and can expand as a Taylor series to $O(\delta T)^2$ near $T = T_c$ and $z = 1$ as (N and V are held fixed)

$$E(T, z) = E(T_c, 1) + \left(\frac{\delta E}{\delta T} \right)_{T_c, 1} \delta T + \left(\frac{\delta E}{\delta z} \right)_{T_c, 1} \delta z + \frac{1}{2} \left(\frac{\delta^2 E}{\delta T^2} \right)_{T_c, 1} (\delta T)^2.$$

(9.1.34)

We do not need any further terms since δz is $O(\delta T)^2$ from Eq.(9.1.32). The slope of the specific heat at $T = T_c$, is clearly $\delta^2 E/\delta(\delta T)^2$ at $T = T_c$ and $z = 1$ and using Eqs.(9.1.32) and (9.1.34), the slope can be easily shown to be

$$\left(\frac{dC_v}{dT} \right)_{T \to T_c+} = \frac{Nk_B}{T_c} \left[\frac{45}{8} \frac{\xi_{5/2}(1)}{\xi_{3/2}(1)} - \frac{27}{4} \left(\frac{\xi_{3/2}(1)\Gamma(3/2)2}{\pi} \right) \right]$$

$$\cong -0.776 \frac{Nk_B}{T_c}.$$

(9.1.35)

For T approaching T_c from below, Eq.(9.1.12), shows the slope to be

$$\left(\frac{dC_v}{dT} \right)_{T \to T_c-} = \frac{2.89 Nk_B}{T_c}.$$

(9.1.36)

The specific heat, thus, has a maximum at $T = T_c$ and a slope discontinuity of $3.66 Nk_B/T$. The behaviour of C_v as a function of T is shown in Fig.9.2.

It should he emphasized that the above picture is valid for a non-interacting Bose gas and any interaction, however weak, may alter the picture drastically.

We now comment on the procedure followed in Eq.(9.1.3), when the number of particles residing in the ground state, i.e. $p = 0$, were singled out. We could have split off any finite number of terms without changing the value of the integral, i.e., we could equally well have written

$$N = N_0 + N_1 + V \left(\frac{2m}{h^2} \right)^{3/2} \int \frac{\epsilon^{1/2} d\epsilon}{e^{\beta(\epsilon - \mu)} - 1}$$

where

$$N_1 = \frac{1}{z} e^{\beta \epsilon_1} - 1^{-1} \quad \text{and} \quad N_0 = \frac{z}{1 - z}.$$

If this is as reasonable as removing N_0 alone from the sum, then we do not have a consistent procedure for establishing that particles have definitely condensed in the $p = 0$ state. We consider the gas to be in a box of volume V which we ultimately take to be infinity. Now as $z \to 1$, $N_1 \to (e^{\beta \epsilon_1} - 1)^{-1}$ and since the energies of a particle in a box are proportional to $V^{-2/3}$, it is clear that for large V, $N_1 \sim V^{-2/3}$. Note that N_1 $(z = 1)$ is the maximum value of $N_1(z)$ under the restriction $z \leq 1$ and hence what is true for N_1 $(z = 1)$ is true for all z. Thus the corresponding particle density $N_1/V \sim V^{1/3}$ and hence

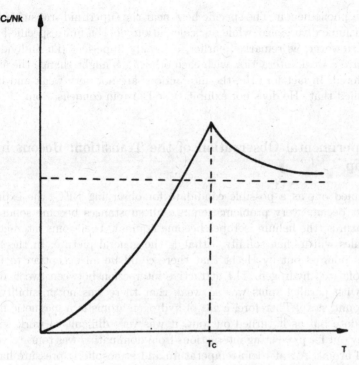

Figure 9.2: Specific heat around the condensation point T.

must vanish in the thermodynamic limit. On the other hand, in a finite volume V the chemical potential adjusts itself to provide a finite limit for

$$\lim_{\substack{z \to 1 \\ V \to \infty}} \frac{z}{V} \cdot \frac{1}{1-z}$$

and thus only N_0 among all possible energy states needs to be retained.

We note that at the beginning of the section, we laid stress on the fact that N, the total number of particles is conserved. This is essential for the condensation and hence in a system with no conservation of numbers (photon gas or a phonon gas) it is not possible to have Bose condensation. For a relativistic Bose gas, the discussion of the condensation must take the antiparticles into account since the number conservation will involve particles and antiparticles simultaneously.

Can this condensation described above be observed experimentally? The only Bose system which remains a liquid at very low temperatures where quantum effects predominate is liquid helium. If one puts the parameters appropriate to ^{4}He in Eq.(9.1.8), then T_c turns out to be 3.14 K. It is known that ^{4}He does undergo a phase transition to a superfluid state at $T \sim 2.17$ K and it is to a certain extent an open question as to whether that transition is related to the

condensation phenomenon. The specific heat near the superfluid transition has
a near logarithmic divergence, while our calculation yields a finite specific heat
at $T = T_c$. However, as remarked earlier, a weakly imperfect (i.e. individual
molecules have a weak interaction with each other) gas might change the situation considered. In fact for ^{4}He, the interactions are not very weak and it is
well established that ^{4}He does not exhibit Bose-Einstein condensation.

9.1.4 Experimental Observation of the Transition: Bosons in a Trap

With ^{4}He ruled out as a possible candidate for observing BEC, the experimental quest became very problematic since all substances become solids at
low temperatures (the helium isotopes become liquids). The atoms interact to
form molecules which then solidify – that is the general picture. In the late
fifties it was pointed out by Hecht that there could be an exception to this
rule: spin polarized hydrogen. The attractive interaction between two hydrogen atoms with parallel spins was so weak that there was no possibility of
forming a bound state. Therefore, a gas of hydrogen atoms in a magnetic field
was a candidate but as it turned out, it was very difficult to work with.
The only way out for preventing interactions from dominating was to use a very
dilute gas of atoms. Air at room temperature and atmospheric pressure has a
molecular density of about $10^{19}/\text{cm}^3$. The dilute gas which will be genuinely
weakly interacting would have a much smaller number density. If one reduces
the density by a few orders of magnitude to $10^{15}/\text{cm}^3$, the temperature for
seeing quantum phenomena would be about 10^{-5} deg. Kelvin or less. A huge
amount of experimental work was necessary to produce temperatures as low
as these: laser cooling followed by evaporative cooling is the process generally
followed. For cooling to occur and condensation take place the gas of atoms
must be confined in a small region and hence the particles have to be trapped.
This is done by engineering a harmonic potential, generally with a combination
of laser beams and magnetic fields, and these magneto-optical traps (MOT)
serve the dual purpose of trapping and cooling. Our immediate concern is that,
the calculation done carried out in subsection (A) has to be modified to take
into account the trapping potential.

The immense experimental problems that one encounters with obtaining a
gas of non-interacting ultra cold atoms were finally overcome in the last decade
of the previous century when a gas of extremely dilute alkali metal atoms was
cooled to temperatures of the order of nanokelvins. The atoms remained in
a gaseous state which was a metastable state at those low temperatures and
showed clear evidence of the condensation. The confining of these atoms in a
given region of space was carried out by using a simple harmonic potential
as a trap. Thus, we have in the actual experiment a set of non interacting
atoms placed in an external oscillator potential. Thus the number of particles
in a quantum state characterized by quantum numbers n_1, n_2 and n_3 is (as

opposed to Eq.(9.1.1))

$$N_{n_1,n_2,n_3} = \frac{1}{e^{[(n_1+n_2+n_3+\frac{3}{2})\hbar\omega - \mu]/k_B T} - 1} \tag{9.1.37}$$

where $\epsilon = \left(n_1 + n_2 + n_3 + \frac{3}{2}\right)\hbar\omega$ is the energy of the single particle. The integers n_1, n_2 and n_3 range from zero to infinity. The chemical potential μ is restricted to be less than or equal to $\frac{3}{2}\hbar\omega$. The transition will occur at a finite temperature T_c if μ acquires the value $\frac{3}{2}\hbar\omega$ at that temperature. The total number of particles of the system is given by

$$N = \sum_{n_1,n_2,n_3} N_{n_1,n_2,n_3} = \sum_{l=1}^{\infty} \sum_{n_1,n_2,n_3} z'^l e^{-l(n_1+n_2+n_3)\hbar\omega/k_B T} \tag{9.1.38}$$

where $z' = z.e^{-3\hbar\omega/2k_B T} = e^{(\mu-3\hbar\omega/2)/k_B T} \leq 1$.

The number of particles in the excited states is given by

$$N_{\text{ex}} = \sum_{l=1}^{\infty} \sum_{[n_1,n_2,n_3 \text{ not all zero}]} z'^l e^{-l(n_1+n_2+n_3)\hbar\omega/k_B T} \tag{9.1.39}$$

At $T = T_c$, $z' = 1$ and we have the maximum possible occupation of the excited states

$$N_{\text{ex}}^{\max} = \sum_{l=1}^{\infty} \sum_{[n_1,n_2,n_3 \text{ not all zero}]} e^{-l(n_1+n_2+n_3)\hbar\omega/k_B T}. \tag{9.1.40}$$

We will carry out the evaluation of the sum over n_1, n_2, n_3 in two different ways: first by considering n_1, n_2, n_3 as continuous variables (this is true for $\hbar\omega/k_B T_c \ll 1$) and then by treating them as discrete variables.

We define $n_1 + n_2 + n_3 = n$ and ask how many different arrangements of n_1, n_2 and n_3 can lead to the same value of n. For given n_1 and n_2, n_3 can range from 0 to $n - n_1 - n_2$ and hence the degeneracy factor $\int \int \int dn_1 dn_2 dn_3$ (keeping n fixed) becomes $\int \int \int dn_1 dn_2 d(n - n_1 - n_2)$. For a fixed n_1, the integral over n_2 gives $\int \left[(n-n_1)^2 - \left(\frac{n-n_1}{2}\right)^2\right] dn_1 = \frac{1}{2}\int_0^n (n-n_1)^2 dn_1 = \frac{1}{6}n^3$. The number of states between n and $n + dn$ is $\frac{1}{6}3n^2 dn = \frac{1}{2}n^2 dn$. Thus,

$$
\begin{aligned}
N_{ex}^{\max} &= \frac{1}{2}\sum_{l=1}^{\infty}\int e^{-nl\hbar\omega/k_B T_c} n^2 dn \\
&= \frac{1}{2}\left(\frac{k_B T_c}{\hbar\omega}\right)^3 \sum_{l=1}^{\infty}\frac{1}{l^3}\Gamma(3) \\
&= \left(\frac{k_B T_c}{\hbar\omega}\right)^3 \xi(3). \tag{9.1.41}
\end{aligned}
$$

The maximum population of the excited states at T_c will trigger a transition if this number becomes equal to N, the total number of bosons available. If we consider $T < T_C$, then the number in the excited states will decrease and will be less than N. The remaining number $N_0 = N - N_{ex}^{max}$ will settle in the ground state. The transition temperature is, consequently, given by

$$\left(\frac{k_B T_C}{\hbar\omega}\right)^3 \xi(3) = N \tag{9.1.42}$$

For $T < T_C$, the number of particles in the ground state is

$$\begin{aligned}
N_0 &= N - N_{ex}^{max}(T) = \left(\frac{k_B T_c}{\hbar\omega}\right)^3 \xi(3) - \left(\frac{k_B T}{\hbar\omega}\right)^3 \xi(3) \\
&= N\left[1 - \left(\frac{T}{T_C}\right)^3\right].
\end{aligned} \tag{9.1.43}$$

If we compare Eq.(9.1.43) with Eq.(9.1.9), we see how the trapped gas has different quantitative features compared to the homogeneous gas.

As explained earlier the above arguments for continuously varying n_i are valid for $\hbar\omega \ll k_B T$. If we do not take this limit, we need to evaluate the sum

$$\sum_{n_1, n_2, n_3 \text{ not all zero}} e^{-\beta_c l \hbar\omega(n_1 + n_2 + n_3)}.$$

We need to consider three cases:

(i) Two of n_1, n_2 and n_3 are zero (three possibilities), (ii) One of n_1, n_2, n_3 is zero (three possibilities) and (iii) None of n_1, n_2, n_3 is zero.

Thus, we see

$$N_{ex}^{max} = \sum_{l=1}^{\infty}\left[3\sum_{n_1=1}^{\infty} e^{-\beta_c l \hbar\omega n_1} + 3\left(\sum_{n_1=1}^{\infty} e^{-\beta_c l \hbar\omega n_1}\right)^2 + \left(\sum_{n_1=1}^{\infty} e^{-\beta_c l \hbar\omega n_1}\right)^3\right].$$

For condensation to occur at T=T_c,

$$N = \left(\frac{k_B T_C}{\hbar\omega}\right)^3 \xi(3) + \frac{3}{2}\left(\frac{k_B T_C}{\hbar\omega}\right)^2 \xi(2) + \dots. \tag{9.1.44}$$

If $k_B T_c/\hbar\omega \gg 1$, then the first term on the right hand side is the complete story and we have the answer of Eq.(9.1.41). This more careful calculation shows the existence of finite size effects. We can solve Eq.(9.1.44) iteratively for T_c by writing

$$\begin{aligned}
\left(\frac{k_B T_C}{\hbar\omega}\right)^3 &= \frac{N}{\xi(3)} - \frac{3}{2}\left(\frac{k_B T_C}{\hbar\omega}\right)^2 \frac{\xi(2)}{\xi(3)} + \dots \\
&= \frac{N}{\xi(3)} = -\frac{3}{2}\left[\frac{N}{\xi(3)}\right]^{2/3} \frac{\xi(2)}{\xi(3)} \\
&= \frac{N}{\xi(3)}\left[1 - \frac{3}{2}\xi(2)\left[\frac{\xi(3)}{N}\right]^{1/3}\right]
\end{aligned}$$

or

$$\frac{k_B T_C}{\hbar \omega} = \left(\frac{N}{\xi(3)} \right)^{1/3} \left\{ 1 - \frac{\xi(2)}{2} \left[\frac{\xi(3)}{N} \right]^{1/3} + \dots \right\}. \qquad (9.1.45)$$

The second term in the brackets on the right hand side of the above equation refer to the finite size effects. If N is of the order of Avogadro number, then the term is indeed negligible. However, for $N \sim O(10^6)$ or less, we see a significant correction in the condensation temperature. As is usual, the effect of finite size is to lower the transition temperature.

9.1.5   Interaction Effects in BEC

In this subsection we take account of the two-body interactions between the atoms in an approximate formalism. These interactions do not alter the fact that condensate has formed, i.e., at very low temperatures almost all the particles have settled in the lowest available state but still the interactions have a clear experimentally observable signature and consequently they are very important.

We begin by asking, what should be the interaction potential between two particles. The effective interaction between the two particles of mass m each, at low energies is seen to be approximately constant and is given by $U_0 = 4\pi \hbar^2 a/m$ in the momentum space, where 'a' is the scattering length. In coordinate space this corresponds to a contact interaction $U_0 \delta(\mathbf{r_1} - \mathbf{r_2})$, where $\mathbf{r_1}$ and $\mathbf{r_2}$ are the position vectors of the two particles. The Hamiltonian of the two particle system can then be written as

$$H = \sum_{i=1}^{N} \left[\frac{p_i^2}{2m} + V(\mathbf{r_i}) \right] + U_0. \sum_{\text{pairs } \langle ij \rangle} \delta(\mathbf{r_i} - \mathbf{r_j}). \qquad (9.1.46)$$

Since we are specifically discussing the condensed state at $T = 0$, all the N-atoms are taken to be in the same single particle state $\phi(\mathbf{r_i})$ and the wavefunction for the system is taken to be

$$\psi(\mathbf{r_1}, \mathbf{r_2}, \dots \mathbf{r_N}) = \prod_{i=1}^{N} \phi(\mathbf{r_i}) \qquad (9.1.47)$$

Remembering that $\mathbf{p} = -i\hbar \nabla$ we find the energy E of the state by constructing $\int \psi^* H \psi d^3 r_1 \dots d_N^r$. This gives

$$E = N \int d^3 r \left[\frac{\hbar^2}{2m} \mid \nabla \phi(\mathbf{r}) \mid^2 + V(\mathbf{r}) \mid \phi(\mathbf{r}) \mid^2 + \frac{N-1}{2} U_0 \mid \phi(\mathbf{r}) \mid^4 \right]. \qquad (9.1.48)$$

The N in front of the right hand side of the above equation can be absorbed by defining $\psi(\mathbf{r}) = N^{1/2}\phi(\mathbf{r})$. Considering $N \gg 1$, we have

$$E = \int d^3r \left[\frac{\hbar^2}{2m} \mid \nabla\psi(\mathbf{r}) \mid^2 + V(\mathbf{r}) \mid \psi(\mathbf{r}) \mid^2 + \frac{1}{2}U_0 \mid \psi(\mathbf{r}) \mid^4 \right]. \qquad (9.1.49)$$

with the constraint

$$\int \mid \psi(\mathbf{r}) \mid^2 d^3r = N \qquad (9.1.50)$$

The function $\psi(\mathbf{r})$ can then be found by optimising E in Eq.(9.1.49), subject to the constraint of Eq.(9.1.50). The constraint can be handled by a Lagrange multiplier μ, so that $\delta E - \mu\delta N = 0$ and μ becomes the chemical potential. The procedure is equivalent to minimising $E - \mu N$ with respect to ψ^* or ψ at fixed μ. Minimising with respect to ψ^*,

$$-\frac{\hbar^2}{2m}\nabla^2\psi(\mathbf{r}) + V(\mathbf{r})\psi(\mathbf{r}) + U_0 \mid \psi(\mathbf{r}) \mid^2 \psi(\mathbf{r}) = \mu\psi(\mathbf{r}) \qquad (9.1.51)$$

which is the time-independent Gross-Pitaevski equation. We restrict ourselves to an isotropic potential here, so that $V(\mathbf{r}) = \frac{1}{2}m\omega^2r^2$ and use the trial wave function

$$\psi(\mathbf{r}) = \frac{N^{1/2}}{\pi^{3/4}b^{3/2}} \cdot \exp\left(-\frac{r^2}{2b^2}\right) \qquad (9.1.52)$$

which is the simply the ground state wave-function for the simple harmonic oscillator in the absence of any interactions. It is the unknown width b of the wave function that is supposed to take account of interactions. Carrying out the necessary integrations of Eq.(9.1.49), we get

$$E = \frac{3}{4}N\hbar\omega \left(\frac{a_0{}^2}{b^2} + \frac{b^2}{a_0{}^2} \right) + \frac{N^2U_0}{2(2\pi)^{3/2}b^3} \qquad (9.1.53)$$

where $a_0 = \sqrt{\hbar/m\omega}$ is the oscillator length scale. Expressing U_0 in terms of the scattering length a, we can write (in terms of $x = a/a_0$)

$$E = \frac{3}{4}N\hbar\omega \left[\frac{1}{x^2} + x^2 + \frac{4}{3}\frac{1}{(2\pi)^{1/2}}\frac{Na}{a_0}\frac{1}{x^3} \right] \qquad (9.1.54)$$

The minimum of E as a function of x is now to be found and the energy of the condensate to be obtained. In Fig.9.3 we show typical E vs. x plots. It should be noted that a "metastable" minimum can also exist for negative scattering lengths (attractive interactions) provided that Na/a_0 is not high enough. The position of the minimum in E can be obtained from

$$x_m{}^5 - x_m - \frac{2}{\sqrt{2\pi}} \cdot \frac{Na}{a_0} = 0 \qquad (9.1.55)$$

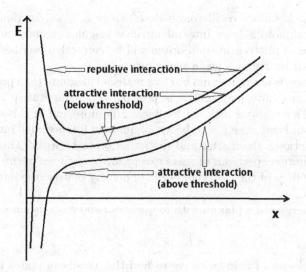

Figure 9.3: Plot of E vs. x in Eq.(9.1.54) for different values of the scattering length embodying the type of inter-particle interaction – repulsive or attractive. For attractive interaction, below a threshold value of the scattering length [Eq.(9.1.56)], condensate can form.

The minimum will cease to be an extremum if the second derivative vanishes as well, which happens at the critical value of

$$\frac{N \mid a \mid}{a_0} = \frac{2(2\pi)^{1/2}}{5^{9/4}} \simeq 0.67 \tag{9.1.56}$$

It is easy to see that if the particle number is large enough to overwhelm the kinetic energy term [the second term on the right hand side of Eq.(9.1.54)], then the minimum x goes as $x_m \propto N^{1/5}$ and the energy per particle $E/N \propto N^{2/5}$. This can actually be inferred from experiments, giving credence to the fact that interactions are important and simultaneously, the particles are a part of the condensate.

9.2 Two Important Phenomena

9.2.1 Black Body Radiation and Stefan-Boltzmann law

The radiation in thermal equilibrium is called black body radiation and the radiation may be regarded as a gas consisting of photons. The equations of the radiation field are governed by Maxwell's equations which are linear and hence the electromagnetic oscillations which constitute radiation can be written as a set of uncoupled harmonic oscillators and the energy quanta of these oscillators are the photons. Since the governing equations are linear and the description in terms of simple harmonic oscillators is exact, there is no interaction among the photons. The photons have integral intrinsic angular momentum (spin) and hence the gas of photons in equilibrium will be correctly described by the Bose-Einstein distribution at a given temperature.

If the radiation is not in vacuum but in a material medium, then properties of the medium come into play and it is possible that constitutive relations entering Maxwell's equations may not be linear any more and in this case, for the condition of an ideal Bose gas to be approximately fulfilled, one must have the interaction between the matter and the radiation to be weak. This is true all across the radiation spectrum in gases except for frequencies in the vicinity of the absorption lines of the material. For high density matter, the interaction may not be weak.

Since the energy ϵ of a photon with frequency ω and momentum k is given by

$$\epsilon = \hbar\omega = \hbar c k \tag{9.2.1}$$

where c is the velocity of light, we have to find the density of states in terms of k or ω. We also need to find the chemical potential μ. If the particle number is not conserved the chemical potential has got to be zero. This is because the average number of particles in the system will be determined by the condition of equilibrium, namely the condition minimizing the free energy F with respect to all the variables in the system. This leads to $\frac{\partial F}{\partial N} = 0$ but since $\mu = (\frac{\partial F}{\partial N})_{V,T}$ we have the condition of vanishing chemical potential. Thus the number of photons with energy $\hbar\omega$ is

$$N(\omega) = (e^{\hbar\omega/k_B T} - 1)^{-1}. \tag{9.2.2}$$

We now need to specify the number of states with momentum between k and $k + dk$. To do so we note that the spatial and temporal dependence of the electric field corresponding to the photon will have the plane waveform determined by $e^{i(\mathbf{k}\cdot\mathbf{r} - \omega t)}$. Now periodic boundary conditions on the electric field in a cube of size L will force the usual quantization

$$\mathbf{k} = \frac{2\pi}{L} \tag{9.2.3}$$

where $\mathbf{n} = n_x \hat{i} + n_y \hat{j} + n_z \hat{k}$, n_x, n_y and n_z being integers and thus the number of states in the phase space volume $V d^3 k$ follow the free-particle argument of Chapter 4 exactly, and can be written as $V/(2\pi)^3 . 4\pi k^2 dk$. Now the spin degeneracy has to be included and although $s = 1$ would lead to $2s + 1 = 3$, we note that the electromagnetic oscillations are transverse and hence there can be only two polarization directions (the two transverse directions). Thus the spin degeneracy is 2 and the number of photons with energy between $\hbar\omega$ and $\hbar(\omega + d\omega)$ can be written as

$$N(\omega)d\omega = \frac{8\pi V c}{(2\pi)^3} \frac{\omega^2 d\omega}{e^{\hbar\omega\beta} - 1}. \tag{9.2.4}$$

The total energy of the radiation at a temperature T is

$$
\begin{aligned}
E &= \int_0^\infty \hbar\omega N(\omega)d\omega = \frac{8\pi V\hbar}{(2\pi)^3 c^3}\int_0^\infty \frac{\omega^3 d\omega}{e^{\beta\hbar\omega}-1} \\
&= \frac{8\pi V\hbar}{(2\pi)^3(\beta\hbar)^4 c^3}\int_0^\infty \frac{x^3 dx}{e^x - 1} \qquad (\text{where } x = \beta\hbar\omega) \\
&= 8\pi\left(\frac{1}{hc}\right)^3 (k_B T)^4 \sum_{n=0}^\infty \frac{1}{(n+1)^4}\Gamma(4) \\
&= 48\pi\left(\frac{1}{hc}\right)^3 V(k_B T)^4\xi(4) = \frac{48\pi^5}{90}V.k_B T\left(\frac{k_B T}{hc}\right)^3 \\
&= \frac{8}{15}\pi^2 VkT\left(\frac{\pi k_B T}{hc}\right)^3 \qquad\qquad\qquad (9.2.5)
\end{aligned}
$$

This T^4-dependence of the radiated energy on temperature is known as the Stefan-Boltzmann law. The specific heat at constant volume is consequently,

$$
C_V = \left(\frac{\partial E}{\partial T}\right)_V = \frac{32}{15}\pi^2\left(\frac{\pi k_B T}{hc}\right)^3 \qquad (9.2.6)
$$

What we would like to verify experimentally is the T^4 - law of Eq.(9.2.5) or more importantly the actual distribution $E(\omega, T)$ which we define by

$$
E(T)/V = \int_0^\infty E(\omega, T)d\omega \qquad (9.2.7)
$$

leading to

$$
E(\omega, T) = \frac{\hbar}{\pi^2 c^3}\cdot\frac{\omega^3}{e^{\beta\hbar\omega}-1}. \qquad (9.2.8)
$$

To test the above relation we imagine a small hole through which the black body radiation communicates with the outside world. Now, the amount of energy of frequency ω coming out per unit area per unit time in a direction θ to the normal is the amount of energy of frequency ω contained in a volume element of height $c.\cos\theta$ and base of area unity, while the probability of coming out in the solid angle $d\Omega = 2\pi\sin\theta d\theta$ is $d\Omega/4\pi$ (isotropy). Hence the amount of energy coming out per unit area per unit time is obtained by integrating over the hemisphere and hence

$$
\begin{aligned}
I(\omega, T) &= \int_{\text{hemisphere}} \frac{d\Omega}{4\pi}c.\cos\theta E(\omega, T) \\
&= \frac{c}{4}E(\omega, T). \qquad\qquad\qquad (9.2.9)
\end{aligned}
$$

while the total energy emitted, $I(T)$, is given by

$$I(T) = \int_0^\infty d\omega I(\omega, T) = \sigma T^4 \qquad (9.2.10)$$

where

$$\sigma = \frac{\pi^2 k^4}{60(\hbar c)^3} \qquad (9.2.11)$$

is called the Stefan-Boltzmann constant.

The measurement of both $I(\omega, T)$ and $I(T)$ can be carried out and provides an excellent verification of the Bose distribution for photons. The shape of spectral curves of $I(\omega, T)$ vs ω at different temperatures is shown in Fig.9.3.

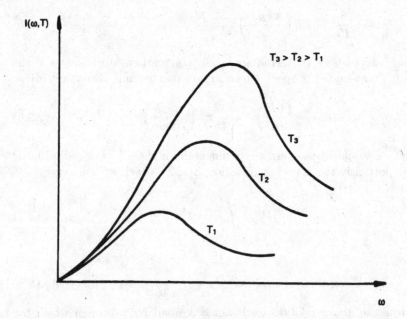

Figure 9.4: Superfluidity of 4He.

Note that there is a characteristic maximum in the spectrum $I(\omega, T)$ and as the temperature increases, the maximum shifts to higher frequencies. This is known as Wien's displacement law.

Finally, consider a body surrounded by black body radiation. The amount of radiation energy incident per unit time on a surface area dS at an angle θ to a chosen direction is clearly $cE(\omega, T) \cos \theta dS d\omega$. If the absorptive power of the body is $A(\theta, \omega)$ at this angle and frequency then the amount of energy absorbed per unit time is $cA(\theta, \omega)E(\omega, T) \cos \theta dS d\omega$. We consider fluorescence and scattering to be absent. Also, radiation is merely absorbed by the body: it does not pass through. If the emissivity of the surface is $J(\theta, \omega)$, then the amount of energy emitted per unit time is $J(\theta, \omega)dSd\omega$ and energy balance yields

$$J(\theta, \omega)/A(\omega, \theta) = cE(\omega, T)\cos\theta \qquad (9.2.12)$$

independent of the properties of the body: this is Kirchoff's law. For a black body $A(\omega, \theta) = 1$ and the actual emissive power is independent of the material property.

9.2.2 Casimir Effect

In this Section, we focus on another aspect of electromagnetic radiation. We imagine a situation in which we have nothing but electromagnetic radiation – it is electromagnetic radiation at $T = 0$ which arises from an esoteric quantum effect called vacuum fluctuations. This refers to a quantum fluctuation which results in the formation of a virtual particle-antiparticle pair from empty space and which subsequently annihilates giving off radiation. The pair formation is caused by energy fluctuation allowed by Heisenberg's time-energy uncertainty relation.

There is a certain amount of radiation associated with this radiation that comes from summing over the modes of wavenumber $\mathbf{k}$, with the energy of each mode being $\frac{1}{2}\hbar ck$. The k-values are continuously distributed. The half comes from the fact that this is the zero-point energy of the quantum oscillator associated with the electromagnetic field (for quantum oscillators see Eq.(4.5.6)). We now introduce two infinitely extended parallel conducting plates separated by a distance L in the z-direction, along which the allowed k-values are as given by Eq.(9.2.3), i.e., $k_z = \frac{n\pi}{L}$ where n is an integer and the total energy will be obtained by taking a discrete sum over integers n and an integration over the continuously distributed wave numbers k_x and k_y. The energy associated with each k is $2 \times \frac{1}{2}\hbar ck = \hbar c\sqrt{k_x^2 + k_y^2 + \frac{n^2\pi^2}{L^2}}$, the factor of 2 coming from the fact that as light is transversely polarized in three dimensions, there will be two polarization states for each k. The total energy per unit plate area can then be written as

$$
\begin{aligned}
\frac{E}{A} &= \hbar c \sum_n \int \frac{d^2p}{h^2}\sqrt{k_x^2 + k_y^2 + \frac{n^2\pi^2}{L^2}} \\
&= \hbar c \sum_n \int \frac{d^2k}{2\pi^2}\sqrt{k_x^2 + k_y^2 + \frac{n^2\pi^2}{L^2}} \\
&= \frac{\hbar c}{6\pi} \sum_n \left[\left(\Lambda^2 + \frac{n^2\pi^2}{L^2}\right)^{3/2} - \frac{n^3\pi^3}{L^3}\right] \\
&= \frac{\pi^2\hbar c}{6L^3} \sum_n (\Lambda^3 - n^3) \\
&= \frac{\pi^2\hbar c}{6L^3}\left(\Lambda^3 N - \sum_n n^3\right) \qquad (9.2.13)
\end{aligned}
$$

where Λ is the maximum value of the two dimensional wavenumber $k = \sqrt{k_x{}^2 + k_y{}^2}$, and N is an upper limit on the sum over n. We can write $\Lambda^3 N = \frac{1}{4}X^4 = \int_0^X y^3 dy$, where X is some large number and thus Eq.(9.2.13) becomes

$$\frac{E}{A} = \frac{\pi^2 \hbar c}{6L^3} \left[\int_0^X y^3 dy - \sum_{n=1}^N n^3 \right]. \tag{9.2.14}$$

With $X \to \infty$ as well as $N \to \infty$, we can now use the Euler-McLaurin sum formula to write

$$\frac{E}{A} = -\frac{\pi^2 \hbar c}{720 L^3}. \tag{9.2.15}$$

The force between the two plates come from a negative derivative with respect to L, and we get an attractive force between the plates given by

$$F = -\frac{\pi^2 \hbar c}{240 L^4}. \tag{9.2.16}$$

This result was obtained by Casimir in 1948, and the unambiguous experimental confirmation came about fifty years later. At finite temperatures the thermal fluctuations suppress the quantum fluctuations (which are responsible for the effect) and hence the magnitude of the force gets reduced.

9.3 Superfluidity of Helium-4

9.3.1 General Characteristics

Helium has several remarkable properties. It remains a liquid even at the lowest temperatures under normal pressures. This is because of the inert gas nature which causes the interaction between helium atoms to be weak. In spite of this weak interaction the other inert gases do solidify. Helium, however, because of its very low mass has a high degree of zero-point kinetic energy and the combination of low mass and weak interatomic forces keeps it a liquid at $T \cong 0$. Thus the two isotopes of helium - 4He and 3He are the only liquids which can show quantum features-all other liquids freeze into solids before the temperature can become low enough for the De Broglie wavelength to become of the order of the atomic spacing.

4He is known to liquify at $T \cong 5K$. If this liquid is further cooled, then a further change occurs at $T = T_\lambda = 2.17K$. Below T_λ, the liquid boils without bubbling-its thermal conductivity becomes infinite! In addition to the apparently infinite thermal conductivity, its viscosity, under certain conditions, also appears to be zero. In particular, if it flows through a narrow capillary at speeds less than a critical velocity v_c, then it shows no viscosity (this narrow capillary is called a superleak). The narrower the capillary, higher the critical velocity. However, if the viscosity is measured by the rotating cylinder technique (a rotating can placed in a liquid bath experiences a drag due to the viscosity of

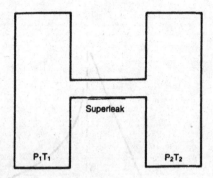

Figure 9.5: Flow caused by temperature gradient.

the liquid), the measured value goes down with decreasing temperature. This behaviour of viscosity is explained by a model which assumes that below T_λ, liquid helium is an intimate mixture of two liquids-a normal fluid which exhibits viscosity and a superfluid which does not exhibit any viscosity. The superleak allows only the superfluid to flow.

A further assumption that the superfluid carries no entropy helps explain the mechano-caloric effect. If two containers of 4He, A and B, are maintained at the same temperature but different pressures and connected by a superleak (see Fig.9.4), then the temperature of the container at the lower pressure falls and a temperature difference ΔT is maintained which is related to the pressure difference ΔP as $\Delta T = \Delta P/\rho s$, where s is the specific entropy. Conversely, if A and B are at the same pressure but different temperatures, then on being connected by a superleak a pressure difference ΔP develops, given by $\Delta P = \rho s \Delta T$ as stated before. If one of the containers (the one at lower temperature) is a thin tube, then the pressure head causes a fountain of liquid helium to emerge.

Yet another interesting property of superfluid helium is its ability to creep up the side of the beaker in which it is kept. For all liquids Van der Waals forces of attraction between beaker and liquid molecules cause the formation of a layer of liquid whose thickness decreases with height. In all liquids, except He in its superfluid phase, the layer formation is inhibited by the small but finite temperature difference between the wall and the liquid. If the wall is at a higher temperature, the layer forms droplets and falls back into the liquid. For $T < T_\lambda$, the infinite thermal conductivity prevents any temperature difference and hence the liquid layer creeps up the wall and escapes from the beaker. The above properties clearly show that for $T < T_\lambda$, the liquid is in different thermodynamic state and this state is called He II. The transition at $T = T_\lambda$ is second order (no latent heat) and is called the lambda transition due to the behaviour of the specific heat, which becomes infinitely large at $T = T_\lambda$, and the specific heat vs temperature curve has the appearance of the letter lambda (λ) (see Fig.(9.6)).

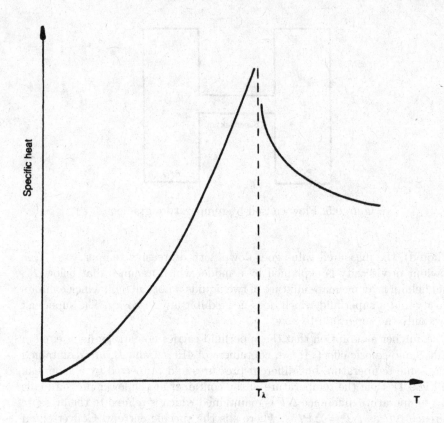

Figure 9.6: Specific heat near the superfluid transition.

It should be noted that 3He does not exhibit the above features unless the temperature is lowered to milli-kelvins. The prime difference between 3He and 4He being one of statistics (3He is half integral spin, while, 4He has spin zero) we expect that the superfluid property of 4He is due to the fact that it obeys Bose statistics. The main purpose of this section is to bring out the connection between superfluidity and the Bose distribution.

To conclude this qualitative introduction to the properties of He II, we note that the picture of He II as being composed of two kinds of fluids leads to a prediction about wave propagation. If we call ρ_n and ρ_S the densities of the normal and superfluid components of He II, then the actual density $\rho = \rho_S + \rho_n$. Generally wave propagation implies oscillations in the density ρ (sound waves), but in this case we can have out of phase oscillations in ρ_S and ρ_n so that ρ_n need not vary. However, the oscillations in ρ_S and ρ_n will propagate in the form of waves and these oscillations are called second sound. Since the total entropy of He II is determined by ρ_n, the oscillation in ρ_n will manifest itself as oscillations in entropy and consequently as oscillations in temperature. The prediction and subsequent experimental verification of the existence of these temperature oscillations was one of the principal achievements of the "two-fluid" theory.

Information regarding the energy spectrum (i.e. relation between energy and momentum) of a collection of helium atoms can be had from the study of low temperature specific heat. It is experimentally observed that for very low temperatures, the specific heat is proportional to T^3 (the specific heat of a phonon gas), while for relatively higher temperatures there is an additional contribution having the form $e^{-\Delta/k_B T}$, where Δ is a temperature independent energy scale. The very low temperature properties are determined by the lowest momentum values and the T^3-law in this regime suggests that for very small p, the dispersion is phonon-like, i.e.

$$E = \hbar\omega = \hbar c k = cp \quad (p \text{ small}). \tag{9.3.1}$$

The exponential behaviour at higher temperature suggest that for higher momenta (which are sampled at higher temperatures), the dispersion relation has the form

$$E = \hbar\omega = \Delta + \frac{(p - p_0)^2}{2m} \quad (p \cong p_0) \tag{9.3.2}$$

(that this leads to the desired specific heat is the subject of one of the problems). The total specific heat is the sum of those obtained separately on the basis of Eqs.(9.3.1) and (9.3.2). Comparing with the experimentally measured specific heat, Landau could find the values of the fitting parameters c, Δ, p_0 and m^*. The value of c thus obtained is the velocity of sound at low temperatures and the experimental confirmation of this value vindicates the essential correctness of Eqs.(9.3.1) and (9.3.2). However, the above dispersion [Eq.(9.3.2)] yields only parts of the E vs. p curve, as shown in Fig.9.6. The full curve cannot be obtained theoretically, although it can be inferred from scattering measurements by a relationship which we now discuss.

9.3.2 The Energy Spectrum

We begin by assuming that the fluid has an equilibrium density $\bar{\rho}$ when its velocity is zero. The excitations cause a density fluctuation $\delta\rho$ and provide a small velocity v. The energy can be written to quadratic order as

$$E = \bar{E} + \int \psi(\mathbf{r})\delta\rho d^3r + \frac{1}{2}\int \phi(\mathbf{r} - \mathbf{r}')\delta\rho(\mathbf{r})\delta\rho(\mathbf{r}')d^3r d^3r' + \frac{1}{2}\int \bar{\rho}v^2(\mathbf{r})d^3r \tag{9.3.3}$$

where $\psi(\mathbf{r})$ and $\phi(\mathbf{r}, \mathbf{r}')$ are to be thought of as coefficients for expansion in powers of $\delta\rho(\mathbf{r})$.

The fact that the unperturbed fluid is isotropic and homogeneous makes $\phi(\mathbf{r}, \mathbf{r}')$ a function of $(\mathbf{r} - \mathbf{r}')$ alone and also makes $\psi(\mathbf{r})$ a constant. The constraint of the total mass remaining unchanged implies $\int \delta\rho(\mathbf{r})d^3r = 0$ and thus we have

$$E = \bar{E} + \frac{\bar{\rho}}{2}\int v^2(\mathbf{r})d^3 + \frac{1}{2}\int \phi(\mathbf{r} - \mathbf{r}')\delta\rho(\mathbf{r})\delta\rho(\mathbf{r}')d^3r d^3r' \tag{9.3.4}$$

where $\bar{E}$ is determined by $\bar{\rho}$ alone. The continuity equation provides a constraint in the form

$$\frac{\partial \rho}{\partial t} + \nabla \cdot (\rho \mathbf{V}) = 0 \qquad (9.3.5)$$

where in equilibrium, the velocity $\mathbf{V} = 0$ and the density ρ is a constant $\bar{\rho}$. Small fluctuations from equilibrium result in a density variation $\delta\rho$ and a velocity variation $\mathbf{v}$, where

$$\frac{\partial}{\partial t}\delta\rho + \bar{\rho}\nabla \cdot \mathbf{v} = 0. \qquad (9.3.6)$$

To incorporate the constraint of Eq.(9.3.6) in Eq.(9.3.4), we write down the energy expression in momentum space by using the Fourier transforms

$$\mathbf{v}(\mathbf{r}) = \int \mathbf{v}(\mathbf{k})e^{i\mathbf{k}.\mathbf{r}}\frac{d^3k}{(2\pi)^3} \qquad (9.3.7)$$

$$\delta\rho(\mathbf{r}) = \int \delta\rho(\mathbf{k})e^{i\mathbf{k}.\mathbf{r}}\frac{d^3k}{(2\pi)^3} \qquad (9.3.8)$$

and

$$\phi(\mathbf{r} - \mathbf{r}') = \int \delta\rho(\mathbf{k})e^{i\mathbf{k}.(\mathbf{r}-\mathbf{r}')}\frac{d^3k}{(2\pi)^3}. \qquad (9.3.9)$$

The energy becomes

$$E = \bar{E} + \frac{\bar{\rho}}{2}\sum_k \mathbf{v}(\mathbf{k}) \cdot \mathbf{v}(-\mathbf{k}) + \frac{1}{2}\sum_k \phi(\mathbf{k})\delta\rho(\mathbf{k})\delta\rho(-\mathbf{k}). \qquad (9.3.10)$$

In momentum space, Eq.(9.3.6) becomes

$$\frac{d}{dt}\delta\rho(\mathbf{k}) = -i\mathbf{k}\bar{\rho} \cdot \mathbf{v}(\mathbf{k}). \qquad (9.3.11)$$

Substituting for $\mathbf{v}(\mathbf{k})$ in Eq.(9.3.10), bearing in mind that small oscillations in a liquid are longitudinal [$\mathbf{v}(\mathbf{k})$ is parallel to $\mathbf{k}$], we have

$$E - \bar{E} = \frac{1}{2} \cdot \sum_k \frac{\left|\delta\dot{\rho}(k)\right|^2}{\bar{\rho}k^2} + \frac{1}{2}\sum_k \phi(k)|\delta\rho(k)|^2 \qquad (9.3.12)$$

where use has been made of the fact that $\delta\rho$ is real implies $\delta\rho(\mathbf{k})^* = \delta\rho(-\mathbf{k})$ and similarly for $\mathbf{v}(\mathbf{k})$. The right hand side of Eq.(9.3.12) corresponds to a set of non-interacting oscillators (one for each k value) with frequency $\Omega(\mathbf{k})$ given by

$$\Omega^2(\mathbf{k}) = \bar{\rho}.\phi(\mathbf{k}).k^2 \qquad (9.3.13)$$

and accordingly the energy spectrum for each $\mathbf{k}$ value is given by

$$E(\mathbf{k}) = \left(n + \frac{1}{2}\right)\hbar\Omega(\mathbf{k}) \qquad (9.3.14)$$

where n is an integer or zero. The ground state energy of the system is given by $n = 0$ and hence

$$E_{\text{ground}} = \bar{E} + \sum_k \frac{1}{2}\hbar\Omega(\mathbf{k}). \qquad (9.3.15)$$

The second term on the right hand side is the zero-point energy and we have

$$\frac{1}{2}\hbar\Omega(\mathbf{k}) = \frac{1}{2\bar{\rho}k^2}\left\langle\left|\delta\dot{\rho}(\mathbf{k})\right|^2\right\rangle + \frac{1}{2}\phi(\mathbf{k})\langle|\delta\rho(\mathbf{k})|^2\rangle = \phi(\mathbf{k})\langle|\delta\rho(\mathbf{k})|^2\rangle \qquad (9.3.16)$$

where we have made use of the virial theorem (angular brackets denote expectation values) for the simple harmonic oscillator which asserts that the two expectation values in the first equality of Eq.(9.3.16) are equal. The excitation energy $\epsilon(\mathbf{k})$ is clearly $\hbar\Omega(\mathbf{k})$ and hence

$$\begin{aligned}
\epsilon(\mathbf{p}) &= \epsilon(\mathbf{k}) = \hbar\Omega(\mathbf{k}) = \hbar\frac{\Omega(\mathbf{k})^2}{\Omega(\mathbf{k})} \\
&= \frac{\hbar^2\bar{\rho}\phi(\mathbf{k})k^2}{2\phi(\mathbf{k})\langle|\delta\rho(\mathbf{k})|^2\rangle} = \frac{\hbar^2\bar{\rho}k^2}{2\langle|\delta\rho(\mathbf{k})|^2\rangle} \\
&= \frac{p^2}{2S(\mathbf{p})}
\end{aligned} \qquad (9.3.17)$$

where $\mathbf{p} = \hbar\mathbf{k}$ is the momentum and $S(\mathbf{p}) = \langle|\delta\rho(\mathbf{k})|^2\rangle/\bar{\rho}$ is the Fourier transform of the density correlation function or the structure factor.

$$S(\mathbf{r} - \mathbf{r}') = \langle\delta\rho(\mathbf{r})\delta\rho(\mathbf{r}')\rangle/\bar{\rho}. \qquad (9.3.18)$$

As we have seen before (see Chapter 5), the structure factor can be experimentally determined from neutron scattering and thus the excitation curve can be experimentally obtained using Eq.(9.3.17). The curve shown in Fig.9.7 does have a linear behaviour near $\mathbf{p} = 0$ and a minimum at $\mathbf{p}_0$: features which were present in Landau's theory shown in Fig.9.6.

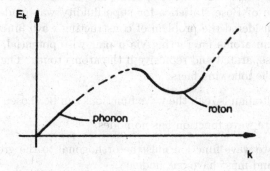

Figure 9.7: The spectrum according to Landau.

9.3.3   Occurrence of Superfluidity

An excitation curve of the kind shown in Fig.9.7 does lead to superfluidity.

This can be understood by considering He-II flowing down a capillary with velocity V. The kinetic energy $E = 1/2MV^2$. For the fluid to be normal there will be momentum exchange with the capillary walls and hence excitations will be created in the fluid. In a superfluid such excitations cannot be formed and consequently there is no momentum or energy transfer and hence no viscosity. If an excitation of momentum $\mathbf{p}$ is created then the fluid velocity changes to $\mathbf{V}$, given by

$$\mathbf{V}' = \mathbf{V} - \mathbf{p}/M. \tag{9.3.19}$$

The energy of the fluid is

$$E = \frac{1}{2}MV'^2 = \frac{1}{2}M(V - p/M)^2 \cong \frac{1}{2}MV^2 - \mathbf{V}.\mathbf{p} = E - \mathbf{V}.\mathbf{p}. \tag{9.3.20}$$

If the excitation of momentum $\mathbf{p}$ has an energy $\epsilon(\mathbf{p})$ then the energy of the fluid must have decreased by $\epsilon(\mathbf{p})$. If the maximum value of $\mathbf{V}.\mathbf{p}$ which is pV, is less than $\epsilon(\mathbf{p})$ then the excitation cannot be formed and hence we have a critical velocity V_c given by

$$V_c = \epsilon(\mathbf{p})/|\mathbf{p}_{\min}|. \tag{9.3.21}$$

If the excitation curve gives a non-zero minimum value of $\epsilon(\mathbf{p})/p$ then we expect that no excitations can be formed in the system for flow velocities lower than V_c and superfluidity will be observed. Clearly if $\epsilon(\mathbf{p})/p$ vanishes at the origin i.e., as $p = |\mathbf{p}| \to 0$, we will never get a finite V_c. Hence if superfluidity is to be observed, the low lying excitations must be phonons – the quanta for sound waves. The above semi-phenomenological discussion of superfluidity does not invoke the question of statistics at any point. The lesson from the above discussion is that the low-p excited states can only be sound waves for superfluidity and this restriction on the excited states ought to be a consequence of the Bose statistics: this is the final issue that we want to address.

The importance of Bose statistics for superfluidity was made plausible by Feynman who considered the problem of constructing wave functions for a collection of N helium atoms interacting via a pair wise potential, attractive when the atoms are separated and repulsive if the atoms touch. The argument proceeds by noting the following facts:

1. Since the Hamiltonian is real, the wave functions can be chosen to be real.

2. The ground state wave function has no nodes.

3. The excited state wave function must he orthogonal to the ground state wave function and must have one node.

The important point to understand is why Bose statistics forces the low lying excitations to be phonons as shown in Eq. (9.3.1)

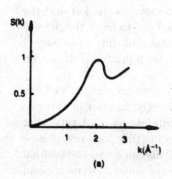
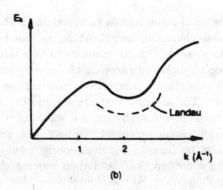

Figure 9.8: (a) Experimental structure for $S(k)$. (b) Using $S(k)$ to obtain E_k. Solid curve agrees qualitatively with Landau's curve, but is off in magnitude.

As a starting hypothesis we can picture the ground state wave function to be constant if the atoms are separated and zero if they touch. We denote the function by $\phi(r_1, r_2 \ldots r_N)$. If the first excited state wave function is $\psi(r_1, r_2 \ldots r_N)$ then $\int d^3r_1 \ldots d^3r_N \phi(r_1 \ldots r_N)\psi(r_1 \ldots r_N)$ has to be equal to zero. If we call a given distribution of r_1, r_2, ...r_N a configuration, then we note that over approximately half the configurations ψ must be positive (we call these configurations A) and over the other half it must be negative (we call these B). If ψ is not a phonon state then the change from A to B cannot alter the density and if it has to be a low energy state then the change must involve large distances that is large scale movements of atoms must be necessary to change A to B.

Let us divide the system into two volumes left and right: if there are more atoms to the left we have the configuration A (let us say) and if there are more to the right, we have B (Fig. 9.8). Now, to convert A to B, we would need to move atoms from left to right (large scale displacements) but in the process the density will change to the reverse (high on left and low on right) (Fig.9.8) Thus these wave functions are phonon-like. If the density is not to vary, then

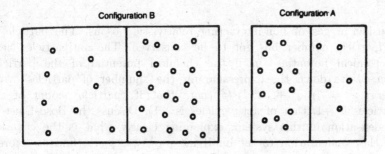

Figure 9.9: Excited state configuration.

for an A atom moved to the right, a B atom has to move to the left and the small changes in neighbouring positions etc. are ironed out to leave the density unchanged. This too appears to be a large displacement and thus a low energy non-phonon state keeping the density unchanged has been produced. However, this conclusion is erroneous due to Bose statistics.

The atoms are identical and obey Bose statistics, so an exchange of two atoms does not change the wave function. Consequently, instead of the two large distance changes, we first imagine interchanging atoms of A type that takes us close to the final configuration in the B-form (this does not change the wave function) and then moving them a small distance to get the final configuration. So the change from A to B requires very small movements and costs a lot of energy. The exchange would change the sign of the wavefunction if the identical particles were fermions and the above argument fails. Thus the low-lying states cannot be non-phonon like as a consequence of He atoms obeying Bose statistics.

9.4   Problems for Chapter 9

1. Consider a BE gas of particles whose energy(ϵ)-momentum(p) relation is $\epsilon = \alpha p^S$, where α is a constant and s is a variable index (usual case $s = 2$) . In a D-dimensional space, find the boundary in the $s - D$ plane which demarcates the non-condensation region from that in which BE condensation occurs at a finite temperature.

2. For a BE gas in a gravitational field g, show that the condensation sets in at

$$T_c \approx T_c^{(0)} \left[1 + \frac{8}{9} \frac{1}{\Gamma(3/2)} \left(\frac{m\pi g L}{k_B T_c^{(0)}} \right)^{1/2} \right]$$

where $T_c^{(0)}$ is the temperature at which condensation occurs in the absence and L is the length of the container in the z-direction (direction of gravity).

3. Consider a gas of non-interacting relativistic bosons. The particle + antiparticle number has got to be conserved. The antiparticles have a chemical potential $-\mu$ if the chemical potential of the particles is μ. Write down the expression for the "number of particles" with energy $\epsilon = (p^2 c^2 + m^2 c^4)^{1/2}$ (note that if "particle" count of the particles is $+1$, that of antiparticles is -1). Discuss the Bose Einstein condensation in this system, explaining clearly what is the criterion for the condensation to set in. If inc $mc^2 << k_B T_c$, find an expression for the critical temperature T_c in terms of the fundamental constants.

4. The Landau model treats superfluid helium at low temperatures ($\leq 1K$) as a weakly excited quantum system in which deviations from the ground state ($T = 0K$) are described in terms of a "gas of elementary excitations" of the system against a quiescent background. The gas of excitations corresponds to the normal fluid and the background to the superfluid. Landau suggested the excitation spectrum shown in Fig.9.6 on phenomenological grounds. For $p \ll p_0$ the spectrum is linear and hence gives a specific heat $\propto T^3$ at very low temperatures. For $p \approx p_0$, Landau called the excitations rotons and gave them the frequency spectrum

$$\epsilon_{(p)} = \Delta + (p - p_0)^2 / 2\mu.$$

The rotons are bosons and their number indefinite. Show that the specific heat of a roton gas is

$$C_r = R \left\{ \frac{3}{4} + \frac{\Delta}{k_B T} + \left(\frac{\Delta}{k_B T} \right)^2 \right\} e^{-\Delta / k_B T}$$

5. The most visible "super" quality of the fluidity of liquid 4He is its ability to crawl up walls and out of a containing vessel. The reason is that it coats the wall with a thick film. At height z above the free liquid surface (in equilibrium with its own vapour), the film adhering to a vertical wall has a thickness proportional to $z^{-1/2}$ Derive this result by the following steps:

   a) Suppose the temperature is such that the superfluid fraction is f. Consider an element of the adhering film at height z, of thickness d, vertical dimension Δz and unit width in the other direction. Treat the superfluid component in this volume element as a free Bose gas. Write down its energy, with contribution from potential energy due to gravity and kinetic energy due to its being confined to a thickness d.

   b) Minimize the energy with respect to d.

10

Superconductivity

10.1 Introduction

In the last four chapters we have dealt with fermions and bosons. In each instance that we have taken, we have dealt separately with either a collection of fermions or a collection of bosons. In this chapter we introduce a different type of problem – a problem where we start with a collection of fermions but end up with a state which can be characterized as a condensation of bosonic objects formed from two bound electrons of opposite spins. At the heart of the phenomenon is the possibility of the formation of this bound state. But then one needs an attractive interaction between two fermions. Our attempt in this mainly qualitative chapter will be to focus mainly on this attraction. We begin by recalling the salient experimentally observed features of a superconductor.

In this chapter, we will give a fairly phenomenological picture of conventional superconductivity. Generally metals which are not very good conductors at room temperatures become superconductors when the temperature is reduced to $5 - 10K$. We give a qualitative picture of these superconductors in view of the interest generated of late in superconductivity at considerably higher temperatures ($\sim 100K$) in certain ceramic compounds. The mechanism of that superconductivity (see Sec. 10.5) is supposed to be different from the mechanism that we will be discussing in the next section. The characteristic properties of a superconductor like tin are:

i) It behaves as if there is no D.C. electrical resistivity. Current, if set up in superconductors, does not seem to decay.

ii) It behaves as a perfect diamagnet. A sample in thermal equilibrium in an applied magnetic field carries surface currents. These currents give rise to an additional field that exactly cancels the field in the interior of the specimen.

iii) The superconductor behaves as if there was an energy gap of width 2Δ centred around the Fermi level in the set of allowed one-electron levels. The gap increases as the temperature decreases and levels off to a maximum value of $\Delta(0)$ at very low temperatures.

The transition to a superconducting state is sharp in bulk specimen. Below a critical temperature T_c the superconducting properties appear, the most dramatic of which is the apparent vanishing of the D.C. resistance. Above T_c, the resistivity has the form $\rho(T) = \rho_0 + BT^5$, the constant term appearing from impurity and defect scattering and the term in T^5 arising from phonon scattering. Below T_c, these mechanisms are ineffective in dissipating the current and the resistivity abruptly drops to zero. The superconductivity can be destroyed by large magnetic fields, if the current exceeds a critical value (the size of the critical current is determined by the geometry of the sample) or by applying an A.C. electric field at sufficiently high frequencies.

Consider a superconductor at a temperature below the critical temperature T_c. As a magnetic field is applied a certain amount of energy is expended to establish the magnetic field of the screening currents so that the magnetic field inside the specimen is cancelled out. If the applied field is large enough it will be energetically favourable for the system to revert back to the normal state, allowing the field to penetrate. Although the normal state has a higher free energy than the superconducting state below T_c in zero field the increase in free energy will be compensated for by the decrease in magnetic energy as the screening currents disappear. The manner in which field penetration occurs can depend upon the material for the same geometry and it has been seen that materials can behave in two distinct ways:

1. Below a critical field $H_C(T)$ that increases as T decreases further below T_c, there is no penetration at all. For $H > H_c$, the entire sample reverts to the normal state and the field penetrates perfectly. These are called the type-I materials.

2. Below a lower critical field $H_{c1}(T)$, there is no penetration of the field at all, while above an upper critical field $H_{c2}(T)$, the sample reverts back to the normal state with full penetration of the field. For $H_{c1} < H < H_{c2}$, there is partial penetration of the flux, and the sample develops a rather complicated microscopic structure of both normal and superconducting regions, known as the mixed state. In the mixed state (Abrikosov state), the flux penetrates the sample in the form of thin filaments of flux. Within each filament the field is high and the material not superconducting. Outside the core of the filaments, the material remains superconducting. Circulating around each filament is a vortex of screening current. These are calledtype-II materials.

At low temperatures, the specific heat of a metal is of the form $AT + BT^3$, where the linear term is due to electronic excitations and the cubic term is the lattice contribution. For superconductors below T_c, this form is significantly altered. The linear term is absent and instead replaced by a term whose temperature dependence is of the form $e^{-\Delta/kT}$. This is the characteristic behaviour of a system whose excited states are separated from the ground state by an energy of order Δ. Experiments show that the energy gap is of the order kT_c.

Further evidence of the existence of a gap in the energy spectrum can be had from tunneling experiments. The conduction electrons in a superconductor and a normal metal can be brought into thermal equilibrium by placing them in close contact such that there is just a thin insulating layer separating them. In thermal equilibrium, the chemical potential of the two metals will be equal and if a potential difference is now applied, the chemical potential of one will be raised with respect to the other and if both metals are in the normal state, a current will immediately flow. This is the tunneling current. However, if one of the metals is normal and the other superconducting, the potential difference has to be raised to at least before a tunneling current begins to flow.

Acoustic attenuation and A.C. electrical conductivity also provide evidence for the existence of the gap. In the next section we will give, following Weisskopf, a qualitative description of the phenomenon of superconductivity.

10.2 Pairing Theory: Qualitative

Consider the metal as a cubic lattice of positive ions filled with a degenerate gas of free electrons at zero temperature. The lattice distance d is of the order of a few Bohr radii (Bohr radius a_B is of the order of $\hbar^2/me^2$). If ϵ_A is an atomic energy of a few electron volts, then clearly

$$\epsilon_A \sim \frac{e^2}{d} = \frac{\hbar^2}{md} \cdot \frac{me^2}{\hbar^2} = \frac{\hbar^2}{m} \cdot \frac{d}{d^2 a_B} = \frac{\hbar^2}{md^2} f \sim \frac{\hbar^2}{md^2}. \tag{10.2.1}$$

where $f = d/a_B$ is a number of order unity though always somewhat greater than unity. The maximum momentum and hence velocity of the electrons is possessed by those electrons at the top of the Fermi surface

$$p_F \sim \frac{\hbar}{d}$$

$$v_F \sim \frac{\hbar}{md}. \tag{10.2.2}$$

The mechanics of the lattice can be described as that of independent oscillators of mass M and frequency given by the Debye frequency ω_D. The potential energy of the displacement (typically d) is

$$\frac{1}{2} M \omega_D^2 \left(\frac{d}{2}\right)^2 \sim \epsilon_A \sim \frac{\hbar^2}{m(d/2)^2}$$

$$\omega_D = b \frac{\hbar}{d^2} \cdot \frac{1}{\sqrt{mM}} \sim \sqrt{\frac{m}{M}} \frac{\epsilon_A}{\hbar}. \tag{10.2.3}$$

The number b is 8 in our estimate and indeed lies between 5 and 10 for most metals.

We now describe the effect of the motion of the electron on the lattice. Consider an electron at the Fermi surface. It spends a time $\tau \sim d\nu_F$ near the ion. During this time the momentum transferred to the ion is approximately

$$\Delta p \sim \tau . \frac{e^2}{d^2} \sim \frac{e^2}{d\nu_F}. \tag{10.2.4}$$

This makes the ion travel towards the electron and the displacement suffered can be estimated as

$$\begin{aligned} \delta &\sim \frac{\Delta p}{M\omega_D} \\ &\sim \frac{e^2}{d\nu_F} \frac{1}{M} \sqrt{\frac{mM}{\hbar}} d^2 \\ &= \sqrt{\frac{m}{M}} \frac{d}{a_B} . d \\ &\sim \sqrt{\frac{m}{M}} d. \end{aligned} \tag{10.2.5}$$

The ions therefore are displaced a distance δ towards the path of the electron. They do not remain in this displaced position for ever but return after a time of the order of ω_D^{-1}. It should be noted that we are at $T = 0$ and the displacement δ does not spread over the lattice in the form of sound waves. The displacement is a part of the quantum state of the electron in a lattice, different from that of the free electron. In moving through the lattice, the electron behaves like a quasi-particle. The fact that the ion will return after a time ω_D^{-1} implies that the length of the region over which there are displaced ions associated with an electron is

$$l \sim \nu\tau \sim \frac{\nu}{\omega_D} \sim \sqrt{\frac{M}{m}} . d \; . \tag{10.2.6}$$

The displacement of the ion towards the electron implies that in this tube of length l and diameter d the potential is different from that we would find in an undeformed lattice. The potential in the absence of the deformation is e^2/d and since the separation has decreased by a distance δ, we must have a decrease in the potential by the amount

$$\Delta U = -\frac{e^2}{d^2}\delta \sim -\frac{e^2}{d}\sqrt{\frac{m}{M}}. \tag{10.2.7}$$

A second electron entering this tube sees a screening of the field of the first electron and hence feels an effective attraction. Only the electrons near the Fermi surface can make use of such a weak interaction-the electrons below the Fermi surface are "frozen" in that they cannot easily change their quantum states. In order to feel the effect of the potential tube fully, the electrons must meet "head-on", i.e. to say the closest distance of approach should be equal to or less than d, since d is of the order of wavelength of the electrons at the Fermi surface. This means the electrons must be in a relative S-wave state. In a higher angular momentum state, the impact parameter will be higher and the electrons would miss their mutual potential tubes.

The mutual potential then has the form (relative to the usual background)

$$V(r) \approx -\sqrt{\frac{m}{M}}\frac{e^2}{d}\delta_{LO} \quad \text{for} \ r < 1$$
$$\approx 0 \quad \text{for} \ r > 1 \quad\quad (10.2.8)$$

where δ_{LO} is the δ-function that decrees that only the S-wave state is important. The two electrons must have opposite spins to satisfy Pauli principle in the S-wave state. Thus, we have the result that two electrons of opposite momenta and spins feel a weak attraction for each other at the Fermi surface.

It is this attraction that gives rise to bound states on top of the Fermi distribution. If we visualize the potential of Eq.(10.2.8) to be a trough (in $D = 3$) of depth U and radial extension l, then indeed there would be bound states with energy

$$\epsilon \approx -U + \frac{\pi^2\hbar^2}{2ml^2} \quad\quad (10.2.9)$$

if $U \gg \hbar^2/ml^2$. Although the potential of Eq.(10.2.8) would satisfy this criterion, it is not obvious that a bound state would exist. This is because the two electrons are at the Fermi surface and have a relative kinetic energy of $2\epsilon_F$. Since $\epsilon_F \gg U$, it can be argued that they would be in the continuum and hence unbound. However, it should be noted that because of the Pauli principle, this state would be the lowest possible state and hence the above argument may not be valid. The effect, as it turns out, is that a bound state is formed although the binding energy is much weaker than that expected on the basis of Eq.(10.2.9). The potential of Eq.(10.2.8) decreases the energy of a pair of electrons at the Fermi surface from $2\epsilon_F$ to $(2\epsilon_F - \Delta)$ if the total momentum is zero and we are in the S-wave state (see Sec.10.4 for the relevant demonstration using the Schrodinger equation), with

$$\Delta = 2\hbar\omega_D.e^{-\xi.2\pi\epsilon_F/e^2 d} \quad\quad ...(10.2.10)$$

where ξ is a numerical constant of order unity.

The pair of electrons is called a Cooper pair and Δ is their binding energy. If we define,

$$\gamma = e^{-\xi.2\pi\epsilon_F d/e^2} \quad\quad (10.2.10)$$

then $\gamma \sim 10^{-2}$ to 10^{-3} $(\epsilon_F d/e^2 \sim 1)$ and we can write

$$\frac{\Delta}{\epsilon_F} = 2\sqrt{m/M}\gamma \quad\quad (10.2.11)$$

making Δ roughly 10^{-4} to 10^{-5} times smaller than atomic energies. Thus, $\Delta \sim 10^{-4} eV$ which corresponds to a temperature of a few degrees Kelvin.

Since the wavefunction of a Cooper pair is an S-wave, the motion it describes is the periodic back and forth movement of the pair covering a relative, distance ρ between them. For distances $r > \rho$, the wavefunction goes to zero much faster than r^{-1}. This is quite analogous to the motion of two nucleons in a deuteron or the electron and positron in the ground state of positronium. If the momentum spread of the wavefunction around the Fermi momentum (i.e., approximately the range of momentum of the free particle states which have to be superposed to give the wavefunction of the Cooper pair) is Δp then the energy spread would be

$$\Delta \epsilon \approx \Delta p . p_F / m \qquad (10.2.12)$$

and since $\Delta \epsilon$ is of the order of the binding energy Δ,

$$\frac{\Delta p}{p_F} \sim m\Delta / p_F^2 \sim \Delta / \epsilon_F \ll 1. \qquad (10.2.13)$$

Hence the momentum spread is very small and thus for $r \ll \rho$, the pair-wavefunction is almost $\sin(\rho_F \, r/h)$. For $r > \rho$, the wavefunction vanishes due to interference of the waves with different p. The "size" ρ of the pair can then be estimated as

$$\rho \sim \frac{\hbar}{\Delta p} \sim \frac{\hbar p_F}{m\Delta} \sim \frac{p_F^2}{m\Delta} \sim \frac{\epsilon_F}{\Delta} . d \gg d \qquad (10.2.14)$$

The wavefunction is thus tremendously spread out which is a characteristic of the low binding energy. The large extension of the wavefunction is responsible for the absence of the influence of electrostatic repulsion on the binding. For distances somewhat larger than d, the electron interactions are screened out by the electron gas. Since $\rho \gg d$, the effect of the repulsion which is confined to distances smaller than a few d, changes the wavefunction only near the centre and the extension of the binding potential being $l \gg d$, the binding energy is largely unaffected. The length ρ is called the coherence length and can be of the order of 10^{-4}cm.

We now turn to the electron gas as a whole. Cooper pairs are formed only among those electrons that lie in the uppermost part of the Fermi distribution, not lower than Δ below the Fermi-surface in terms of energy, or Δp below p_F in momentum. Electrons lying lower than that cannot make use of the empty states to form a Cooper pair wavefunction. All Cooper pairs have momentum $p = 0$ and spin $S = 0$. That puts them in the same quantum state. This does not violate any statistics since the two-electron system obeys Bose statistics. It is this issue of ending with a collection of bosons after starting with a gas of fermions that makes this chapter different from the previous ones on of quantum statistics, as has been remarked at the very beginning of this chapter.

The number n' of Cooper pairs can be estimated from the number of electrons n (number per unit volume) which would be of $O\left(\frac{1}{d^3}\right)$

$$n' = n\frac{\Delta}{\epsilon_F} \sim \frac{\Delta}{d^3\epsilon_F} \sim \frac{1}{d^2\rho}. \qquad (10.2.15)$$

The distance D between Cooper pairs is

$$D \sim \left(\frac{1}{n'}\right)^{1/3} \sim d\left(\frac{\rho}{d}\right)^{1/3} \sim (30 \text{ to } 40)d. \qquad (10.2.16)$$

The size of a Cooper pair is much larger than that and hence the Cooper pairs must necessarily overlap. However, it is important to note that two Cooper pair wave functions are still orthogonal. The Cooper pair wavefunction is also orthogonal to the wave functions of the free electrons which are more than Δ, below the Fermi level. The interaction between the electrons discussed above, thus changes the Fermi distribution to an ensemble of Cooper pairs among the n' electrons in the Δ-layer and of free electrons below the Δ-layer.

We next demonstrate that a finite amount of energy of the order of Δ is required to reach the first excited state of this ensemble. For a free-electron Fermi gas, the excited states are infinitely close to the ground state: they are the states in which one or more electrons are moved into a level slightly higher than the Fermi level. These excited states carry electric current. The existence of a finite energy required to reach the first excited state in the present case can be seen from the fact that if a free electron is to be taken outside the Fermi surface, it needs to be given an energy of $O(\Delta)$ and if an electron from the Δ-layer is to be taken across the Fermi level, the pair has to be broken requiring an energy of order Δ. To complete this argument, one has to understand what prevents a simple Cooper pair from changing its motion from the momentum state $p = 0$ to a very small momentum p' without breaking up. If $p'^2/2m \ll \Delta$, then this would represent an excited state of the ensemble with energy much lower than Δ. However, this is impossible since it would require that the free electron states making up the Cooper pair change to contain pairs of free electron levels for which the total momentum is p'. Since all electrons in the Δ-layer are paired, the change of motion of one pair would affect all the other pairs. The only way in which one Cooper pair can change its momentum from $p = 0$ to p' is by having all the Cooper pairs change simultaneously and that would cost an energy much larger than Δ.

One must now consider how an ensemble of electrons with zero total momentum $p = 0$ ever produces a current and even a supercurrent without any resistance at all. It might seem that in a state $p = 0$, there is no net motion of charge. The reason that a current can be carried is that in the presence of a magnetic field, the velocity is no longer proportional to the momentum. The velocity is obtained from the kinetic energy by differentiation with respect to the momentum p and since the kinetic energy in the presence of a magnetic field, characterized by a vector potential, is $\frac{1}{2m}\left(\mathbf{p} - \frac{e\mathbf{A}}{c}\right)^2$ we find for the velocity,

$$v = \frac{1}{m}(\mathbf{p} - e\mathbf{A}/c). \tag{10.2.17}$$

The current density produced by the ensemble of electrons is

$$
\begin{aligned}
\mathbf{j} &= e\sum \mathbf{v}_i = \frac{e}{m}\sum \mathbf{p}_i - \frac{e^2}{mc}n\mathbf{A} \\
&= -\frac{e^2 n}{mc}\mathbf{A} \tag{10.2.18}
\end{aligned}
$$

since $\sum \mathbf{p}_i = $ total momentum $= 0$ and n is the total number of electrons per unit volume. We can write

$$\mathbf{j} = \frac{e^2 n}{mc}\mathbf{A} = -\frac{\mathbf{A}}{\lambda^2} \tag{10.2.19}$$

where

$$\lambda = \left(\frac{mc^2}{ne^2}\right)^{1/2} = \left(\frac{mc^2 d}{e^2}\right)^{1/2} d = \left(\frac{d}{a_B}\right)^{1/2}\frac{\hbar c}{e^2}d. \tag{10.2.20}$$

Now, the fine structure constant $e^2/\hbar c = 1/137$ and hence the length λ is of the order of hundred times the lattice distance.

The current is supported because of the energy gap. The current above is just the induction current set up when the magnetic field is switched on. The kinetic energy (T) increases when the field is switched on

$$
\begin{aligned}
T &= \frac{1}{2m}\sum_i \left(\mathbf{p}_i - \frac{e\mathbf{A}}{c}\right)^2 \\
&= \frac{1}{2m}\sum \mathbf{p}_i^2 + \frac{e^2 n}{2mc^2}A^2 \tag{10.2.21}
\end{aligned}
$$

where ($\sum \mathbf{p}_i.\mathbf{A} = 0$ since $\sum \mathbf{p}_i = 0$). If the metal were non-superconducting, this increase of kinetic energy would be compensated by changing the momentum of each electron by $e\mathbf{A}/c$ so that the kinetic energy reverts to its original value within a short relaxation time. In the superconductor, the Cooper pairs at the crust prevent such a redistribution-momentum redistribution costs energy and hence the situation with altered kinetic energy has to persist and thus the current does not decay.

However, it must be pointed out that a gap in the spectrum is not the sole criterion for persistent current. Had that been so, insulators would have shown superconducting behaviour as well. The reason an insulator does not behave in a similar fashion is because the free particle kinetic energy expression is no longer correct at the top of a band. If we consider a one-dimensional situation for simplicity, then $T(p)$ is like $p^2/2m$ for the lower part of the band. Near the upper end of the band the function $T(p)$ has a point of inflection and the tangent is horizontal at the upper end, i.e., $dT/dp = 0$ at $p = \pm p_{\max}$. Now, in the absence of the magnetic field

$$j = e \sum \frac{dT}{dp} = e \int_{-p_{\max}}^{p_{\max}} \frac{dT}{dp} \cdot g(p) dp = 0 \qquad (10.2.22)$$

since $g(p)$ the density of states is an even function and dT/dp is an odd function of p.

In the presence of a magnetic field, we have the kinetic energy as some function of $(\mathbf{p} - e\mathbf{A}/c)$ and expanding as a power series of the potential A, which is usually small

$$T(p - eA/c) = T(p) - eA/c \frac{dT}{dp} + \dots$$

The current is

$$
\begin{aligned}
j &= e \int_{-p_{\max}}^{p_{\max}} \frac{dT}{dp} g(p) dp \\
&= e \int_{-p_{\max}}^{p_{\max}} \frac{dT(p)}{dp} g(p) - \frac{e^2 A}{c} \int_{-p_{max}}^{p_{max}} \frac{d^2 T}{dp^2} g(p) dp \\
&= -\frac{e^2 A g}{c} \int_{-p_{\max}}^{p_{\max}} \frac{d^2 T}{dp^2} dp \\
&= \left(-\frac{e^2 A g}{c} \frac{dT}{dp} \right)_{p_{\max}}^{p_{\max}} \\
&= 0 \qquad\qquad\qquad\qquad (10.2.23)
\end{aligned}
$$

We have used the fact that a one-dimensional density of states is p-independent and that at the top of the band $dT/dp=0$. Thus, there is no current in the presence of a magnetic field.

We now show that the magnetic field does not penetrate into the super-conductor. The current being given by Eq.(10.2.19), we have the magnetic field as being given by

$$\nabla \times \mathbf{B} = \mathbf{j} = -\frac{\mathbf{A}}{\lambda^2}$$

Another curl operation leads to

$$\nabla^2 \mathbf{B} = \mathbf{B}/\lambda^2$$

showing that the field dies down exponentially from the surface.

Consider an infinite superconductor with a plane surface. In a rectangular coordinate system, the space $x > 0$ is filled with superconducting metal, the plane defined by $x = 0$ is the surface. Let us imagine that there is a magnetic field B_0 for $x > 0$ in the y-direction. Inside the material

$$B_y = B_0 e^{-x/\lambda} \qquad (x > 0)$$

and $B_x = B_z = 0$. The corresponding vector potential is

$$A_z = -\lambda B_0 e^{-x/\lambda} \qquad (x > 0)$$

with $A_x = A_y = 0$

$$j_z = -A_z/\lambda^2 = \frac{B_0}{\lambda} e^{-x/\lambda} \qquad (x > 0)$$

with $j_x = j_y = 0$. Thus there is a current in the z-direction but confined within a depth λ from the top. The total current flowing across a rectangle of depth λ and width S in the y-direction is clearly $B_0/\lambda.\lambda S = SB_0$ (we have replaced $e^{-x/\lambda}$ by 1 for $x < \lambda$ and by zero for $x > \lambda$.

Finally consider a superconducting wire with its axis in the z-direction, with a circular magnetic field B_0 around the wire, parallel to its surface and perpendicular to the z-axis. We conclude from the preceding discussion that there will be a current j in the wire flowing in the z-direction of strength

$$k = 2\pi R B_0.$$

The current is restricted to a depth λ from the surface of the wire, This current produces a circular magnetic field B_0 at the radius R and the situation is self-consistent and stable. The current produces the field B_0 which, in turn, causes the same current to appear in the "frozen" electron gas, kept in the ground state by the Cooper pair crust at the top of the Fermi surface, so long as $kT < \Delta$. The electrons at a depth greater than λ from the surface do not move as there is no magnetic field there, while the electrons within a depth λ move keeping the total momentum zero. This state of affairs does not continue for all values of B_0. As the magnetic field is increased, the kinetic energy of the electrons increases, and if this increment can break the Cooper pairs, then the electrons will redistribute and get rid of the excess energy by losing the kinetic energy, i.e. stopping the current. If we consider a unit area on the surface of the wire, then the additional kinetic energy in a volume $V = \lambda.1$ is

$$\Delta T = V \frac{ne^2}{2mc^2} A^2 = V \frac{A^2}{2\lambda^2}.$$

The current will be stable if ΔT is less than the total energy G of the binding of Cooper pairs

$$G = V \Delta n' = V.\Delta.n.\Delta/\epsilon_F$$

where n' is the number of Cooper pairs per unit volume. The critical magnetic field B_{cr} is given by $B_{cr} \sim A_{cr}/\lambda$ and hence

$$\frac{V}{2}B_{cr}^2 \sim 2n\Delta^2/\epsilon_F$$
$$B_{cr}^2 \sim 2n\Delta^2/\epsilon_F. \tag{10.2.24}$$

With $\Delta/\epsilon_F \sim 10^{-4}$ and $\epsilon_F \sim eV, B_{cr}$ turns out to be around $300G$. A superconductor with a critical field of this order is termed a type-I superconductor.

This essentially concludes the brief survey of superconductivity. We have not attempted to touch upon the topics of flux quantization or Josephson tunneling. Our focus has been on the pairing mechanism of conventional superconductivity. Accordingly, in the two remaining sections of this chapter we will attempt to quantify the attractive interaction and demonstrate that in the presence of the attraction, two electrons near the Fermi surface will always form a bound state.

10.3   Origin of the Attractive Interaction

In this section, we attempt to quantify the heuristic argument about the attractive interaction between the two electrons. To do so, we work in terms of the dielectric constant and obtain the interaction between two electrons in momentum space in the form

$$V(k,\omega) = \frac{e^2}{\epsilon(k,\omega)k^2} \tag{10.3.1}$$

where $\epsilon(k,\omega)$ is the momentum and frequency dependent dielectric constant. The effect that we want to incorporate is the strong reaction of the lattice to the electron and the screening of the electron charge by the ion cloud. It is this screening which is capable of overcompensating the electron charge. A second electron coming along may now see a positive charge and feel attracted.

To calculate $\epsilon(k,\omega)$, we consider an external charge distribution $\delta\rho$, having the Fourier transform

$$\delta\rho(r) = \delta\rho \cos(k.r)e^{i\omega t} + \text{c.c} . \tag{10.3.2}$$

If the screening charge formed in response to $\delta\rho$ is $\rho(r)$ with the Fourier Transform

$$\rho(r) = \rho \cos(k.r)e^{i\omega t} + \text{c.c} . \tag{10.3.3}$$

then the dielectric constant $\epsilon(k,\omega)$ is

$$\epsilon(k,\omega) = \frac{\delta\rho}{\rho + \delta\rho}. \tag{10.3.4}$$

We need to calculate the response charge ρ, which will have contributions from ions as well as electrons denoted by ρ_i and ρ_e respectively. Clearly

$$\rho = \rho_i + \rho_e$$

The total charge sets up the potential V, which satisfies

$$\nabla^2 V = -(\delta\rho + \rho_i + \rho_e). \tag{10.3.5}$$

In the resulting electric field $\mathbf{E} = -\nabla V$, the ions move according to the equation of motion

$$M\dot{\mathbf{J}}_i = nZe^2\mathbf{E} \tag{10.3.6}$$

where J_i is the ionic current, n the number density of ions and Z the ionic charge. The ionic current satisfies the continuity equation

$$\frac{\partial \rho_i}{\partial t} + \nabla.\mathbf{J}_i = 0. \tag{10.3.7}$$

Combining Eqs.(10.3.5)-(10.3.7), we get

$$\frac{\partial \rho_i}{\partial t} = \frac{nZe^2}{M}\nabla^2 V = -\frac{nZe^2}{M}(\delta\rho + \rho_i + \rho_e)$$

In Fourier space,

$$\omega^2[\rho_i(\mathbf{k},\omega)] = \omega_i^2[\delta\rho(\mathbf{k},\omega) + \rho_i(\mathbf{k},\omega) + \rho_e(\mathbf{k},\omega)] \tag{10.3.8}$$

where $\omega_i^2 = \frac{nZe^2}{M}$. (Typically, $\omega_i \approx 10^{13}/S$)

One now needs to know the response of the electrons alone. The characteristic frequencies involved with electrons are i) the plasma frequency $\omega_e = \omega_i\sqrt{M/m}$ and ii) the frequency associated with the Fermi energy $\omega_F = \epsilon_F/h$. Both ω_e and ω_F are much greater than ω_i and hence for frequencies ω close to ω_i, the electron gas is virtually static. We calculate the response in what is known as the Thomas-Fermi approximation. We recall from Chapter 7 that the Fermi energy of an electron gas is proportional to $n^{2/3}$, where n is the number density. We now assume that in an external potential $V(r)$, the Fermi energy is raised to $\epsilon_F + eV(r)$ and the resulting density $n_e(r)$ is proportional to $[\epsilon_F + eV(r)]^{3/2}$. We write $n_e(r) = ne + \rho_e(r)$ and expanding in $\rho_e(r)$ and $eV(r)$

$$\frac{\rho_e(r)}{-ne} = \frac{3}{2}\frac{eV}{\epsilon_F}. \tag{10.3.9}$$

Using Eq.(10.3.5) in momentum space

$$\rho_e(k) = -\frac{k_S^2}{k^2}(\rho + \delta\rho) \tag{10.3.10}$$

where

$$k_S^2 = \frac{3}{2}\frac{ne^2}{\epsilon_F}.$$

Combining Eqs.(10.3.8) and (10.3.10),

$$\rho = \rho_i + \rho_e = \left(\frac{\omega_i^2}{\omega^2} - \frac{k_S^2}{k^2} \right)(\rho + \delta\rho). \qquad (10.3.11)$$

This leads to

$$\epsilon(k, \omega) = \frac{\omega^2(k_S^2 + k^2) - \omega_i^2 k^2}{\omega^2 k^2}$$

Writing,

$$\omega_k^2 = \omega_i^2 \frac{k^2}{k^2 + k_S^2}$$

$$\frac{1}{\epsilon(k, \omega)} = \frac{\omega^2 k^2}{(k_S^2 + k_S^2)[\omega^2 - \omega_k^2]}$$

$$= \frac{k^2}{k^2 + k_S^2} \left[1 + \frac{\omega_k^2}{\omega^2 - \omega_k^2} \right]. \qquad (10.3.12)$$

For $\omega \leq \omega_k$, $\epsilon(k, \omega)$ can become negative signaling an attraction.

10.4 Cooper Instability

In this section, we demonstrate using the Schrodinger equation, the formation of the bound state of two electrons at the Fermi surface through a weak attractive interaction. We begin with a degenerate gas of N free electrons at $T = 0$, with the Fermi energy given by ϵ_F and the Fermi momentum by k_F. Two electrons at the Fermi surface are taken to interact with a weak attractive interaction. Their centre of mass is at rest. The two-particle Schrodinger equation takes the form

$$-\frac{h^2}{2m}(\nabla_1^2 + \nabla_2^2)\psi + V(r_1 - r_2)\psi = E\psi. \qquad (10.4.1)$$

We will measure the energy relative to the Fermi energy and hence write $E = 2\epsilon_F + \epsilon_b$. Further, it will be convenient to work in the momentum space and hence we introduce the momentum space wave function $\phi(k)$ as

$$\psi(r_1, r_2) = \int \phi(k) e^{i\bar{k} \cdot (\bar{r}_1 - \bar{r}_2)} \frac{d^3 k}{(2\pi)^3}. \qquad (10.4.2)$$

Translation invariance implies that $\psi(\bar{r}_1 - \bar{r}_2)$ is a function of $(\bar{r}_1 - \bar{r}_2)$ above. The filled Fermi sphere immediately puts the following vital restriction on $\phi(k)$

$$\phi(k) = 0 \quad \text{for } |k| < k_f. \qquad (10.4.3)$$

The states inside the Fermi sphere are not available to the two electrons because of the exclusion principle. In momentum space, Eq.(10.3.1) becomes

$$\frac{h^2}{2m}(k^2 + k^2)\phi(k) + \sum_{k'} V_{kk'}\phi(k') = (2\epsilon_F + \epsilon_b)\phi(k)$$

or,

$$\left(\frac{h^2}{m}k^2 - 2\epsilon_F - \epsilon_b\right)\phi(k) = -\sum_{k'} V_{kk'}\phi(\bar{k}') \tag{10.4.4}$$

We now have to take a form for $V_{kk'}$. As seen qualitatively in Sec.10.2 the attractive interaction has a long range in coordinate space. Accordingly, it is short ranged in momentum space and exists over a range $\hbar\omega_D$ (ω_D being the Debye frequency ν_m of Sec.4.5) above the Fermi surface.

In that range, we take it to be a constant V_0, so that

$$V_{kk'} = -V_0 \quad \text{if } k < k',$$

such that

$$\epsilon_F < \frac{h^2 k'^2}{2m} < \epsilon_F + \hbar\omega_D. \tag{10.4.5}$$

The negative sign comes from the fact that it is an attraction. We can now write Eq.(10.4.4) as

$$\phi(k) = \frac{V_0 C}{\frac{h^2 k^2}{m} - 2\epsilon_F - \epsilon_b}. \tag{10.4.6}$$

where C is a constant given by

$$C = \sum_{k'} \phi(\bar{k}') \tag{10.4.7}$$

the summation being over the range where $V_{k,k'} \neq 0$. From Eqs.(10.4.6) and (10.4.7)

$$C = \sum_k \phi(k) = V_0 C \sum_k \frac{1}{\frac{h^2 k^2}{m} - 2\epsilon_F - \epsilon_b}. \tag{10.4.8}$$

The sum over k brings in a phase space factor $\frac{4\pi k^2 dk}{(2\pi)^3} = N(\epsilon)d\epsilon$ where $N(\epsilon)$ is the density of one electron level for a given spin. In terms of the energy variable measured from the Fermi level $\left(\epsilon = \frac{h^2 k^2}{2m} - \epsilon_F\right)$, Eq.(10.4.8) becomes

$$1 = V_0 \int_0^{\hbar\omega_D} \frac{N(\epsilon)d\epsilon}{2\epsilon - \epsilon_b}. \tag{10.4.9}$$

Over the narrow range of integration shown in the above equation, $N(\epsilon)$ is approximately constant and equal to $N(0)$: the density of states at the Fermi level. Thus,

$$1 = \frac{N(0)V_0}{2}\ln\left(1 - \frac{2\hbar\omega_D}{\epsilon_b}\right) \tag{10.4.10}$$

leading to

$$\epsilon_b = \frac{-2\hbar\omega_D \cdot \exp\left[-\frac{2}{N(0)V_0}\right]}{1 - \exp\left[-\frac{2}{N(0)V_0}\right]} \tag{10.4.11}$$

which in the weak coupling limit $N(\omega)V_0 \ll 1$, becomes

$$\epsilon_b \approx -2\hbar\omega_D e^{-\frac{2}{N(0)V_0}}. \tag{10.4.12}$$

The negative sign indicates that the electrons form a bound state. The bound state is formed regardless of how small the potential V_0 is. This is the unusual feature which comes from the blocking off of a certain region of momentum space by the filled Fermi sphere. It should be noted that Eq.(10.4.12) cannot be obtained in perturbation theory.

10.5   A Different Pairing: High Temperature Superconductivity in Cuprates

Till 1986, the highest temperature upto which superconductivity was seen to survive was only about 30 degree Kelvin. As explained in previous Sections, electrons near the Fermi surface could form pairs despite the strong Coulomb repulsion between two electrons, because of the coupling to the vibrations of the lattice that could induce a weak attraction, and at the Fermi surface, however weak the attraction, it could lead to a bound state. It was widely believed that the mechanism could not keep the superconductivity going beyond 30 degree Kelvin (MgB_2 with the T_c at 39 degree Kelvin was an exception). The materials which became superconducting at low enough temperatures were all metals at normal temperatures. But then came 1986 and copper oxide (cuprate) based materials were found to become superconductors at temperatures much higher than 30 degree Kelvin. The first group of such compounds $La_{2-x}\,Ba_x\,Cu\,O_4$ had T_c at around 40 degree Kelvin, which rose to around 100 degree Kelvin $Y\,Ba_2\,Cu_3\,0_{6+x}$, and to 150 degree Kelvin for mercury based cuprates. Surprisingly enough, these materials at room temperatures were such poor conductors that they could hardly be recognized as metals. In fact, if the composition was varied slightly (the x of $La_{2-x}\,Ba_x\,Cu\,O_4$) then the material became an antiferromagnetic insulator. Magnetism comes from the strong repulsive interactions (Coulomb repulsion) between electrons, and superconductivity is the result of weak attraction between electrons and hence, superconductivity and magnetism are completely at variance with each other. Superconductivity in these materials was consequently a very big surprise.

Over the last quarter of a century the properties of cuprates have been studied with a great deal of precision, and it has become quite clear that the usual quantum theory of electronic properties of solids which can account for most metals and superconductors becomes quite inadequate for this class of materials. They form what in general can be termed "highly correlated electron systems". One of the important facts is that, at temperatures much above the transition temperature T_c the conductivity in cuprates is almost two orders smaller than in simple metals and shows a frequency and temperature dependence which is not in agreement with the usuals theory of metals. This has led to the usual description of these oxides as "strange metals" or "bad metals" (it should be noted that "strange" behaviour has also been found in other highly correlated electron system which are not linked to high temperature superconductivity). The copper oxides also exhibit a tremendously complicated behaviour above T_c. This regime is called a "pseudo-gap" regime and is characterized by a reduction of the electron density of states but no obvious broken symmetry.

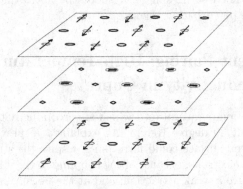

Figure 10.1: A schematic representation of two 2D cuprate layers separated by an inert buffer layer.

The crystal structure of the cuprates exhibiting high-temperature superconductivity shows a layered structure with two-dimensional copper oxide layers separated by ionic, electronically inert buffer layers (see Fig.(10.1)). The highly correlated electrons in the undoped system (i.e., $x = 0$) give rise to an antiferromagnetic behaviour. The copper oxide planes can be 'doped' by changing the makeup of the interleaved "charge reservoirs" layers so that the electrons are removed (hole-doped) or added (electron doped). Hole doping rapidly suppresses superconductivity and at a critical doping superconductivity sets in. The transition temperature reaches a maximum at a critical doping, then declines and vanishes for a doping exceeding a certain maximum limit. A very simplified typical phase diagram is as shown in Fig.(10.2) (it should be borne in mind that it is possible in certain situations for the superconducting state to exist in presence of the antiferromagnetism).

The fact that strong electron repulsions are responsible for the undoped system to be an insulator points to the fact that the appropriate model for describing these systems is the Hubbard model. In Section 7.7 we studied the tight-binding model, which is a lattice model, where the ability of an electron to hop from one site to another is captured. In the Hubbard model one introduces a strong repulsion between electrons by stipulating that if two electrons (spin up and spin down) occupy the same site, then it cost an energy $U > 0$. Finally, one needs the chemical potential which will allow us to control the number of particles (see discussions on the grand canonical ensemble in Chapter 3). The

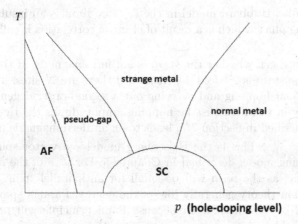

Figure 10.2: Typical phase diagram with hole-doping in high-T_c cuprate super-conductors (AF: antiferromagnetic, SC: superconducting).

Hamiltonian for the Hubbard model then has the structure

$$H = H_{\text{tight binding}} + U \sum_j n_{j\uparrow} n_{j\downarrow} - \mu \sum_j (n_{j\uparrow} + n_{j\downarrow}). \qquad (10.5.1)$$

In the above, $n_{j\uparrow}$ and $n_{j\downarrow}$ are the number operators for electrons with spin-up and spin-down respectively. If the filling is one electron per site then the lattice is half-filled. Studies of the Hubbard model often deal with the half-filled case.

The simplest situation to deal with is the one-site model. The hopping term is absent and we have four possibilities.

1. no electrons and hence $E = 0$.

2. one electron and hence $E = -\mu$ (two possibilities for spin-up and spin-down).

3. two electrons and hence $E = U - 2\mu$.

The partition function at temperature T is

$$Z = 1 + 2e^{\beta\mu} + e^{2\beta\mu - \beta U}. \qquad (10.5.2)$$

We can get the average occupation number $\rho = \langle N \rangle$ as

$$\rho = k_B T \frac{\partial}{\partial \mu} \ln Z = \frac{2(e^{2\beta\mu} + e^{2\beta\mu - \beta U})}{1 + 2e^{2\beta\mu} + e^{2\beta\mu - \beta U}}. \qquad (10.5.3)$$

A plot of ρ vs. μ for different values of U and β can be instructive. In particular, if $U = 4$ and $\beta = 2$, then one finds that μ has to jump when the filling ρ passes through 1. The jump in μ shows the existence of a gap and hence the half-filled Hubbard model in the $U \to \infty$ limit is an insulator. This kind of insulating phase which is a result of electron correlation is called a Mott insulator.

We need to check whether the state is antiferromagnetic. If the hopping vanishes then the state is 2^N-fold degenerate, if there are N sites. Turning on a nearest neighbour hopping and carrying out a second-order degenerate perturbation theory in t (where t is the hopping amplitude for the tight-binding Hamiltonian described in Section 7.7) leads to an antiferromagneric interaction $J\vec{S_i}.\vec{S_j}$ with $J = \frac{4t^2}{U}$. This is the Heisenberg model – a vector spin generalization of the Ising model described in Chapter 5. For $J > 0$, the ordering is antiferromagnetic, as the term will be small for antiparallel spins. A second order perturbation theory generally lowers the energy. Parallel spins will not be able to avail of this mechanism because Pauli principle will prevent the use of virtual doubly occupied states in writing down the sum over states for the second order energy shift in perturbation theory. Thus the Hubbard model cansupport an insulating antiferromagnetic state.

It should be noted that for the strongly correlated electron systems, electrons are more closely associated with the atomic orbitals and hence the Coulomb energy is much larger than the kinetic energy. The dynamic screening of charge fluctuations which led to the effective dielectric constant of Eq.(10.3.12) will no longer be effective. One has to look for a different mechanism. One possibility (investigated by a number of authors but certainly not the last word on the subject) is the possibility of interaction being mediated by spin fluctuations. In the presence hole-doping, the long range antiferromagnetic order quickly disappears but spin fluctuations are going to remain strong. A given electron spin $\vec{S}$ at position $\vec{r}$ and time t polarizes the spin of the surrounding electrons via the exchange interactions and thus acts like a local magnetic field at $(\vec{r},t)$. A spin at $\vec{S}$ at $(\vec{r'},t')$ will respond to this local field through a dynamic spin susceptibility $\chi(\vec{r} - \vec{r'}, \vec{t} - \vec{t'})$ and this will lead to an effective Zeeman energy contributing a negative term proportional to $\vec{S}(\vec{r},t).\vec{S}(\vec{r'},t')$ and thus effectively introducing an attractive interaction between the electrons via their spins – a paramagnon exchange induced pairing interaction.

Rather early on, in fact before the first high temperature superconductor was first reported in 1986, the possible pairing interactions in the Hubbard model had been investigated. The most striking feature of these investigations was that the paramagnon mediated exchange interaction was anisotropic and the singlet d-wave pairing ($d_{x^2-y^2}$ and $d_{3z^2-r^2}$) was not favoured. This is in complete contrast to the isotropic interaction we considered in Section 10.4. It is worth noticing that in the conventional superconductor the Coulomb pairing interaction (screened by electron-lattice interaction) is retarded in time, while for the unconventional superconductors the pairing interactions in the Hubbard model is spatially nonlocal. The pairing interaction peaks at a large momentum transfer characteristic of a nearest neighbour antiferromagnetic or spin-density wave correlation. Interestingly enough, if one considers the unrealistic weak coupling regime $U \ll t$, a similar picture emerges. The structure of the superconducting energy gap follows a variant of the conventional theory (BCS). One finds a momentum dependent sign changing order parameter where $\Delta(\vec{k})$ and $\Delta(\vec{k} + \vec{Q})$ have opposite signs. For small $\vec{Q}$ interactions are pair-breaking and they promote pair formation at large $\vec{Q}$. If there are antiferromagnetic correlations, the pairing interaction has a peak if $\vec{Q} = \vec{Q}_{AF}$, where $\vec{Q}_{AF}$ is the antiferromagnetic ordering vector. While the situation is still unsettled, it seems that the Hubbard model will hold the key for understanding the high temperature superconductivity of the cuprates.

11

Phase Transitions

11.1   Introduction

We came across two models exhibiting second-order phase transitions, transitions which occur without requiring any latent heat, in Chapter 5. One was the model of real gases given by Van der Waals equation where in the $P - V$ diagram there were coexisting regions (i.e. coexisting liquid and gas phases) for sufficiently low temperatures but at a critical temperature $T = T_c$, the coexisting region shrunk to zero and above T_c, there was a single phase (gaseous) region alone. Setting $P = P_c$ and $V = V_c$, i.e., the pressure and volume corresponding to the point at which the coexisting region shrinks to zero, if one lowers the temperature from above T to below T_c, then a transition occurs from a one-phase region to a two-phase region-this transition requires no latent heat. Based on Van der Waals equation, properties near the critical point were examined in Chapter 5. The result was an example of what is known as a mean field theory. We found characteristically universal critical behaviour: the universality was so strong that the results were independent of the dimensionality of space. Starting with Van der Waals equation, one cannot but help get the mean field behaviour since the equation itself can be considered an equation of state in the mean field approximation. A more interesting behaviour is found from the Ising model where we consider classical spins at each site of a regular lattice. The spins can point up or down relative to an external field, i.e. they can take the values ± 1. At the site $\mathbf{n}$ (where $\mathbf{n}$ is a D-dimensional vector), the spin is denoted by $S(\mathbf{n})$ and the energy of interaction between two spins $S(\mathbf{n})$ and $S(\mathbf{m}+\mathbf{n})$ at two different sites is given by $-J(\mathbf{n})S(\mathbf{m})S(\mathbf{m}+\mathbf{n})$. The main task is the evaluation of the partition function

$$Z = \sum_{\{S(\mathbf{m})\}} \exp\left(\beta \sum_{m,n} J(\mathbf{n})S(\mathbf{m})S(\mathbf{m}+\mathbf{n})\right) = \sum_{\{S(\mathbf{m})\}} e^A \qquad (11.1.1)$$

where the quantity $\beta \sum_{m,n} J(\mathbf{n})S(\mathbf{m})S(\mathbf{m}+\mathbf{n})$ is known as the action. The relation to the "action" as known in classical or quantum mechanics can be found in Section 11.4. If the interaction is of the nearest neighbour variety, then $J(\mathbf{n})$ is clearly constant but its value can depend on the direction of the coupling since the coupling can be in any one of the D-directions available in the D-dimensional lattice. In different spatial dimensions, the action A can consequently be written for different dimensions as:

1. For $D = 1$,

$$A = \beta J \sum_i S(i)S(i+1) \tag{11.1.2}$$

    (i labels the one-dimensional lattice sites.)

2. For $D = 2$,

$$A = \beta \sum_{i,j} J_1 S(i,j)S(i+1,j) + J_2 S(i,j)S(i,j+1). \tag{11.1.3}$$

    Note that in two dimensions, we require two integers i and j to label the spatial sites.

3. For $D = 3$,

$$\begin{aligned} A \;=\; & \beta \sum_{i,j,k} [J_1 S(i,j,k)S(i+1,j,k) + J_2 S(i,j,k)S(i,j+1,k) \\ & + \; J_3 S(i,j,k)S(i,j,k+1)] \end{aligned} \tag{11.1.4}$$

and so on.

In Chapter 5, we have already dealt with the situation in $D = 1$ by constructing the transfer matrix and obtaining the different thermodynamic quantities. We now give a somewhat different demonstration of the fact that there can be no long-range order in the one-dimensional model at any finite temperature. To do so, we imagine a completely ordered state at some $T \neq 0$ and investigate the effect of flipping some of the spins. The flipping which costs the least energy is shown in Fig.(11.1).

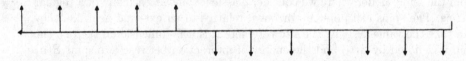

Figure 11.1: Flipping a chunk of spins in a 1D Ising chain.

The change ΔE in energy caused by the flipping shown in Fig.(11.1) is clearly $2J$. However, whether the flipped state is thermodynamically favourable or not is determined by the change ΔF in the free energy which is given by

$\Delta F = \Delta E - T\Delta S$, To estimate ΔS, we note that the length of the domain of overturned spins can be anything between the lattice spacing a and the total length L of the system. Consequently the change in entropy ΔS is of the order of $\ln L$ and thus

$$\Delta F = \Delta E - T\Delta S = 2J - T.\ln\ L < 0 \qquad (11.1.5)$$

i.e., for all $T \neq 0$ it is possible to find a L such that the above inequality holds. Hence it is thermodynamically favourable to over-turn the spins and consequently the system cannot order except at $T = 0$.

Since the above model does not show any ordered behaviour, it is natural to ask whether the one dimensional Ising model can have an ordered state at $T \neq 0$ if the interaction $J(n)$ is long ranged. This would mean that the action of Eq.(11.1.2) now has the form

$$A = \beta \sum_{i,n} J(n)S(i)S(i+n) \qquad (11.1.6)$$

and $J(n) \propto n^{-\alpha}$. If we now flip the spins starting from an ordered configuration as shown in Fig.(11 .1), then the energy expended would be

$$\Delta E \propto \sum_{i,j} \frac{1}{|i-j|^\alpha} \propto \int \frac{dxdy}{|x-y|^\alpha}. \qquad (11.1.7)$$

If we want the energy barrier to overcome the disordering tendency of the entropy term then ΔE must be at least of $O(\ln\ L)$. Clearly the integral in Eq.(11.1.7) is $O(\ln\ L)$ if $\alpha = 2$. For $\alpha > 2$, ΔE is a finite number and hence the entropy term will always dominate. Consequently long-range interactions falling off faster than n^{-2} cannot cause the Ising model to have an ordered state at $T \neq 0$. On the other hand, interactions which are slower than n^{-2} give a ΔE larger than $O(\ln\ L)$ and hence will cause a transition to an ordered state for $T \neq 0$.

We now turn to the two-dimensional model and try to construct an argument similar to the one for the short ranged one-dimensional model. Once again we imagine a perfectly aligned system and cause a defect in it by over-turning some spins. The flipping with minimum energy expenditure is shown in Fig.(11.2).

The flipping shown in Fig.(11.2) which creates the smallest possible domain of overturned spin requires an amount of energy $\Delta E = 8J$. Keeping this energy the same, we cannot have any larger domains. It is easy to see that if larger domains of overturned spins are created then a larger amount of energy is expended. If spins are overturned in a region of size L, the energy expended would be proportional to L. This is the major difference from the one-dimensional case. There we could create arbitrarily long regions of over-turned spins without extra expenditure of energy. The change in entropy is always of $O(\ln\ L)$ where L is the domain size. For large L it is possible to

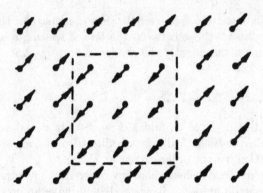

Figure 11.2: Formation of a region of upturned spins in 2D.

have the ordered state favoured at low enough temperatures. Hence it is possible for the nearest neighbour Ising model to exhibit a phase transition in two dimensions. The above argument is not rigorous, it is to be viewed as a plausibility argument. Needless to say that two-dimensional models with long-range interactions will also exhibit phase transitions at a finite temperature.

11.2   Transfer Matrix for the 2D Ising Model

11.2.1   The Conversion to a Quantum Problem

The nearest neighbour two-dimensional Ising model is described by the action of Eq.(11.1.3). Defining $\beta_1 = \beta J_1$ and $\beta_2 = \beta J_2$, our task is to calculate the partition function

$$Z = \sum_{\{S(i,j)\}} \exp\left[\sum_{ij}\{\beta_1 S(i,j)S(i+1,j) + \beta_2 S(i,j)S(i,j+1)\}\right]. \quad (11.2.1)$$

The transfer matrix that we are seeking will be from a given row to the next row. For the one-dimensional problem in Chapter 5, the transfer matrix took us from one site to the next. In this two-dimensional case it is from one row to the next and we can easily imagine that for the three-dimensional case it will be from one plane to the next.

In what follows we will not try to find the transfer matrix for the general case of arbitrary β_1 and β_2. Instead, we will concentrate on a limit which allows the transfer matrix operator to have the form $e^{-H\tau}$, which becomes $1 - H\tau$ in the limit $\tau \ll 1$. The couplings β_1 and β_2 will have to be adjusted to obtain this limit. Since we expect the model to have a phase transition, one would find a relation between β_1 and β_2 which would give the transition point. At the transition point, we expect long-range order, which would imply an infinite correlation length and we would expect all dependence on lattice spacings to

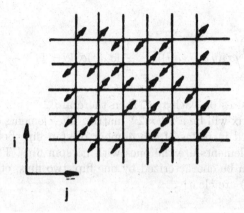

Figure 11.3: Two dimensional Ising lattice.

vanish. The latter may be taken as the signal for the transition point. Thermodynamic quantities, if calculated, would show a non-analytic behaviour at the transition point. We will not be able to calculate the spontaneous magnetization, however, since that would require the presence of an external field which would have to be set equal to zero after the derivative of the free energy with respect to the field is taken.

We start with the restricted goal of finding the transfer matrix by noting that apart from constants, the action of Eq.(11.2.1) can be written as

$$
\begin{aligned}
A &= -\sum_{ij}[\beta_1 S(i,j)S(i+1,j)+\beta_2 S(i,j+1)S(i,j+1)] \\
&= \sum_{i,j}\left\{\frac{\beta_1}{2}[S(i,j)-S(i+1,j)]^2-\beta_2 S(i,j)S(i,j+1)\right\} \\
&\quad + \text{ constants}.
\end{aligned}
$$

(11.2.2)

We now focus on a particular i and write

$$
A = \sum_i L(i,i+1)
$$

(11.2.3)

where

$$
\begin{aligned}
L(i,i+1) &= \sum_j\left\{\frac{\beta_1}{2}[S(i,j)-S(i+1,j)]^2\right. \\
&\quad \left. -\frac{\beta_2}{2}[S(i,j)S(i,j+1)+S(i+1,j)S(i+1,j+1)]\right\}.
\end{aligned}
$$

(11.2.4)

Introducing the notation

$$
S(i,j)=\sigma_3(j), \quad S(i+1,j)=s_3(j)
$$

(11.2.5)

we write

$$L(i, i+1) = \sum_j \left\{ \frac{\beta_1}{2} [\sigma_3(j) - s_3(j)]^2 - \frac{\beta_2}{2} [\sigma_3(j)\sigma_3(j+1) + s_3(j)s_3(j+1)] \right\}.$$

$$(11.2.6)$$

If there are N sites on each row, then there are 2^N spin configurations and the transfer matrix will be a $2^N \times 2^N$ matrix. The elements of the transfer matrix can be labelled in terms of the number of spin flips between the two rows. The diagonal elements are the ones with no spin flips. The various off-diagonal elements can be characterized by one flip, two flips, etc. The matrix elements of $L(i, i+1)$ are clearly:

$$L(i, i+1)_{(0 \text{ flip})} = -\beta_2 \sum_j \sigma_3(j)\sigma_3(j+1) \qquad (11.2.7)$$

where $\sigma_3 = s_3$ for no flip. For one flip,

$$L(i, i+1)_{(1 \text{ flip})} = 2\beta_1 - \frac{\beta_2}{2} \sum_j [\sigma_3(j)\sigma_3(j+1) + s_3(j+1)]. \qquad (11.2.8)$$

The j-value at which the flip occurs leads to $\sigma_3(j) - s_3(j) = \pm 2$. For n flips,

$$L(i, i+1)_{(n \text{ flips})} = 2n\beta_1 - \frac{\beta_2}{2} \sum_j [\sigma_3(j)\sigma_3(j+1) + s_3(j+1)]. \qquad (11.2.9)$$

Since $T(i, i+1) = e^{-L(i,i+1)}$, the corresponding matrix elements for T are

$$T(i, i+1)_{(0 \text{ flip})} = \exp\left[\beta_2 \sum_j \sigma_3(j)\sigma_3(j+1) \right] \qquad (11.2.10)$$

$$T(i, i+1)_{(1 \text{ flip})} = \exp\left[-2\beta_1 + \frac{\beta_2}{2} \left\{ \sum_j \sigma_3(j)\sigma_3(j+1) \right. \right.$$

$$\left. \left. + \sum_j s_3(j)s_3(j+1) \right\} \right] \qquad (11.2.11)$$

$$T(i, i+1)_{(n \text{ flips})} = \exp\left[-2n\beta_1 + \frac{\beta_2}{2} \left\{ \sum_j \sigma_3(j)\sigma_3(j+1) \right. \right.$$

$$\left. \left. + \sum_j s_3(j)s_3(j+1) \right\} \right]. \qquad (11.2.12)$$

At this point, we bring in the simplification of writing $T = e^{-H\tau} \sim 1 - H\tau$, which would lead to the following relations between matrix elements:

$$T_{(0 \text{ flip})} = 1 - \tau H_{(0 \text{ flip})} \qquad (11.2.13)$$

$$T_{(1 \text{ flip})} = -\tau H_{(1 \text{ flip})} \qquad (11.2.14)$$

$$T_{(n \text{ flips})} = -\tau H_{(n \text{ flips})}. \qquad (11.2.15)$$

Note that the operator 1 cannot have matrix elements between states with flipped spins and hence does not appear in Eqs.(11.2.14) and (11.2.15). Comparing Eqs.(11.2.10) and (11.2.13), we conclude that for the form of the latter to hold, we must have $\beta_2 = O(\tau)$ in the former. Similarly comparison of Eqs.(11.2.11) and (11.2.14) leads to the conclusion that $e^{-2\beta_1} = O(\tau)$ and thus we can write

$$\beta_2 = \lambda e^{-2\beta_1} \qquad (11.2.16)$$

where λ is a number independent of τ. If we now choose

$$e^{-2\beta_1} = \tau \qquad (11.2.17)$$

then it follows that

$$\beta_2 = \lambda \tau. \qquad (11.2.18)$$

If we now compare Eqs.(11.2.12) and (11.2.15), we find that the matrix element of H for n flips must be of the order τ^{n-1} and hence for $n \geq 2$ the matrix elements are negligible in the limit $\tau \to 0$. Thus, in this limit the only relevant matrix elements of H are those with zero flips (diagonal elements) and those with a single flip. The matrix elements are

$$H_{(0 \text{ flip})} = -\lambda \sum_k \sigma_3(j)\sigma_3(j+1) \qquad (11.2.19)$$

$$H_{(1 \text{ flip})} = -1. \qquad (11.2.20)$$

Clearly the representation of H can be in terms of a 2×2 matrix and the obvious candidate for the spin flipping matrix is the Pauli matrix $\sigma_1 = \begin{pmatrix} 0 & 1 \\ 1 & 0 \end{pmatrix}$. As for the diagonal elements, we can use the Pauli matrix $\sigma_3 = \begin{pmatrix} 1 & 0 \\ 0 & -1 \end{pmatrix}$ at each site j and thus obtain the Hamiltonian operator

$$H = -[\sum_j \sigma_1(j) + \lambda \sum_j \sigma_3(j)\sigma_3(j+1)] \qquad (11.2.21)$$

which will reproduce correctly all the matrix elements of Eqs.(11.2.19) and (11.2.20).

The eigenvalue spectrum of the transfer matrix is thus to be obtained from the spectrum of the quantum mechanical Hamiltonian of Eq.(11.2.21). It should be noted that the quantum Hamiltonian is not a single particle one but a many particle Hamiltonian for spins sitting on a one-dimensional lattice. To bring out explicitly the fermionic character of the above Hamiltonian it is convenient to consider an equivalent Hamiltonian $\bar{H}$, defined as

$$\bar{H} = -[\sum_n \sigma_3(n) + \lambda \sum_n \sigma_1(n)\sigma_1(n+1)] \tag{11.2.22}$$

which is obtained from H by the orthogonal transformation that takes $\sigma_1 \to \sigma_3$ and $\sigma_3 \to \sigma_1$.

11.2.2   The Transition Temperature

We begin by noting that the Pauli matrices for raising and lowering spins can be constructed as

$$\sigma^{\pm}(n) = \frac{1}{2}[\sigma_1(n) \pm i\sigma_2(n)] \tag{11.2.23}$$

The matrices $\sigma^{\pm}(n)$ anticommute on the same site and $(\sigma^{\pm})^2 = 1$. In this they behave like fermions. But true fermion operators have anticommutation properties on separate lattice sites as well. With this in mind, we construct the following operators for a lattice where n runs from $-N$ to N in steps of unity:

$$C(n) = \prod_{j=-N}^{n-1} [e^{i\pi\sigma^+(j)\sigma^-(j)}]\sigma^-(n) \tag{11.2.24}$$

and

$$C^+(n) = \sigma^+(n) \prod_{j=-N}^{n-1} [e^{-i\pi\sigma^+(j)\sigma^-(j)}]. \tag{11.2.25}$$

Making use of the results

$$\sigma^-(n)\sigma^+(n) = \frac{1}{2}[1 - \sigma_3(n)] \tag{11.2.26}$$

$$\sigma^-(n)\sigma^+(n) = \frac{1}{2}[1 + \sigma_3(n)] \tag{11.2.27}$$

$$e^{i\pi\sigma_3/2} = i\sigma_3 \tag{11.2.28}$$

we find

$$C(n) = \prod_{j=-N}^{n-1} [-\sigma_3(j)]\sigma^-(n) \tag{11.2.29}$$

and

$$C^+(n) = \sigma^+(n) \prod_{j=-N}^{n-1} [-\sigma_3(j)] \tag{11.2.30}$$

from which it is straightforward to verify that

$$\{C(n), C^+(m)\} = \delta_{mn} \tag{11.2.31}$$
$$\{C(n), C(m)\} = 0 \tag{11.2.32}$$

where $\{A, B\} = AB + BA$ is the anticommutator. The $C(n)$ and $C^+(n)$ are thus true fermionic operators and we next have to express the Hamiltonian of Eq.(11.2.22) in terms of them. From Eq.(11.2.27)

$$\sigma_3(n) = 2\sigma^+(n)\sigma^-(n) - 1 \tag{11.2.33}$$

and thence using Eqs.(11.2.29) and (11.2.30), we have $\sigma^+(n)\sigma^-(n) = C^+(n)C(n)$ so that

$$\sigma_3(n) = 2C^+(n)C(n) - 1. \tag{11.2.34}$$

The coupling term is tackled by writing

$$
\begin{aligned}
\sigma_1(n)\sigma_1(n+1) &= [\sigma^+(n) + \sigma^-(n)][\sigma^+(n+1) + \sigma^-(n+1)] \\
&= \sigma^+(n)\sigma^+(n+1) + \sigma^+(n)\sigma^+(n+1) \\
&+ \sigma^-(n)\sigma^+(n+1)\sigma^+(n)\sigma^-(n+1)
\end{aligned}
\tag{11.2.35}
$$

Now,

$$
\begin{aligned}
C^+(n)C(n+1) &= \sigma^+(n)[-\sigma_3(n)]\sigma^-(n+1) \\
&= \sigma^+(n)\sigma^-(n+1)
\end{aligned}
\tag{11.2.36}
$$

since $\sigma^+(n)\sigma_3(n) = -\sigma^+(n)$. Similarly, one finds

$$
\begin{aligned}
C(n)C^+(n+1) &= -\sigma^-(n)\sigma^+(n+1) \\
C^+(n)C^+(n+1) &= \sigma^+(n)\sigma^+(n+1) \\
C(n)C(n+1) &= -\sigma^-(n)\sigma^-(n+1)
\end{aligned}
\tag{11.2.37}
$$

leading to (dropping constant terms)

$$\bar{H} = -2\sum_n C^+(n)C(n) - \lambda \sum_n [C^+(n)][C^+(n+1) + C(n+1)]. \tag{11.2.38}$$

Since $\bar{H}$ is quadratic, the model will be exactly solvable. It is convenient to work in momentum space by defining the operators

$$a_k = \left(\frac{1}{2N+1}\right)^{1/2} \sum_{n=-N}^{N} e^{ikn}C(n) \tag{11.2.39}$$

where,

$$k = 0, \pm\frac{2\pi}{2N+1}, \pm\frac{4\pi}{2N+1}, \ldots, \frac{2\pi N}{2N+1}.$$

We have, in choosing k, made a commitment to a boundary condition linking a_{-N} to a_N which, however, is not vital for our present purpose. The operators a_k have the correct fermionic anticommutation relations, namely,

$$
\begin{aligned}
\{a_k^+, a_{k'}\} &= \delta_{k,k'} \\
\{a_k, a_{k'}\} &= \{a_k^+, a_{k'}^+\} = 0.
\end{aligned}
\tag{11.2.40}
$$

Inverting Eq.(11.2.39), we can write

$$C(n) = \left(\frac{1}{2N+1}\right)^{1/2} \sum_k e^{ikn} a_k \qquad (11.2.41)$$

and now it is possible to write each term of $\bar{H}$ in terms of a_k and a_k^+ to obtain,

$$\bar{H} = -2\sum_k (1 + \lambda \cos k) a_k^+ a_k - \lambda \sum_k (e^{-ik} a_k^+ a_{-k}^+ - e^{ik} a_k a_{-k}). \qquad (11.2.42)$$

If we restrict the sum over k to only $k > 0$ by explicitly writing out the terms with $k < 0$ and carrying out the transformation $k = -k$ for them, we find

$$\bar{H} = -2\sum_{k>0}(1 + \lambda \cos k)(a_k^+ a_k + a_{-k}^+ a_{-k}) + 2i\lambda \sum_{k>0} \sin k(a_k^+ a_{-k}^+ + a_k a_{-k}).$$
$$(11.2.43)$$

The diagonalization of $\bar{H}$ is done by a canonical transformation (Bogoliubov Valatin transformation used in superconductivity). The idea is to obtain a set of operators η_k such that $\bar{H}$ is written in terms of them as

$$\bar{H} = \sum_k \eta_k^+ \eta_k + \text{constant}. \qquad (11.2.44)$$

We define

$$\begin{aligned} \eta_k &= u_k a_k + i v_k a_{-k}^+ \\ \eta_k^+ &= u_k a_k^+ - i v_k a_{-k} \end{aligned} \qquad (11.2.45)$$

where u_k and v_k depend only on the magnitude of k and are real quantities. The requirement that the transformation is canonical, i.e.,

$$\begin{aligned} \{\eta_k, \eta_{k'}^+\} &= \delta_{kk'} \\ \{\eta_k, \eta_{k'}\} &= \{\eta_k^+, \eta_{k'}^+\} = 0 \end{aligned} \qquad (11.2.46)$$

leads immediately to

$$u_k^2 + v_k^2 = 1. \qquad (11.2.47)$$

Demanding that $\bar{H}$ has the form of Eq. (11.2.31), requires

$$4(1 + \lambda \cos k)u_k v_k + 2\lambda \sin k(u_k^2 - v_k^2) = 0. \qquad (11.2.48)$$

Straightforward algebraic manipulation now yields

$$\Lambda_k = 2\sqrt{1 + 2\lambda \cos k + \lambda^2} \qquad (11.2.49)$$

for the energy spectrum of the particles whose creation operator is η_k^+. The minimum value of the excitation energy is obtained for $k = \pm\pi$ and is given by

$$A_k = 2|1 - \lambda| \qquad (11.2.50)$$

At this point it is instructive to introduce the dimensions of the various quantities. In the above, the dimensionless wavenumber is k. In the vicinity of $k = \pi$, we introduce the dimensional wavenumber p (also the momentum if $\hbar = 1$) and write

$$k = \pi + p.a \tag{11.2.51}$$

where a is a length equal to the lattice spacing. We are interested in the range $p.a \ll 1$. The dimensional energy $(\hbar = c = 1)$ will similarly be defined as $E_k = \Lambda_k/2a$ and we have from Eq.(11.2.49) for $pa < 1$

$$
\begin{aligned}
E_p &= \frac{\Lambda_{\pi+pa}}{2a} \\
&= \frac{1}{a}\sqrt{1 + 2\lambda \cos(\pi + pa) + \lambda^2} \\
&= \frac{1}{a}\sqrt{1 - 2\lambda \cos pa + \lambda^2} \\
&= \frac{1}{a}\sqrt{(1-\lambda)^2 + p^2 a^2} \\
&= \sqrt{\frac{(1-\lambda)^2}{a^2} + p^2}.
\end{aligned}
\tag{11.2.52}
$$

The energy spectrum of the η-particles is the relativistic energy of a particle of momentum p. The mass of the particle depends on the lattice spacing and thus the spectrum is dependent on the lattice variable a except at $\lambda = 1$, when we have the energy spectrum given by $E_p = |p|$ independent of the lattice spacing. In accordance with our discussion at the beginning of this section the critical point (the phase transition point) is then $\lambda = 1$. To see how this compares with the exact result obtained by Kramers and Wannier, we note that $\lambda = 1$ implies [see Eqs.(11.2.17) and (11.2.18)]

$$\beta_2 = e^{-2\beta_1}. \tag{11.2.53}$$

The exact result for the transition curve is

$$\sinh(2\beta_2)\sinh(2\beta_1) = 1. \tag{11.2.54}$$

If $\beta_1 \to \infty$ and $\beta_2 \to 0$ (the limits considered by us for $\tau \to 0$, then $\sinh 2\beta_1 \to \frac{1}{2}e^{2\beta_1}$ and $\sinh 2\beta_2 \to 2\beta_2$ and Eq.(11.2.54) reduce to Eq.(11.2.53). Thus, the identification of $\lambda = 1$ as the transition point is indeed correct.

11.2.3   The Specific Heat

Although, we have identified the transition point, we have not yet found the largest eigenvalue of the transfer matrix. To do so, it is best to return to Eq.(11.2.43) and work near $\lambda = 1$ and $k \cong \pi$. We will write $k = \pi + K$, where $K \ll 1$ and replace $\sin k = \sin(\pi + K) = -\sin K$ by $-K$. We will work to the lowest (in this case the first) order in $\lambda - 1$ and K. Consequently, the Hamiltonian of Eq.(11.2.43) becomes

$$\bar{H} = 2(\lambda-1) \sum_{K>0} (a_K^+ a_K + a_{-K} a_{-K}) - 2i\lambda \sum_{K>0} K(a_K^+ a_{-K}^+ + a_K a_{-K}). \quad (11.2.55)$$

Introducing a fermion field $\phi(x)$ [instead of the fermionic variables $C(n)$ on the lattice], by the relation

$$\phi(x) = \frac{1}{\sqrt{L}} \sum_K a_K e^{-iKx} \quad (11.2.56)$$

we have

$$\int_0^L \phi(x)\nabla\phi(x)dx = -\sum_K iK a_K a_{-K} = -2i \sum_{K>0} K a_K a_{-K}$$

and to within additive constants the Hamiltonian of Eq.(11.2.55) can be written as

$$\bar{H} = \int dx H(x) \quad (11.2.57)$$

where

$$H(x) = (\phi\nabla\phi - \phi^+\nabla\phi^+) + 2(\lambda - 1)\phi^+\phi. \quad (11.2.58)$$

The Hamiltonian density of the Dirac equation for a particle of mass m in one dimension is

$$H_D(x) = -i\psi^+ \alpha\nabla\psi + m\psi^+ \beta\psi \quad (11.2.59)$$

where the Dirac matrices α and β need to satisfy

$$\begin{aligned} \alpha\beta + \beta\alpha &= 0, \\ \alpha^2 = \beta^2 &= 1 \end{aligned} \quad (11.2.60)$$

which can be satisfied in a 2×2 representation by the Pauli matrices given by

$$\alpha = \sigma_2 = \begin{pmatrix} 0 & -i \\ i & 0 \end{pmatrix}$$

$$\beta = \sigma_3 = \begin{pmatrix} 1 & 0 \\ 0 & -1 \end{pmatrix} \quad (11.2.61)$$

making the wavefunction ψ a two-component object. The similarity between Eqs.(11.2.58) and (11.2.59) is obvious but they cannot be identical unless we introduce a two-component object involving the ϕ's. To do so we imagine a second Ising lattice superposed on the first but not interacting with it. The fields on the two lattices are $\phi_1(x)$ and $\phi_2(x)$ and we define the connection between ψ and ϕ by writing

$$\psi = \begin{pmatrix} \phi_1 + i\phi_2 \\ \phi_1^+ + i\phi_2^+ \end{pmatrix} \tag{11.2.62}$$

where the fields ϕ_1 and ϕ_2 are supposed to anticommute, so that

$$\{\phi_i(x), \phi_j^+(x')\} = \delta_{ij}\delta(x - x')$$
$$\{\phi_i(x), \phi_j(x')\} = 0. \tag{11.2.63}$$

The equivalence is now complete since

$$-i\psi^+\alpha\nabla\psi = \psi^+ \begin{pmatrix} 0 & -1 \\ 1 & 0 \end{pmatrix} \nabla\psi$$
$$= (\phi_1\nabla\phi_1 - \phi_1^+\nabla\phi_1^+) + (\phi_2\nabla\phi_2 - \phi_2^+\nabla\phi_2^+)$$
$$+ i\nabla(\phi_1\phi_2 + \phi_2^+\phi_1^+)$$

The last term is a total derivative and cannot contribute to $\bar{H}$ on integration. Similarly, the mass term is

$$\psi^+\beta\psi = \psi^+ \begin{pmatrix} 1 & 0 \\ 0 & -1 \end{pmatrix} \psi$$
$$= [\phi_1^+, \phi_1] + [\phi_2^+, \phi_2] + i\{\phi_1^+, \phi_2\} - i\{\phi_2^+, \phi_1\}$$
$$= 2\phi_1^+\phi_1 + 2\phi_2^+\phi_2 - 2\delta(0)$$

since the anticommutators vanish and the commutators have been evaluated with Eq.(11.2.63). The delta function is to be smeared out over a lattice spacing and hence not dangerous. Discarding it as another constant, we have the identification

$$\int H_D(x)dx = \int [H(\phi_1) + H(\phi_2)]dx$$

where the mass m of the Dirac equation is given by $m = (\lambda - 1)$. The ground state energy is found by filling up all the negative energy single-particle states of the Dirac equation and this determines the largest eigenvalue of the transfer matrix. The ground state energy is clearly

$$E_G = -\sum_K (m^2 + K^2)^{1/2} = -\frac{L}{\pi}\int_0^{K_c} dK\sqrt{m^2 + K^2} = -\frac{L}{\pi}\int_0^{K_c} dk\sqrt{(\lambda - 1)^2 + K^2}$$
$$\tag{11.2.64}$$

where the cutoff has to come since our approximation has to break down at large values of K. To make the connection with thermodynamics clear, we need to bring in the temperature variable. To do so we note that $\lambda = e^{2\beta_1}\beta_2$ is a function of temperature, i.e., $\lambda = f(T)$ since $\beta_1 = \beta J_1$ and $\beta_2 = \beta J_2$ are determined entirely by $\beta = (k_B T)^{-1}$ for a given J_1 and J_2 and the $\lambda - 1$ defines the critical temperature T_c such that $f(T_c) = 1$. Hence $\lambda - 1 = f(T) - f(T_c) = A_1(T - T_c) + A_2(T - T_c)^2$ in a Taylor expansion. If $\lambda \cong 1$, we can retain the first term alone and thus $\lambda - 1 = A_1(T - T_c)$ in our approximation, where A_1 is a constant of order unity. The largest eigenvalue of the transfer matrix e^{E_G} will dominate in the thermodynamic limit when the lattice size becomes very large. Consequently the partition function $Z = e^{-2E_G L}$ ($2L$ for the two non-interacting lattices) and the free energy F per unit length is (in units of $k_B T$)

$$F = -\frac{1}{2L} \ln Z = E_G = -\frac{L}{\pi} \int_0^{K_c} \sqrt{A_1^2 (T - T_c)^2 + K^2} dK. \qquad (11.2.65)$$

The entropy is obtained as $S = -\frac{\partial F}{\partial T}$ and thus

$$S = A_1(T - T_c) \frac{L}{\pi} \int_0^{K_c} \frac{dK}{\sqrt{A_1^2 (T - T_c)^2 + K^2}}.$$

Further differentiation yields the specific heat as

$$C = T \frac{\partial S}{\partial T} = TA_1 \frac{L}{\pi} \left[\int_0^K \frac{dK}{\sqrt{A_1^2 (T - T_c)^2 + K^2}} \right.$$
$$\left. - A_1(T - T_c)^2 \int_0^K \frac{dK}{[A_1^2 (T - T_c)^2 + K^2]^{3/2}} \right]. \qquad (11.2.66)$$

We want the asymptotic behaviour of C as $T \to T_c$.

Note that both the integrals in Eq.(11.2.66) diverge at $K = 0$ if $T = T_c$. However the second integral is multiplied by $(T - T_c)^2$ which cancels the $(T - T_c)^2$ divergence of the integral. The critical behaviour is contained in the first term from which we find for $T \cong T_c$

$$C \sim C_0 \ln (T - T_c)^{-1} \qquad (11.2.67)$$

showing that the specific heat is singular at the transition point and that the singularity is logarithmic – a result first established by Onsager on the basis of his complete solution.

11.3   Planar Model and Heisenberg Model

The discussion of the nearest neighbour Ising model shows that it does not order at any non-zero temperature for $D = 1$, but does so for $D = 2$. The question we can ask is whether the model defined on lattices with D between 1 and 2 will show a finite temperature transition or not. To answer this question, we note that in any dimension D, if we begin with ordered spins and then overturn a few to form a domain of 'wrong' spins, then the size of the domain will be given by L^{D-1}, where L is the typical length scale in the problem. If $2J$ is the amount of energy required to turn a spin, then the energy cost of having a disordered domain is of the order JL^{D-1}. For $D > 1$, it costs a large amount of energy to create a large domain of disorder – in particular, the larger the domain of wrong spins, more the energy expenditure to create it (it is only in $D = 1$ that arbitrarily large domains can be produced with no extra cost in energy). The entropy change can at most be $O(\ln L)$ and hence for $D > 1$, the energy change dominates, making it possible for the Ising model to exhibit a finite temperature phase transition for all $D > 1$. The case $D = 1$ is accordingly called the lower critical dimension, i.e., a dimension above which the phase transition occurs at some non-zero temperature.

The Ising spins can only point up or down. However, magnetic moments in general are vectors and hence we can, in general, consider $\mathbf{S}(\mathbf{m})$ to be a vector $\mathbf{S}$ at the lattice point $\mathbf{m}$. The energy of the interaction (nearest neighbour interaction and isotropy being assumed) is accordingly

$$E = -J \sum_m \mathbf{S}(\mathbf{m}).\mathbf{S}(\mathbf{m}+1) \qquad (11.3.1)$$

(ferromagnetic if $J > 0$ and antiferromagnetic if $J < 0$).

If $\mathbf{S}$ is a three-dimensional vector, we have the Heisenberg model and if $\mathbf{S}$ is a two-dimensional vector, then we have the planar model. The most important feature of the vector spin models is the alteration in the lower critical dimension from the $D = 1$ for the Ising case.

The alteration in the lower critical dimension is best seen by considering domains of upturned spin at any finite temperature assuming that at $T = 0$ the spins are all aligned. If the upturned spin domain is of size L, then we have to consider the boundary between the spins pointing one way and the spins pointing the other way. This boundary is not sharp since the spins can be canted and the transition from up-spins to down-spins occurs in a region of size of order L. In a D-dimensional space, the sharp interface for the Ising model implies an energy cost of $O(L^{D-1})$. The blurred interface with an extension of $O(L)$ will cost a lower amount of energy-namely of $O(L^{D-2})$ since the turning of spins is over a domain of length L and in that length L, the energy cost is negligible. Thus the effective dimension for calculating the energy change is reduced by one. Thus to form a macroscopic domain of disordered spin, the energy cost is $O(L^{D-2})$. The entropy gained is $O(\ln L)$ and consequently the free energy change on disordering is

$$\Delta F = O(L^{D-2}) - TO(\ln L) \qquad (11.3.2)$$

Clearly for $D < 2$, it is not possible to have a positive ΔF at any finite temperature and hence the disordering will lower the free energy. Consequently, there is no long-range order at any nonzero T for $D < 2$. The lower critical dimension is thus $D = 2$. To investigate the situation at $D = 2$, one needs a more sophisticated argument. We simply mention here that it was shown by Mermin, Wagner and Hohenberg that at any nonzero T, the planar model does not order in $D = 2$. However, there is a particular kind of ordering for the planar model in $D = 2$, which was found by Kosterlitz and Thouless.

11.4   Statistical Mechanics and Field Theory

In Sec.11.2, we found that the two-dimensional Ising model was equivalent to a many-body quantum problem in one spatial dimension which is a field theory in $1 + 1$ dimensions, i.e. one spatial dimension and time. Had we started with the one-dimensional Ising model, the equivalent quantum Hamiltonian would be a single particle Hamiltonian, i.e., the one-dimensional statistical mechanics problem is equivalent to a problem in quantum mechanics. We anticipate from the above, the general result, that a $(D+1)$-dimensional statistical mechanical problem would be having a quantum field theory counterpart in D spatial dimensions and time.

To illustrate the point, we consider a self-coupled scalar field in D spatial dimensions with the Lagrangian

$$L = \int d^D x \left[\frac{1}{2} \left(\frac{\partial \phi}{\partial t} \right)^2 - \frac{1}{2} \left(\vec{\nabla} \phi \right)^2 - \frac{1}{2} m^2 \phi^2 - \frac{\lambda}{4} \phi^4 \right] \qquad (11.4.1)$$

A formal expression for a transition amplitude from time t_i to t_f can be written down as

$$Z = \int D[\phi] \exp \left(\int_{t_i}^{t_f} L dt / \hbar \right) \qquad (11.4.2)$$

We now consider the imaginary time variable

$$t' = it \qquad (11.4.3)$$

and introduce a lattice spacing a in the spatial direction and a spacing τ in the temporal direction t'. Defining the Euclidean action $S_E = -S = -i \int L dt'$ and labelling the temporal direction as 0 and all spatial dimensions as $l = 1, 2, ..., D$, we find

$$\begin{aligned}
S_E &= \int d^D x dt' \left[\frac{1}{2} \left(\frac{\partial \phi}{\partial t'} \right)^2 + \frac{1}{2} \left(\vec{\nabla} \phi \right)^2 + \frac{m^2}{2} \phi^2 + \frac{\lambda}{4} \phi^4 \right] \\
&= \sum_{n_i} \left\{ \frac{a^D}{2\tau} [\Delta_0 \phi(n_i)^2] + \frac{\tau}{2} a^{D-2} \sum_l [\Delta_l \phi(n_i)]^2 \right. \\
&\quad + \left. \frac{\tau}{2} a^D m^2 \phi^2(n_i) + \frac{\tau}{4} \lambda a^D \phi^4(n_i) \right\}
\end{aligned} \qquad (11.4.4)$$

where

$$\Delta_\mu \phi(\mathbf{n}) = \phi(\mathbf{n} + \mu) - \phi(\mathbf{n}) \qquad (11.4.5)$$

while $\mathbf{n}$ labels the lattice points ($\mathbf{n}$ has $D + 1$ components).

Defining

$$K_\tau = \frac{a^D}{2\tau}, \quad K = \frac{\tau}{2} a^{D-2}, \quad b = \frac{\tau}{2} a^D m^2 \quad \text{and} \quad U = \frac{\tau}{4} \lambda a^D, \qquad (11.4.6)$$

the Euclidean action can be written as

$$S_E = \sum_{n_i} \left[K_\tau (\Delta_0 \phi(n_i))^2 + K \sum_l (\Delta_l \phi(n_i))^2 + b\phi(n_i)^2 + U\phi(n_i)^4 \right] \quad (11.4.7)$$

and the generating functional Z of the field theory is now

$$Z = \prod_{n_i} \int_{-\infty}^{\infty} d\phi(n_i) e^{-S/\hbar} \qquad (11.4.8)$$

where the functional integration over the ϕ-field has been given the meaning of integrating each site variable $\phi(\mathbf{n})$ over all possible values. From Eq.(11.4.8) it is clear that Z is the partition function of a statistical mechanical system where the variable is $\phi(\mathbf{n})$ defined on lattice points in a $(D+1)$-dimensional lattice and thus a field theory in D spatial dimensions and time leads to a statistical mechanical problem in $(D+1)$-dimensions. Note that $\hbar$ of the quantum problem plays the role of $k_B T$ (temperature) in the statistical mechanical problem.

For the statistical mechanical problem, we can construct the transfer matrix to propagate the field $\phi(n)$ at one time slice to the field $\phi(n)$ on the next time slice. The n in the argument are the D "spatial" variables-the different time values on the two successive slices is taken care of by ϕ and ϕ'. We can define the transfer matrix by (we set $\hbar = 1$).

$$\langle \phi' | T | \phi \rangle = \exp \left[-\sum_{n_i} \left\{ K_\tau [\phi'(n_i) - \phi(n_i)]^2 + \frac{1}{2} K \sum_l [(\Delta_l \phi')^2 + (\Delta_l \phi)^2] \right. \right.$$

$$\left. \left. + \frac{b}{2} [\phi'^2(n_i) + \phi^2(n_i)] + \frac{u}{2} [\phi'^4(n_i) + \phi^4(n_i)] \right\} \right] \qquad (11.4.9)$$

and obtain the partition function as

$$Z = \text{Tr} T^{N+1} \qquad (11.4.10)$$

by introducing complete sets of $|\phi\rangle$ states at every time slicing, imposing periodic boundary conditions and summing over all possible initial conditions. We get an operator expression for T by introducing at every spatial point l, the conjugate field by the relation

$$[\phi(\mathbf{l}), \pi(\mathbf{l}')] = i\delta_{\mathbf{l}\mathbf{l}'} \qquad (11.4.11)$$

which readily yields

$$
\begin{aligned}
T \;=\; & \exp\left[\sum n_i\left\{[-\tfrac{1}{2}K\sum_l(\Delta_l\phi'(n_i))] + b\phi'(n_i)^2 + u\phi'(n_i)^4\right\}\right] \\
& \times \;\exp\left[-\frac{1}{4K_\tau}\pi(n_i)^2\right] \\
& \times \;\exp\left[\sum n_i\left\{[-\tfrac{1}{2}K\sum_l(\Delta_l\phi(n_i))] + b\phi(n_i)^2 + u\phi(n_i)^4\right\}\right].
\end{aligned}
$$

$$(11.4.12)$$

We can formally write $T = e^{-H_S\tau}$ although H_S is not very revealing, except when $\tau \to 0$ and H_S is the canonical Hamiltonian of the original field theory.

The operator T is Hermitian and hence its eigenvalues will be real and can be written as $e^{-E_S\tau}$. If the corresponding eigenfunction is $|i\rangle$, we can write a spectral decomposition of T as

$$
T = \sum_i |i\rangle e^{-E_S\tau}\langle i| \tag{11.4.13}
$$

knowing that the set $\{|i\rangle\}$ is complete. To get the partition function we need T raised to a large power N and clearly

$$
T^N = \sum_i |i\rangle e^{-NE_S\tau}\langle i| \tag{11.4.14}
$$

As $N \to \infty$ the sum will be dominated by the lowest value of E_S, if it is unique and hence

$$
\lim_{N\to\infty} T^N \to |0\rangle e^{-NE_0\tau}\langle 0| = |0\rangle e^{-E_0 L_\tau}\langle 0| \tag{11.4.15}
$$

where L_τ is the total length in the temporal direction. The partition function is thus

$$
Z = e^{-E_0 L_\tau} = e^{-\omega_0 V L_\tau} \tag{11.4.16}
$$

where we have written $E_0 = \omega_0 V$ keeping in mind the extensive nature of E_0 (V is the spatial volume). On the other hand the thermodynamic identity relates Z to the free energy F as

$$
Z = e^{-F} = e^{-fV L_\tau} \tag{11.4.17}
$$

where we have taken the extensive nature of F in $(D+1)$ dimensions into account. Comparing Eqs.(11.4.17) and (11.4.16)

$$
f = \omega_0. \tag{11.4.18}
$$

This yields the first connection between the statistical mechanical problem and the field theory. Further connections are revealed by considering the propagator of the field theory defined as $(t > 0)$

$$\Delta(t, x) = \langle 0|\phi(t, x)\phi(0, 0)|0\rangle \qquad (11.4.19)$$

where $|0\rangle$ is the Heisenberg picture ground state of the Hamiltonian H_s and $\phi(x, t)$ is the field operator in the Heisenberg picture. The corresponding correlation function in the statistical mechanical problem is

$$\Gamma(n_0, n) = \frac{1}{Z} \int \prod_{n'_0, n'} d\phi(n'_0, n')\phi(n_0, n)\phi(0, 0)e^{-S} \qquad (11.4.20)$$

where n_0 is the temporal part of **n** and n the spatial part.

Writing the Heisenberg operator $\phi(x, t)$ in terms of the Schrodinger operator as $\phi(x, t) = e^{iH_s t}\phi(x)e^{-iH_s t}$, we have

$$\Delta(t, x) = \langle 0|\phi(x)e^{-iH_s t}\phi(0)|0\rangle e^{iE_0 t} \qquad (11.4.21)$$

and manipulations of Eq.(11.4.20) in terms of the transfer matrix shows the correspondence

$$\Gamma(n_0, n) = \Delta(-in_0\tau, n). \qquad (11.4.22)$$

Finally, if one is not at $T = T_c$ (the critical point), then the statistical mechanical correlation function falls of exponentially, characterized by a definite correlation length. In particular, we will have

$$\Gamma(n_0, n) \sim e^{-n_0/\xi} \qquad (11.4.23)$$

where ξ is the correlation length. Now from Eq.(11.4.21), we have

$$\begin{aligned}
\Delta(-in_0\tau, 0) &= \langle 0|\phi(0)e^{-H_s n_0 \tau}\phi(0)|0\rangle e^{E_0 n_0 \tau} \\
&= \sum_l |\langle 0|\phi(0)|l\rangle|^2 e^{(E_0 - E_l)n_0\tau} \\
&\sim |\langle 0|\phi(0)|1\rangle|^2 e^{-(E_1 - E_0)n_0\tau}, \qquad (11.4.24)
\end{aligned}$$

if n_0 is very large (we have simply taken the sum to be dominated by the largest term and in the ordered sequence of eigenvalues, E_1 is closest to E_0).

Comparing with Eq.(11.4.23), ξ is simply $(E_1 - E_0)^{-1}$. But this energy difference is the mass of the lightest particle at zero momentum and thus the reciprocal of the correlation length is the mass gap of the field theory. The complete correspondence is as follows:

Statistical Mechanics	Field Theory
$k_B T$ is the scale of action	$\hbar$ is the scale of action
Free energy density	Vacuum energy density
Correlation function	Propagator
Inverse correlation length	Mass gap

Turning now to our analysis of the 2-D Ising model in Sec. 11.2, we note that the energy spectrum was found to be $E_p = \sqrt{\frac{(1-\lambda)^2}{a^2} + p^2}$ and hence the lightest particle at zero momentum (i.e., $p = 0$) has a mass $m = (1 - \lambda)/a$. Thus at $\lambda = 1$, the mass gap vanishes yielding an infinite correlation length, implying that $\lambda = 1$ is indeed the critical point of the theory – a fact which we had established then by arguing the absence of lattice effects at $\lambda = 1$.

11.5 Kac-Hubbard-Stratonovich Transformation

The Kac-Hubbard-Stratanovich transformation is in general a procedure for converting one model into another. It is particularly useful in critical phenomena where it is employed for converting a model defied on a lattice to a field theoretic model. Here, we will consider the simplest model which is the Ising model on a D-dimensional lattice. The Ising spins at each lattice site can take the values ± 1 only and the partition function is given by

$$Z_N(J) = \sum_{\{S_m\}} \exp[\Sigma J_{ij} S_i S_j] \qquad (11.5.1)$$

where S_i is the spin at site i (the site is a point on a D-dimensional lattice in general and requires D coordinates for specification) and J_{ij} is the interaction with the property that $J_{ij} = J_{ji}$ and $J_{ii} = 0$ (we are using units where $k_B T = 1$). The total number of spins is N.

We will demonstrate that Eq.(11.5.1) can be rewritten as

$$Z_N(J) = e^{f_0(J)} \int_{-\infty}^{\infty} \prod_{i=1}^{N} d\bar{S}_i \exp\left[-\Sigma Q_{ij} \bar{S}_i \bar{S}_j - \Sigma W(\bar{S}_i)\right] \qquad (11.5.2)$$

where $f_0(J)$ is a smooth analytic function of J. The integration is over each continuous spin variable $\bar{S}_i$ of which there are as many as there were original sites. The limits for each integration are $-\infty$ and $+\infty$. The weighting function $W(S_i)$ will be

$$-W(S_i) = \ln \ \cosh \ S_i - \frac{1}{2}Q_{ii}S_i^2 = -\frac{1}{2}(Q_{ii} - 1)S_i^2 - \frac{1}{12}S_i^4 + \dots \qquad (11.5.3)$$

where Q_{ii} will be defined as we go along. We note that

$$\frac{1}{2}(S_i + S_j)^2 = \frac{1}{2}S_i^2 + \frac{1}{2}S_j^2 + \frac{1}{2}.2S_iS_j = 1 + S_iS_j \geq 0$$

and that interactions can be rewritten as

$$J_{ij}S_iS_j = (-P_0\delta_{ij} + P_{ij})S_iS_j$$

(the diagonal elements of P_{ij} are P_0 to ensure $J_{ii} = 0$) and hence

$$\sum J_{ij}S_iS_j = -\frac{NP_0}{2} + \frac{1}{2}\sum_{i=1}^{N}\sum_{j=1}^{N} P_{ij}S_iS_j. \qquad (11.5.4)$$

If $[S_1, \ldots, S_N]$ is represented by the row vector S^T and the transposed column vector by S, then the partition function becomes

$$Z_N = \exp\left(-\frac{N P_0}{2}\right) \sum_{\{S_m\}} \exp\left(\frac{1}{2} S^T P S\right). \qquad (11.5.5)$$

At this point consider a quadratic form in N continuous variables Y_i given by

$$Q(y) = \sum_{i=l}^{N} \sum_{j=l}^{N} Q_{ij} Y_i Y_j \qquad (11.5.6)$$

where Q_{ij} is a symmetric, positive definite matrix and can be diagonalized by a proper orthogonal transformation O. In terms of new variables x_i

$$x_i = \sum_j O_{ij} Y_j \qquad (11.5.7)$$

the quadratic form is

$$Q(y) = \sum_{j=1}^{N} \lambda_j x_j^2 \qquad (11.5.8)$$

where λ_j are the real positive eigenvalues of Q. It follows that the determinant of Q is

$$\text{Det}|Q| = \prod_{j=1}^{N} \lambda_j. \qquad (11.5.9)$$

The partition function-like expression

$$I(Q) = \int_{-\infty}^{\infty} d^N Y \exp\left(-\frac{1}{2} Y^T Q Y\right) \qquad (11.5.10)$$

can be evaluated by making the change from y to x, noting that the Jacobian of the transformation is 1 and the resulting integrals are Gaussian, yielding

$$I(Q) = \prod_{j=1}^{N} \sqrt{\frac{2\pi}{\lambda_j}} = \frac{(2\pi)^{N/2}}{\sqrt{\text{Det}|Q|}} \qquad (11.5.11)$$

A shift in the variable y is now made according to

$$Y = \bar{S} + Q^{-1} S. \qquad (11.5.12)$$

The $\bar{S}$ are the continuous variables, while the S represent fixed shift parameters. The transpose of Eq.(11.5.12) is given by

$$Y^T = \bar{S}^T + S^T Q^{-1} \qquad (11.5.13)$$

and the quadratic form of Eq.(11.5.6) becomes

$$
\begin{aligned}
Y^T Q Y &= (\bar{S}^T + S^T Q^{-1})Q(\bar{S} + Q^{-1}S) \\
&= \bar{S}^T Q \bar{S} + \bar{S}^T S + S^T \bar{S} + S^T Q^{-1}S \\
&= \bar{S}^T Q \bar{S} + 2 \sum_{i=1}^{N} S_i \bar{S}_i + S^T Q^{-1}S
\end{aligned}
$$

leading to

$$
I(Q) = \exp\left(-\frac{1}{2}S^T P S\right) \int_{-\infty}^{\infty} d^N \bar{S} \exp\left[-\frac{1}{2}\bar{S}^T Q \bar{S} - \sum_{i=1}^{N} S_i \bar{S}_i\right] \qquad (11.5.14)
$$

where $P = Q^{-1}$. Combining this with Eq.(11.5.11) we find

$$
I(Q) = (2\pi)^{N/2} \sqrt{\mathrm{Det}|P|}. \qquad (11.5.15)
$$

Using the form of $e^{+\frac{1}{2}S^T P S}$ obtained from Eqs.(11.5.14) and (11.5.15), the partition function of Eq.(11.5.5) becomes

$$
\begin{aligned}
Z_N &= \frac{e^{-NP_0/2}}{(2\pi)^{N/2}\sqrt{\mathrm{Det}|P|}} \sum_{S_i} \int_{-\infty}^{\infty} d^N \bar{S} \exp\left[-\frac{1}{2}\bar{S}^T Q \bar{S} - \sum_{i=1}^{\bar{N}} S_i \bar{S}_i\right] \\
&= \frac{e^{-NP_0/2}}{(2\pi)^{N/2}\sqrt{\mathrm{Det}|P|}} \int_{-\infty}^{\infty} d^N \bar{S} \exp\left[-\frac{1}{2}\bar{S}^T Q \bar{S} + \ln \, \cosh \, \bar{S}_i\right] \\
&= \frac{e^{-NP_0/2}}{(2\pi)^{N/2}\sqrt{\mathrm{Det}|P|}} \int_{-\infty}^{\infty} d^N \bar{S} \exp\left[-\sum_{i,j}^{N} Q_{ij} \bar{S}_i \bar{S}_j\right. \\
&\quad\quad \left. - \sum_i \left(\frac{1}{2}Q_{ii}\bar{S}_i^2 - \ln \, \cosh \, \bar{S}_i\right)\right] \qquad (11.5.16)
\end{aligned}
$$

We have thus reduced the partition function to the form of Eq.(11.5.2) and we have effectively transformed the discrete sum over spins to an integration over continuous variables.

11.6   Critical Phenomena: Scaling Laws

In this Section, we will use the terminology of the magnetic transition to introduce the phenomenology. At high temperatures the individual magnetic moments are disordered and point in random directions and the entropy contribution to the free energy dominates. As the temperature is lowered, the short - range interaction, which favours alignment of the individual magnetic moments, begins to dominate and the equilibrium state of the system is one in which there

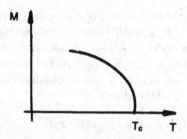

Figure 11.4: Magnetization near T_c.

is overall magnetization-this spontaneous magnetization grows as the temperature is lowered. The temperature at which the spontaneous magnetization first appears is the critical temperature T_c. The growth of the magnetization M, as the temperature is lowered is continuous (Fig.11.4) and in the immediate vicinity of the critical temperature is characterized by a relation of the form

$$M \propto |T_c - T|^\beta. \tag{11.6.1}$$

A relation of this form is called a scaling law and the number β is called a critical exponent. In Chapter 5, we had encountered such a behaviour in both the Van der Waals gas and the approximate solutions of the Ising model in arbitrary dimensions.

As the transition is approached from above, the response of the system becomes stronger and stronger to an infinitesimal external magnetic field h. This response is characterized by the isothermal susceptibility $\chi = \partial M/\partial h$ which grows as the critical point is approached (Fig 11.5).

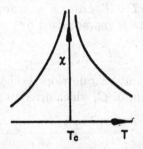

Figure 11.5: Divergence of the susceptibility χ near T_c.

Close to $T = T_c$, whether above or below T_c, the fluctuations in the magnetization are extremely strong and the large susceptibility is characterized by

$$\chi \propto |T - T_c|^{-\gamma} \tag{11.6.2}$$

What lies behind this large susceptibility is the long range correlation that develops in the system as T approaches T_c. To consider the correlation one has to consider the local magnetization function $\phi(x)$ which is the average over the individual magnetic moment in a small volume surrounding the point x.

Here x is to be understood as a D-dimensional vector and this volume is macroscopic with regard to atomic dimensions but microscopic with regard to sample dimensions, so that $\phi(x)$ is a continuous function. We consider the correlation function $G(x_1, x_2)$ defined as

$$G(x_1, x_2) = \langle \phi(x_1)\phi(x_2) \rangle \qquad (11.6.3)$$

where the angular bracket denotes an average over the equilibrium distribution of the field $\phi(x)$. Translational invariance implies that $G(x_1, x_2)$ is a function of $x_1 - x_2 = x_{12}$.

For $T > T_c$, the correlation function dies off exponentially, being characterized by an asymptotic long distance form $G(x_{12}) \sim e^{-x_{12}/\xi}|x_{12}|^{D-2+\eta}$ where ξ is the correlation length and D is the dimensionality of space. As the temperature approaches T_c, the correlation length ξ increases, approaching infinity according to the relation

$$\xi \propto |T - T_c|^{-\nu} \qquad (11.6.4)$$

At T_c, when ξ is infinite, the correlation function decays according to a power law-this implies long-range behaviour, and is characterized by

$$G(x_{12}) = \frac{1}{|x_{12}|^{D-2+\eta}} \qquad (11.6.5)$$

where the exponent η is the anomalous dimension exponent. As we shall establish later, there is a fluctuation dissipation theorem relating the susceptibility to an integral over the correlation function and the long-range nature of $G(x_{12})$ leads to the divergent susceptibility.

The magnetization M at $T = T_c$ grows if an external magnetic field h is applied and thus the growth law is characterized by the relation

$$M \sim h^{1/\delta} \qquad (11.6.6)$$

Another characteristic quality that develops a singularity at the transition point (Fig 11.6) is the specific heat, C, which diverges as

$$C \sim |T - T_c|^{-\alpha} \qquad (11.6.7)$$

The first astounding thing about the exponents α, β, ν and η is that they are universal.

This implies that they do not depend on the nature of the material considered. We must clarify, of course, that they do depend on whether the magnet considered is Ising-like (the magnetization field $\phi(x)$ is scalar) or Heisenberg-like ($\phi(x)$ is a vector). Also, the exponents would differ for the superfluid transition for which $\phi(x)$ is a complex quantity.

The exponents differ with differing spatial dimensionality as well, i.e., whether $D = 3$ or $D = 2$ is being considered. If we consider the dimensionality n

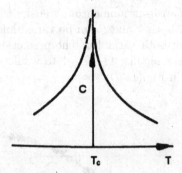

Figure 11.6: Divergence of the specific heat C near T_c.

of the field $\phi(x)$, then $n = 1$, if $\phi(x)$ is a scalar; $n = 2$, if $\phi(x)$ is complex and $n = 3$ if $\phi(x)$ is a three-dimensional vector. Universality means that the exponents introduced above are functions of n and D above. The second remarkable fact is that the exponents are not all independent but connected by the scaling relations

$$\alpha + 2\beta + \gamma = 2 \tag{11.6.8}$$
$$(2 - \eta)\nu = \gamma \tag{11.6.9}$$
$$\beta(1 + \delta) + \alpha = 2 \tag{11.6.10}$$
$$2 - \alpha = D\nu. \tag{11.6.11}$$

Considering, the correlation function $G(x_{12})$ at a temperature different from T_c (i.e. ξ is finite), we note that it has to be a function of two variables x_{12} and ξ. The $\xi = \infty$ form of Eq.(11.6.5) immediately suggests that $G(x_{12}, \xi)$ is a homogeneous function of the two variables, so that we can write

$$G(x_{12}, \xi) = x_{12}^{-(D-2+\eta)} g(x_{12}/\xi) \tag{11.6.12}$$

where the function $g(x)$ has the property that $g(0)$ is a nonzero constant. From the experimental point of view, a more convenient quantity is the Fourier transform $G(k, \xi)$ (here k, like x, has to be understood as a D-dimensional vector) since it can be directly measured by neutron scattering. The Fourier transform is defined as

$$
\begin{aligned}
G(k, \xi) &= \frac{1}{(2\pi)^{D/2}} \int d^D x_{12} G(x_{12}, \xi) e^{ik.x_{12}} \\
&= \frac{1}{(2\pi)^{D/2}} \int d^D x_{12} x_{12}^{-(D-2+\eta)} g\left(\frac{x_{12}}{\xi}\right) e^{ik.x_{12}} \\
&= \xi^{2-\eta} \bar{g}(k\xi) \\
&= k^{-2+\eta} f(k\xi) \tag{11.6.13}
\end{aligned}
$$

The function $f(x)$ has the property that $f(x) \to$ constant as $x \to \infty$. In this limit we also have $\bar{g}(x) \propto x^{-2+\eta}$. The correlation function $G(k, \xi)$ is a function of two variables: the scattering momentum transfer k and the correlation length ξ. In an experiment, both k and ξ can be varied and hence $G(k,\xi)$ can be obtained as a function of both variables. The prediction of Eq.(11.6.13) is that if $G(k,\xi).k^{2-\eta}$ is plotted against $k\xi$, the data would collapse on a single plot. This is demonstrated in Fig.(11.7).

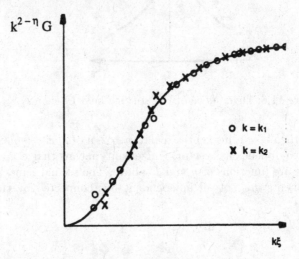

Figure 11.7: Collapse of different plots of the correlation function onto a single plot when the correct scaling variable is used.

The equation of state for a magnet is a relation between M, h and T. Near the critical point it can be cast as a relation between M, h and ξ. From Eqs.(11.1.1) and (11.1.6), we note that

$$M \propto \xi^{-\beta/\nu}, \qquad \text{if } h = 0 \tag{11.6.14}$$

and

$$M \propto h^{1/\delta}, \qquad \text{if } \xi \to \infty. \tag{11.6.15}$$

The consequence of Eqs.(11.6.14) and (11.6.15) is that the equation for M must be a homogeneous function of $\xi^{-\beta/\nu}$ and $h^{1/\delta}$ and thus we can write

$$M = \xi^{-\beta/\nu} m(h\xi^{\beta\delta/\nu}) = t^{\beta} m\left(\frac{h}{t^{\beta\delta}}\right) \tag{11.6.16}$$

where $t = (T_c - T)/T_c$. The function $m(x)$ has the property that $m(0)$ goes to a constant and $m(x) \sim x^{1/\delta}$ when $x \to \infty$. The magnetization M can be measured as a function of two variables t and h and the consequence of Eq.(11.6.16) is that if $Mt^{-\beta}$ is plotted against $h/t^{\beta\delta}$, the data should lie on a single curve. This is illustrated in Fig.11.8.

The immediate task of a theory of critical phenomena is to establish universality, prove the scaling laws [Eqs.(11.6.8) - (11.6.11)] and determine values of the critical exponents as functions of n and D. The tremendous difficulty

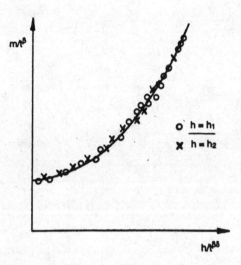

Figure 11.8: Collapse of different curves of the magnetization onto a single curve when the appropriate scaling variable is used.

in doing any of these is the long range nature of the correlations. This is a quintessential many-body problem where perturbation theory refuses to work because of strong correlations. The successful handling of the difficulty required the advent of the renormalization group. However, before discussing that, we will explain the first attempt at solving the problem-the Landau theory-in the next chapter. The amplitudes of individual quantities like magnetization, susceptibility, specific heat etc. are not universal but certain combinations of amplitudes are. A proper theory needs to identify and calculate the universal amplitude ratios as well.

12

Landau Theory and Related Models

12.1 Landau Theory

In any statistical mechanics problem the central concept is that of a partition function. All thermodynamic quantities follow once the partition function is known. Accordingly, one needs to construct the partition function for a system near the critical point. The description of the system is in terms of a field. Near the critical point the local value of the magnetization (the generic term is the order parameter) is the important quantity. There are large fluctuations in the system as it tries to settle down to a state with a spontaneous magnetization and thus the local magnetization is the field variable. The individual magnetic moments when averaged over a small volume about the field point x gives the magnetization (order parameter field $\varphi_\alpha(x)$ ($\alpha = 1, 2 \ldots, n$) and the energy of the system will be a function of φ. From the standard definitions of the partition function

$$Z = \sum_{\text{all possible } \varphi} w(\varphi) e^{-E(\varphi)/k_B T} \tag{12.1.1}$$

where k is the Boltzmann constant and T is the temperature. The factor w is the degeneracy: it is the number of microscopic states associated with a given macroscopic description in terms of $\varphi(x)$. The entropy $S(\varphi) = k_B \ln w(\varphi)$ and thus

$$\begin{aligned} Z &= \sum_{\text{all } \varphi} e^{-[E(\varphi)-TS(\varphi)]/k_B T} \\ &= \sum_{\text{all } \varphi} e^{-F(\varphi)/k_B T} \end{aligned} \tag{12.1.2}$$

where $F(\varphi)$ is the free energy. It is convenient to work in units in which $k_B T = 1$ and we have the result

$$Z = \sum_{\text{all } \varphi} e^{-F(\varphi)}. \tag{12.1.3}$$

This is the same form as obtained after a Hubbard Stratanovich transformation on the Ising model as expressed in Eq.(11.5.15). It is important to realize that it is the free energy functional which enters the calculation of the partition function and to remember that the thermodynamic equilibrium state of a system is the one which minimizes the thermodynamic free energy. The construction of $F(\varphi)$ follows from general principles:

1. Only even functions of φ are admissible since the energy is the same whether magnetic moments are pointing up or down.

2. Since long range properties are of interest the lowest admissible derivative will survive.

3. For an isotropic system there will be $O(n)$ symmetry.

We thus write

$$F(\varphi) = \int d^D x \left[\frac{a}{2} \sum_\alpha \varphi_\alpha^2 + \frac{1}{2} \sum_\alpha \left(\vec{\nabla} \varphi_\alpha \right)^2 + \frac{d}{4} \left(\sum \varphi_\alpha^2 \right)^2 + \ldots \right]. \tag{12.1.4}$$

This particular form of the free energy is known as the Ginzburg-Landau model and serves as the starting point of discussion of phenomena in a wide variety of systems. The simplest evaluation of the partition function follows by evaluating the minimum of $F(\varphi)$ and writing $Z \cong e^{-F_{min}}$. In trying to do so, we note that φ must be independent of space: the spatial dependence always adds to the free energy. Writing $M^2 = \sum_\alpha \varphi_\alpha^2$ we now need to minimize

$$F = V \left[\frac{a}{2} M^2 + \frac{b}{4} M^4 + \ldots \right] \tag{12.1.5}$$

where V is the total volume of the system. If the above is to be a description of the phase transition, then the minimization of F should lead to $M = 0$ for $T > T_c$ and $M \neq 0$ for $T < T_c$. Clearly, the extrema of Eq.(12.2.5) are obtained for

$$M^2 = 0, -a/b \tag{12.1.6}$$

and the corresponding values of F are

$$F_0 = 0, -V \cdot \frac{a^2}{4b}. \tag{12.1.7}$$

It is immediately obvious that the theory will describe critical phenomena if we choose

$$a = a_0(T - T_c)$$
$$b = \text{constant}. \tag{12.1.8}$$

For $T > T_c$, F is minimized for $M = 0$, while for $T < T_c$, F is minimized for

$$M = \left[\frac{a_0}{b}(T_c - T)\right]^{1/2}. \qquad (12.1.9)$$

Comparison with Eq.(11.6.1) shows that

$$\beta = 1/2. \qquad (12.1.10)$$

Since the specific heat C is $-T\frac{\partial^2 F}{\partial T^2}$, we find that $C = 0$ for $T > T_c$ and $C = a_0^2 T_c/2b$ for $T < T_c$. Thus the specific heat has a finite discontinuity at $T = T_c$, but no singularity. Hence, from Eq.(11.6.7),

$$\alpha = 0. \qquad (12.1.11)$$

To calculate the susceptibility, we need to include an external magnetic field H and the quantity to be minimized is

$$F(M, H) = V\left[\frac{a}{2}M^2 + \frac{b}{4}M^4 - MH\right]. \qquad (12.1.12)$$

This leads to

$$aM + bM^3 - H = 0. \qquad (12.1.13)$$

The susceptibility follows as

$$\chi = \left.\frac{\partial M}{\partial H}\right|_{M\to 0} = \frac{1}{a_0}(T - T_c)^{-1} \qquad (12.1.14)$$

for $T > T_c$. For $T < T_c$, we need to consider fluctuations about $M^2 = -a/b$, and we find

$$(a + 3bM^2)\delta M = \delta H \qquad (12.1.15)$$

or

$$
\begin{aligned}
\chi &= \frac{\partial M}{\partial H} = \frac{1}{a + 3bM^2} = -\frac{1}{2a} \\
&= \frac{1}{2a}(T_c - T)^{-1}.
\end{aligned} \qquad (12.1.16)
$$

Once again $\chi \propto |T - T_c|^{-1}$ and hence

$$\gamma = 1. \qquad (12.1.17)$$

Setting $T = T_c$ or $a = 0$ in Eq.(11.6.6) and Eq.(12.1.13), we find

$$\delta = 3 \qquad (12.1.18)$$

which completes the Landau theory of phase transition. Note that the scaling relations (11.6.8a) and (11.6.8c) are satisfied. There is universality: in fact superuniversality in that α, β, γ and δ are independent of everything including n and D.

Though the Landau theory is important as the first nontrivial theory of critical phenomena, it is obviously not quantitatively correct: the problem lies with the ignoring of fluctuations. However, the task of the theory is now clear. One has to evaluate the partition function of Eq.(12.1.1) with the Ginzburg-Landau free energy functional of Eq.(12.1.4). It is more conventional to write $a = m^2$ and $b/4 = \lambda/4!$ and then one is faced with the functional integral

$$Z(H) = \int D[\varphi] e^{-F(\varphi)} \tag{12.1.19}$$

where

$$F(\varphi) = \int d^D x \left[\frac{m^2}{2} \sum_{\alpha=1}^{n} \varphi_\alpha^2 + \frac{1}{2} \sum_\alpha \left(\vec{\nabla} \varphi \right)^2 + \frac{\lambda}{4!} \left(\sum_\alpha \varphi_\alpha^2 \right)^2 - H\varphi_1 \right] \tag{12.1.20}$$

in the presence of an external field H in the one-direction.

We end this section by establishing some thermodynamic identities. First the magnetization M which is defined as the average of $\varphi_1(x)$ is related to $Z(H)$ by

$$\begin{aligned} M &= \lim_{H \to 0} \left\langle \int \varphi_1(x) dx \right\rangle \\ &= \lim_{H \to 0} \frac{1}{Z} \int D[\varphi] \int \varphi_1(x) e^{-F(\varphi)} d^D x \\ &= \lim_{H \to 0} \frac{1}{Z} \frac{\partial Z}{\partial H}. \end{aligned} \tag{12.1.21}$$

The susceptibility is

$$\begin{aligned} \chi &= \frac{\partial M}{\partial H} = -\left(\frac{1}{Z} \frac{\partial Z}{\partial H} \right)^2 + \frac{1}{Z} \frac{\partial^2 Z}{\partial H^2} \\ &= \int \langle \varphi_1(x)\varphi_1(y) \rangle d^D x d^D y - \left[\int \langle \varphi_1(x) \rangle d^D x \right]^2 \\ &= \int \langle \varphi_1(x)\varphi_1(y) \rangle_c d^D y d^D x \end{aligned} \tag{12.1.22}$$

where the subscript 'c' denotes the connected part of the correlation function. Translational invariance implies $G(x, y) = \langle \varphi_1(x)\varphi_1(y) \rangle_c = G(x - y)$ and thus

$$\chi = V \int d^D x_{12} \langle \varphi_1(x_1)\varphi_1(x_2) \rangle_c. \tag{12.1.23}$$

This is the static fluctuation dissipation theorem. The specific heat is found to be

$$C = -T\frac{\partial^2 F}{\partial T^2} = +T\frac{\partial^2}{\partial T^2}(\ln Z)$$

$$= T\left[\frac{1}{Z}\frac{\partial^2 Z}{\partial T^2} - \left(\frac{1}{Z}\frac{\partial Z}{\partial T}\right)^2\right]$$

$$= a_0^2 T \int d^D y \left\langle \sum_{\alpha=1}^n \varphi_\alpha^2(x) \sum_{\beta=1}^n \varphi_\beta^2(y)\right\rangle_c \qquad (12.1.24)$$

where the subscript 'c' means that the factored part $\langle \sum_{\alpha=1}^n \varphi_\alpha^2(x)\rangle^2$ has been subtracted and the correlation function is assumed to have translational invariance. Thus the critical problem in the field theory of phase transition is the calculation of various correlation functions.

12.2 The Gaussian Model

The Landau theory leaves out fluctuations altogether. An exactly solvable model that includes the effect of fluctuations but only upto the quadratic order is called the Gaussian model. The quartic term in the free energy functional of Eq.(12.1.4) is now dropped and the partition function in the Gaussian approximation is Z_G, given by

$$Z_G = \int D[\varphi] \exp\left\{-\int d^D x\left[\frac{a}{2}\sum_{\alpha=1}^n \varphi_\alpha^2 + \frac{1}{2}\sum_{\alpha=1}^n\left(\vec{\nabla}\varphi_\alpha\right)^2\right]\right\}. \qquad (12.2.1)$$

The standard method of evaluating the functional integral is to work in terms of the Fourier components of $\varphi_\alpha(x)$. The functional integral now becomes

$$Z_G = \int \prod_{k,\alpha} d\varphi_\alpha(k) \exp\left\{-\sum_{k,\alpha}\frac{1}{2}(a + k^2)|\varphi_\alpha(k)|^2\right\}$$

$$= \prod_{k,\alpha}(a + k^2)^{-1/2}. \qquad (12.2.2)$$

The thermodynamic free energy is given by

$$F = -k_B T \ln Z_G = \frac{1}{2}k_B T \sum_{k,\alpha}\ln(a + k^2) = \frac{nk_B T}{2}\sum_k \ln(a + k^2). \qquad (12.2.3)$$

The specific heat is to be obtained by differentiating twice with respect to T. This really means differentiation with respect to a, since that is the quantity which changes most as T is varied near T_c (our interest is primarily in the behaviour close to T_c and hence, apart from where it appears in the term a, the temperature can be set equal to T_c everywhere else). We then find

$$C \propto \sum_k \frac{1}{(k^2 + a)^2}$$

$$\propto \int \frac{d^D k}{(2\pi)^D} \cdot \frac{1}{(k^2 + a)^2}$$

$$\propto a^{(D-4)/2}$$

$$\propto (T - T_c)^{-(4-D)/2} \tag{12.2.4}$$

leading to

$$\alpha = \frac{4 - D}{2}. \tag{12.2.5}$$

The correlation function $G(ka)$ is easily evaluated as

$$
\begin{aligned}
G(k, a) &= \langle \varphi_\alpha(k) \varphi_\alpha(-k) \rangle \\
&= \frac{1}{Z_a} \int \prod_{p,\beta} d\varphi_\beta(p) \varphi_\alpha(k) \varphi_\alpha(-k) \exp\left\{ -\sum_{p,\alpha} \frac{1}{2}(a + p^2) \varphi_\alpha(p) \varphi_\alpha(-p) \right\} \\
&= \frac{1}{k^2 + a}.
\end{aligned} \tag{12.2.6}
$$

At $a = 0$, $G(k) \sim k^{-2}$ and thus $\eta = 0$. The susceptibility is obtained from the $k = 0$ value of $G(k, a)$ and thus

$$\chi(a) = G(k = 0, a) = \frac{1}{a} \propto (T - T_c)^{-1}$$

leading to $\gamma = 1$. The inverse Fourier transform of Eq.(12.2.6) clearly shows that $a^{1/2}$ is the inverse correlation length. Thus $\xi^{-1} \propto (T - T_c)^{1/2}$, leading to $\nu = 1/2$. The scaling laws of Eqs.(11.6.8b) and (11.6.8d) are thus seen to be satisfied. The Gaussian model is not defined for $T < T_c$ and hence we cannot make any comments about the value of β in this model.

12.3   The Spherical Limit

We will now discuss a model where the quartic term is included but in a manner that still makes the model exactly solvable. We begin by observing that in Eq.(12.1.4), the first two terms on the right hand side are of order n, but the third term is of order n^2 and hence in the limit $n \to \infty$, the model would make no sense unless we require the coupling constant b to behave as n^{-1}.

Accordingly we write $b = \frac{\mu}{n}$ and rewrite Eq.(12.1.4) as

$$
\begin{aligned}
F(\varphi) &= \int d^D x \left[\frac{a}{2} \sum \varphi_\alpha^2 + \frac{1}{2} \sum \left(\vec{\nabla} \varphi_\alpha \right)^2 + \frac{\mu}{4n} \left(\sum \varphi_\alpha^2 \right)^2 \right] \\
&= \int d^D x \left[\left(\frac{a}{2} + \frac{\mu}{4} \langle \varphi_\alpha^2 \rangle \right) \sum \varphi_\alpha^2 + \frac{1}{2} \sum \left(\vec{\nabla} \varphi_\alpha^2 \right) \right] \\
&\quad + \frac{\mu}{4n} \left(\sum \varphi_\alpha^2 - n \langle \varphi_\alpha^2 \rangle \right) \sum \varphi_\alpha^2 \right]
\end{aligned}
\tag{12.3.1}
$$

In the above $\langle \varphi_\alpha^2 \rangle$ stands for the average value of $\int \varphi_\alpha^2 dx$. On the right hand side of the above equation, the first two terms are clearly of order n, but the third term is $O(1)$ and hence in the limit $n \to \infty$, we can drop the last term and have a model defined by the free energy

$$
F(\varphi) = \int d^D x \left[\frac{a'}{2} \sum \varphi_\alpha^2 + \frac{1}{2} \sum \left(\vec{\nabla} \varphi_\alpha \right)^2 \right]
\tag{12.3.2}
$$

where

$$
a' = a + \frac{\mu}{2} \langle \varphi_\alpha^2 \rangle.
\tag{12.3.3}
$$

The new critical point is obtained when a' goes to zero. From Eq.(12.3.2), it is clear that

$$
\langle \varphi_\alpha(k) \varphi_\alpha(-k) \rangle = \frac{1}{k^2 + a'}
\tag{12.3.4}
$$

(just as in the case of the Gaussian model of the previous Section) and hence

$$
\langle \varphi_\alpha^2 \rangle = \int \frac{d^D k}{(2\pi)^D} \frac{1}{k^2 + a'}
\tag{12.3.5}
$$

With $a = a_0(T - T_c)$, Eq.(12.3.3) becomes

$$
a' = a_0(T - T_c) + \frac{\mu}{2} \int \frac{d^D k}{(2\pi)^D} \cdot \frac{1}{k^2 + a'}.
\tag{12.3.6}
$$

The new transition temperature is T_c', where $a' = 0$. Thus,

$$
0 = a_0(T_c') + \frac{\mu}{2} \int \frac{d^D k}{(2\pi)^D} \cdot \frac{1}{k^2 + a'}.
\tag{12.3.7}
$$

Subtracting Eq.(12.3.7) from Eq.(12.3.6) we have

$$
a' = a_0(T - T_c') + \frac{\mu}{2} \int \frac{d^D k}{(2\pi)^D} \left[\frac{1}{k^2 + a'} - \frac{1}{k^2} \right]
$$

or

$$
\begin{aligned}
a_0(T - T_c') &= a' + \frac{\mu a'}{2} \int \frac{d^D k}{(2\pi)^D} \frac{1}{k^2(k^2 + a')} \\
&= a' + A a'^{\frac{D}{2} - 1}
\end{aligned}
\tag{12.3.8}
$$

where A is a constant independent of a'. Since we are interested in small values of a', the right hand side of Eq.(12.3.8) is dominated by $a'^{\frac{D}{2}-1}$ if $4 > D > 2$, and by a' if $D > 4$. Hence,

$$a' \propto (T - T'_c)^{\frac{2}{D-2}} \quad \text{if} \ 2 < D < 4$$
$$a' \propto T - T'_c \quad \text{if} \ D > 4. \tag{12.3.9}$$

The above manipulations reveal a vital fact: the fluctuation contribution from the quartic term completely dominates the quadratic term for $D < 4$. This makes perturbative calculations around the Gaussian model meaningless for $D < 4$. Just as in the Gaussian model of the previous Section, the correlation length is given by

$$\xi = a^{-1/2} \begin{cases} \propto (T - T_c)^{-1/(D-2)} & \text{for} \ 2 < D < 4 \\ \propto (T - T_c)^{-1/2} & \text{for} \ D > 4 \end{cases} \tag{12.3.10}$$

leading to

$$\nu = \frac{1}{D-2} \quad \text{if} \ 2 < D < 4$$
$$\nu = \frac{1}{2} \quad \text{if} \ D > 4. \tag{12.3.11}$$

The anomalous dimension index η is clearly zero and $\gamma = 2\nu$.

The calculation of specific heat is more difficult. This can be appreciated from Eq.(12.1.24). Using only the quadratic part of the free energy [Eq.(12.3.2)], we can easily evaluate the specific heat. To do this, we write Eq.(12.1.24) as

$$C = a_0^2 T \int \frac{d^D p}{(2\pi)^D} \frac{d^D q}{(2\pi)^D} \left\langle \sum_{\infty=1}^{n} \varphi_\alpha(q)\varphi_\alpha(-q) \sum_{\beta=1}^{n} \varphi_\beta(p)\varphi_\beta(-p) \right\rangle_c$$
$$= n a_0^2 T \int \frac{d^D p}{(2\pi)^D} [\langle \varphi(p)\varphi(-p) \rangle]^2$$
$$\propto \int \frac{d^D p}{(2\pi)^D} \left(\frac{1}{p^2 + a'} \right)^2$$
$$\propto a'^{(D/2)-2} \tag{12.3.12}$$

It would appear on using Eq.(12.3.10) that $\alpha = \frac{4-D}{D-2}$ for $4 > D > 2$. If we check the scaling law $D\nu = 2 - \alpha$ at this point, we would find that it is not satisfied. Thus, the evaluation of the specific heat is a more complicated problem for $n \to \infty$.

To understand the source of the above difficulty, we imagine performing the average in Eq.(12.1.24) by using the full free energy functional of Eq.(12.3.1). This can be done by expanding the part

$$\exp\left\{ \frac{\mu}{4n} \int d^D x \left(\sum \varphi_\alpha^2 - n\langle \varphi_\alpha^2 \rangle \right) \sum \varphi_\alpha^2 \right\} \quad \text{in a power series.}$$

The first term in such an expansion requires the average

$$\frac{\mu}{4n}a_0^2 T \int d^D x d^D y d^D z \left\langle \sum \varphi_\alpha^2(x) \sum \varphi_\alpha^2(z) \left(\sum \varphi_\alpha^2(z) - n\langle \varphi_\alpha^2 \rangle \right) \sum \varphi_\alpha^2(y) \right\rangle.$$

These being Gaussian averages, a little thought will convince the reader that there is a part of the average which is proportional to n^2 and hence the entire term is proportional to n, which is just as big as the zeroth order term found in Eq.(12.3.12). The a' dependence of the term is $(a'^{D/2-2})^2$. If we now imagine considering the higher order terms, a trend will be seen to emerge. At the second order the combinatorial factor yields n^3 while the coupling constant is n^{-2}, giving once again a part proportional to n. The a' dependence is $(a'^{D/2-2})^3$. The result is a geometric series:

$$\begin{aligned} C &\propto a'^{\frac{D}{2}-2} \left[1 - \frac{\mu}{4}a^{\frac{D}{2}-2} + \left(\frac{\mu}{4}a^{\frac{D}{2}-2}\right)^2 \cdots \right] \\ &= a'^{\frac{D}{2}-2} \left(1 + \frac{\mu}{4}.a'^{\frac{D}{2}-2}\right)^{-1} \\ &= \frac{4/\mu}{1 + \frac{4}{\mu}a'^{2-\frac{D}{2}}} \\ &\cong \frac{4}{\mu} \left(1 - \frac{4}{\mu}a'^{2-\frac{D}{2}} + \ldots\right). \end{aligned} \qquad (12.3.13)$$

Using the $T - T_c$ dependence of a' from Eq.(12.3.10), we find

$$\alpha = \frac{D-4}{D-2} \quad \text{for } 4 > D > 2. \qquad (12.3.14)$$

The hyper-scaling law $D\nu = 2 - \alpha$ is now satisfied. The $n \to \infty$ limit is known as the spherical model.

13

The Renormalization Group

13.1 Renormalization Group in Real Space

13.1.1 Introduction

The problem with critical phenomena lies in the very large number of degrees of freedom which makes perturbation theory intractable. As is customary, we will call the free energy F of Eq.(12.1.4), the Hamiltonian H from now and we will see that the constant b has to have a special value for perturbation theory to work, implying that there is a special Hamiltonian for the perturbation theory to be effective. The meaning of that special Hamiltonian will be made clear in this chapter.

To get a preliminary feel for the techniques involved, we follow the early work of Kadanoff and consider a two-dimensional Ising model. The Hamiltonian for this model is $H = -J(\Sigma S_{i,j}S_{i+1,j} + \Sigma S_{i,j}S_{i,j+1})$ and as long range order sets in near the critical point, the spins at different sites get correlated and we can imagine (see Fig.13.1) four spins at the four vertices at a square (a block) acting conjointly to form an effective single spin $\tilde{S}$.

The original spins S have the value $S = \pm 1$ and the new spins are assumed to be properly rescaled to have the value ± 1 once more. The block spins (a block of spins giving a single spin) are assumed to have nearest neighbour interaction like the original spins and the new Hamiltonian is

$$H' = -J_1 \sum \left(\tilde{S}_{i,j}\tilde{S}_{i+1,j} + \tilde{S}_{i,j}\tilde{S}_{i,j+1} \right)$$

where J_1 is the new interaction strength between the block spins. We expect J_1 to be some analytic function of J

$$J_1 = f(J). \tag{13.1.1}$$

We now consider what happens to the correlation length as we construct the block spins. The block spins are twice as far removed as the original spins

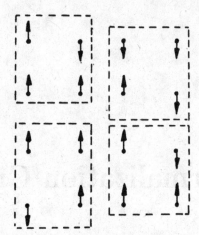

Figure 13.1: Kadanoff's idea of block-spins in a 2D Ising model.

and hence the correlation length (measured in units of lattice spacing) is halved with the formation of the block spins, i.e.

$$\xi\{f(J)\} = \frac{1}{2}\xi(J).\tag{13.1.2}$$

Now imagine doing this block spin construction repeatedly – a J_1 generates a J_2 and so on and eventually it is possible that J does not change any more – we have then reached a fixed point of the transformation given by Eq.(13.1.1) – and the corresponding value J_c of J satisfies

$$J_c = f(J_c).\tag{13.1.3}$$

At $J = J_c$, $\xi(J_c) = \frac{1}{2}.\xi(J_c)$, which implies that ξ is infinite or zero at $J = J_c$. Hence $J = J_c$ corresponds to the critical point, provided ξ is infinity there.

To work near the critical point, we examine $f(J)$ near $J = J_c$ and expand as a Taylor series

$$
\begin{aligned}
f(J) &= f(J_c) + (J - J_c)f'(J_c) + \dots \\
&= f(J_c) + \lambda(J - J_c) + \dots \\
f(J) - f(J_c) &= \lambda(J - J_c) + O[(J - J_c)^2].
\end{aligned}\tag{13.1.4}
$$

Assuming that near the critical point J_c

$$\xi(J) \propto (J - J_c)^{-\nu}\tag{13.1.5}$$

we find

$$\frac{\xi(f(J) - f(J_c))}{\xi(J)} = \left[\frac{(f(J) - f(J_c))}{(J - J_c)}\right]^{-\nu} = .\lambda^{-\nu}.\tag{13.1.6}$$

Clearly this makes sense only if λ is positive. Hence the critical exponent ν can be found if λ is known. To know λ, we must know $f(J)$, which is the central feature of the above scheme.

The philosophy of this approach which is known as the renormalization group transformation is summarized in Eq.(13.1.2). The transformation reduces the correlation length. Repeating the process n times, we would have

$$\xi(J_n) = 2^{-n}\xi(J). \tag{13.1.7}$$

One could always choose an n such that $\xi(J_n)$ is small and conventional perturbation theory can be employed to calculate it.

13.1.2 One-dimensional Ising Model

We now demonstrate how the above picture is implemented for the one dimensional model where the partition function is given by

$$Z = \sum_{\{S_i\}} e^{\beta J \sum_i S_i S_{i+1}} = \sum_{\{S_i\}} e^{K \sum_i S_i S_{i+1}} \tag{13.1.8}$$

with $K = \beta J$. A given spin, say S_5, (see Fig13.2) couples only to S_4 and S_6.

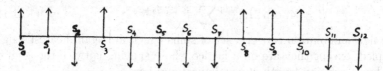

Figure 13.2: One-dimensional spins.

The renormalization group transformation (it should be noted that this operation does not have an inverse and hence is a semi-group) requires increasing the lattice spacing to reduce the degrees of freedom. If we perform the sum over alternate spins $(S_1, S_3, S_5, \dots$ etc) in Eq.(13.1.8), then the lattice spacing will be automatically doubled.

This motivates summing over the alternate spin. We can now write

$$
\begin{aligned}
Z &= \sum_{(S_2, S_4, S_6 \dots)} \sum_{(S_1, S_3 \dots)} \dots e^{K S_1 S_2} e^{K S_2 S_3} e^{K S_3 S_4} e^{K S_4 S_5} \dots \\
&= \sum_{S_2, S_4, S_6} \sum_{S_1} e^{K S_1 S_2} \sum_{S_3} e^{K S_2 S_3 + K S_3 S_4} \sum_{S_5} e^{K S_4 S_5 + K S_5 S_6} \dots.
\end{aligned}
$$

where in the first line above, the dots to the left of the expression $e^{K S_1 S_2}$ imply similar exponentials involving the spins that are to the left of S_1. Consider any one of the sums over an odd spin, say $\sum_{S_3} e^{K(S_2 S_3 + S_3 S_4)}$. The result will depend on S_2 and S_4.

Since there are two independent configurations of S_2 and S_4 (parallel and antiparallel), the result of tracing over S_3 can involve only two undetermined constants. Hence, we anticipate,

$$\sum_{S_3} e^{K(S_2 S_3 + S_3 S_4)} = 2\cosh(S_2 + S_4)K = e^{K_0'} e^{K' S_2 S_4} \tag{13.1.9}$$

If S_2 and S_4 are parallel

$$e^{K_0'} e^{K'} = 2\cosh 2K$$

and if S_2 and S_4 are antiparallel

$$e^{K_0'} e^{-K'} = 2$$

leading to

$$e^{2K'} = \cosh 2K \tag{13.1.10}$$

and

$$e^{2K_0'} = 4\cosh 2K. \tag{13.1.11}$$

We have the arrived at the explicit from for Eq.(13.1.1). The coupling constant $K = \beta J$ transforms to a new coupling constant K' and the partition function is

$$Z = [4\cosh 2K]^{N/4} \sum e^{K'[S_0 S_2 + S_2 S_4 + ...]}. \tag{13.1.12}$$

The new summation is over a set of spins which are again nearest neighbour coupled, except they are on a lattice whose spacing is twice the original lattice. The correlation length is now halved and hence

$$\xi(K') = \frac{1}{2}\xi(K) \tag{13.1.13}$$

which is the explicit form of Eq.(13.1.2).

The new coupling constant K' is related to the old by [see Eq.(13.1.10)]

$$K' = \frac{1}{2}\ln \ \cosh 2K \tag{13.1.14}$$

There are two fixed points of the transformation (obtained by solving $K_c = \frac{1}{2}\ln \ \cosh 2K_c$. One is $K_c = 0$ and the other $K_c = \infty$. The fixed point $K_c = 0$ corresponds to $T = \infty$ (recall that $K = \beta J$) and the fixed point $K_c = \infty$ corresponds to $T = 0$. It is important to understand which is the relevant fixed point that can describe the critical phenomenon.

The correlation length is infinite at the critical point and has to fall off as one moves way from criticality. According to Eq.(13.1.13), if we start at $K \neq K_c$, then each step of the renormalization process makes the correlation length smaller and thus one moves away from the critical point. This indicates that the fixed point that will correspond to a critical point will be an unstable fixed point. In this particular case that we have been discussing, there is only coupling constant K (linked directly to the temperature through $K = \beta J$ and hence the issues are straightforward). In general there will be several coupling constants K_i and then in the space spanned by $\{K_i\}$, there has to be at least one unstable direction and this will be associated with the temperature.

We now test the stability of the fixed points $K_c = 0$ and $K_c = \infty$. For $K_c = 0$, we consider $K = 0 + \delta K$ and write Eq.(13.1.14) as

$$\delta K' = \frac{1}{2} \ln \ (\cosh 2\delta K) = \frac{1}{2} \ln \ \left(1 + \frac{1}{2}(2\delta K)^2 + \right) \approx (\delta K)^2. \qquad (13.1.15)$$

Thus, $\delta K' < \delta K$ and hence $K_c = 0$ is a stable fixed point, although not in the linear stability sense. This is as it should be because the high temperature fixed point is irrelevant.

For $K_c = \infty$, Eq.(13.1.14) yields

$$\delta K' = \delta K \qquad (13.1.16)$$

which gives no obvious information. This is what will later be referred to as the marginal case. To make further progress, we return to Eq.(13.1.14) and write it for $K \gg 1$ as

$$K' = K - \frac{1}{2} \ln 2 + O\left(e^{-2K}\right). \qquad (13.1.17)$$

This shows that on iteration, one actually moves away from $K_c = \infty$ and in that sense it is an unstable fixed point although linear stability analysis did not reveal it. The fact $\xi(K') = \frac{1}{2}\xi(K)$ implies in conjunction with Eq. (13.1.17) is that

$$\xi(k) \propto e^{aK}. \qquad (13.1.18)$$

This leads to $e^{aK - \frac{a}{2} \ln \ 2} = \frac{1}{2}.e^{aK}$, giving $a = 2$.

It follows that $\xi(K) \propto e^{2K}$ and using Eq.(12.1.22) with $\eta = 0$ and $\xi \propto e^{4K} = e^{4J/kT}$, in exact agreement with the result found in Chapter 5.

13.1.3 Two-dimensional Ising Model

We now turn to the two-dimensional Ising model with the coupling constant J in both x and y direction. We want $K = J/k_B T$ exactly as in the one dimensional case. The implementation of the scheme outlined above involves carrying out the sum over all the open circle sites in Fig.13.3.

The particular open circle site we choose for our demonstration is marked zero and the four nearest neighbors are marked 1, 2, 3 and 4.

Showing just the relevant part of the Hamiltonian we can write

$$Z = \sum_{S_1,S_2,S_3,S_4} \sum_{S_0} e^{K(S_0 S_1 + S_0 S_2 + S_0 S_3 + S_0 S_4) + \cdots}. \qquad (13.1.19)$$

The sum over S_0 leads to

$$Z = \sum_{S_1,S_2,S_3,S_4} \left[e^{K(S_1+S_2+S_3+S_4)} + e^{-K(S_1+S_2+S_3+S_4)} \right]. \qquad (13.1.20)$$

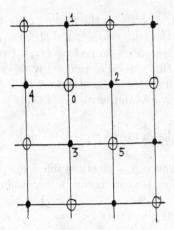

Figure 13.3: Two-dimensional spins.

We recognize the following independent configurations of S_1, S_2, S_3 and S_4:

1. S_1, S_2, S_3, S_4 all pointing up,

2. S_1, S_2, S_3 pointing up and S_4 down,

3. S_1, S_3 up and S_2, S_4 down,

4. S_1, S_2 up and S_3, S_4 down.

Consequently , we expect

$$e^{K(S_1+S_2+S_3+S_4)} + e^{-K(S_1+S_2+S_3+S_4)}$$
$$= \exp\left[K_0 + \frac{K_1}{2}(S_1S_2 + S_2S_3 + S_3S_4 + S_4S_1)\right.$$
$$+ \left. K_2(S_1S_3 + S_2S_4) + K_3S_1S_2S_3S_4\right]. \tag{13.1.21}$$

For $S_1 = S_2 = S_3 = S_4 = 1$ we have

$$2\cosh 4K = e^{K_0+2K_1+2K_2+K_3}. \tag{13.1.22}$$

For $S_1 = S_2 = 1$ and $S_3 = S_4 = -1$, Eq.(13.1.21) yields

$$2 = e^{K_0-2K_2+K_3} \tag{13.1.23}$$

For $S_1 = S_3 = 1$, and $S_2 = S_4 = -1$

$$2 = e^{K_0-2K_1+2K_2+K_3}. \tag{13.1.24}$$

For $S_1 = S_2 = S_3 = 1$ and $S_4 = -1$ we finally have

$$2\cosh 2K = e^{K_0-K_3}. \tag{13.1.25}$$

Using Eqs.(13.1.22) - (13.1.25), we find

$$K_0 = 2(\cosh 2K)^{1/2}(\cosh 4K)^{1/8} \qquad (13.1.26)$$

$$K_1 = \frac{1}{4}\ln \cosh 4K \qquad (13.1.27)$$

$$K_2 = \frac{1}{8}\ln \cosh 4K \qquad (13.1.28)$$

$$K_3 = \frac{1}{8}\ln \cosh 4K - \frac{1}{2}\ln \cosh 2K. \qquad (13.1.29)$$

In writing down Eq.(13.1.27), we note that a contribution to S_1S_2, S_2S_3, S_3S_4, and S_4S_1 also come from the removal of another spin (e.g., removal of the spin marked 5 contributes to S_2S_3 with strength $K_1/2$ and thus the coefficient of S_2S_3 becomes K_1. A similar argument holds for S_1S_2, S_3S_4 and S_4S_1).

We note the problem with the generation of a new Hamiltonian – we started with a nearest neighbour energy expression and now have nearest neighbour, next nearest neighbour and a four spin interaction. At this point if we choose to drop all the new couplings like K_2 and K_3, we would have only Eq.(13.1.27). This recursion is identical to that found in Eq.(13.1.10) for the one-dimensional Ising model and hence will not give a finite temperature transition. To obtained a better result, we improve the theory by taking K_2 into account in an approximate way. Both K_1 and K_2 are positive – they tend to favour the alignment of spins. A possible strategy is to drop K_2 but increase K_1 to K' so that the "alignment tendency remains the same". A crude way of getting K' is to consider all spins aligned. Now in a lattice of $N/2$ spins there are N nearest neighbour bonds and N next nearest neighbour bonds. The total energy would be $Nk_BT(K_1 + K_2)$. We choose K' to get the same total energy. Thus

$$K' = K_1 + K_2 = \frac{3}{8}\ln \cosh 4K. \qquad (13.1.30)$$

The recursion relation now has a new nontrivial fixed point K_c obtained from

$$K_c = \frac{3}{8}\ln \cosh 4K_c \qquad (13.1.31)$$

and is found to be $K_c = 0.50698$.

The flow now splits in a remarkable fashion as shown in Fig.13.4.

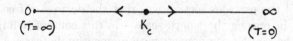

Figure 13.4: The flow.

The fixed point $K = K_c$ is unstable. If we start with $K > K_c$ then the flow goes to $K = \infty$ and starting from $K < K_c$ the flow goes to $K = 0$. In terms of the temperature if $K_c \propto (k_BT_c)^{-1}$, then starting from $T > T_c$, one reaches the fully disordered state $T = \infty$ and starting from $T < T_c$ one reaches the fully ordered state $T = 0$. Thus T_c is the transition temperature. The exact solution of the Ising model gives $K_c = 0.44069$. As we shall see later in this chapter, the information obtained from the derivative of K' at K_c will yield the specific heat exponent α (see the discussion following Eq.(13.3.20). By comparing with Eq.(13.3.21), we note that the exponent $\nu = 0.92$. The new lattice spacing is $\sqrt{2}$ times the old and hence $b = \sqrt{2}$.

13.2   Renormalization Group in Momentum Space

For the models introduced in Chapter 12, the natural implementation of the renormalization group would be in momentum space. Consequently, we try now to carry over the real space picture to the Landau-Ginzburg model of Chapter 12 by writing it in the Fourier space. From now on the free energy functional, which in analogy with other areas of Physics, shall be called a "Hamiltonian" and hence will be denoted by H, thus keeping F for the thermodynamic free energy. We begin by writing Eq.(12.1.4) in momentum space as (by replacing the coefficient a in Eq.(12.1.4) by m^2)

$$
\begin{aligned}
H &= \frac{1}{2} \sum_{\alpha, k < \Lambda} \phi_\alpha(\mathbf{k}) \phi_\alpha(-\mathbf{k})(m^2 + k^2) \\
&+ \frac{\lambda}{4!} \sum_{\alpha, \beta, p_i} \phi_\alpha(\mathbf{p_1}) \phi_\alpha(\mathbf{p_2}) \phi_\beta(\mathbf{p_3}) \phi_\beta(\mathbf{p_4}) \delta^{(D)} \left(\sum_i \mathbf{p_i} \right).
\end{aligned} \tag{13.2.1}
$$

The momentum range is restricted to $k < \Lambda$, where Λ is a large momentum which corresponds to the inverse lattice spacing in the discrete version of the continuum theory. As discussed in detail already, the aim is to calculate the partition function Z and the n-point correlation functions

$$
Z = \sum_{[\phi]} e^{-H(\phi)}, \tag{13.2.2}
$$

$$
G_c^{(2)}(k) = \langle \phi(\mathbf{k}) \phi(-\mathbf{k}) \rangle_c = \frac{1}{Z} \sum_{[\phi]} \phi(\mathbf{k}) \phi(-\mathbf{k}) e^{-H(\phi)} \tag{13.2.3}
$$

and so on.

The lesson of the previous section was that the degrees of freedom have to be reduced by increasing the length scale – in this context it implies that we must remove the Fourier components with momentum between Λ and Λ/b, where $b > 1$. With the degrees of freedom reduced, one now scales the momentum by a factor b, so that the cutoff is restored to its original value Λ. The field ϕ is rescaled as well (corresponding to the spin always being ± 1 in the previous section) so that the gradient term in the Hamiltonian is unaltered. Ensuring that the gradient term remains unaltered, trivial scale transformations are eliminated. These three steps, viz.,

1. Integrating out the degrees of freedom from $k = \Lambda/b$ to Λ.

2. Rescaling momenta $k' = bk$.

3. Rescaling the fields,i.e., $\phi(k) = \zeta_b \phi_\alpha(k') = \zeta_b \phi_\alpha(kb)$

constitute the renormalization group.

Then the process of the renormalization group is embodied in the operation

$$\int \left[\prod_{\alpha=1}^{n} \prod_{\Lambda/b < k' < \Lambda} D\phi_\alpha(k') e^{-H(\phi)} \right]_{\phi_\alpha(k) \to \zeta_b \phi_\alpha(bk)} .$$

The result is a new exponential factor $e^{-H'}$. If we call H' the transformed Hamiltonian $R_b H$, then

$$e^{-R_b H} = \int \left[\prod_{\alpha=1}^{n} \prod_{\Lambda/b < k' < \Lambda} D\phi_\alpha(k') e^{-H} \right]_{\phi_\alpha(k) \to \zeta_b \phi_\alpha(bk)} . \qquad (13.2.4)$$

If we carry out two successive renormalization group transformations, then $\phi_\alpha(k)$ goes to $\zeta_{b'} . \zeta_b \phi_\alpha(b'bk)$. The closure property requires that $R_{b'} R_b = R_{b'b}$ and hence

$$\zeta_{b'b} = \zeta_{b'} \zeta_b. \qquad (13.2.5)$$

This is achieved by the choice

$$\zeta_b = b^y \qquad (13.2.6)$$

We will write y as

$$y = 1 - \frac{\eta}{2} \qquad (13.2.7)$$

and later show that η is indeed the anomalous dimension η of Chapter 11. We now need to know how the partition function and the correlation function transform under the action of R_b.

It should be clear from Eqs.(13.2.2) and (13.2.4) that the partition function is unchanged under the transformation. The correlation function $G_c^{(2)}(k, H)$ transforms from its definition [Eq.(13.2.2)] according to

$$\begin{aligned} G_c^{(2)}(k, H) &= [\zeta_b]^2 G_c^{(2)}(bk, R_b H) \\ &= b^{2-\eta} G_c^{(2)}(bk, R_b H). \qquad (13.2.8) \end{aligned}$$

13.3 Fixed Points and Scaling Laws

We now ask what happens as we repeat the group operation of the previous section. Each step generates a new Hamiltonian and after l steps

$$H_l = (R_b)^l H.$$

A possibility after repeated applications is the emergence of a fixed point Hamiltonian H^*, which does not change on application of R_b, i.e.,

$$R_b H^* = H^*. \tag{13.3.1}$$

The existence of a fixed point cannot be guaranteed from any general considerations.

What changes as the renormalization group operation is performed? It is the coupling constants in the Hamiltonian (in principle, we imagine the Hamiltonian to contain all possible terms consistent with the symmetry) which we denote by K_α that change under the actions of the renormalization group and the fact that H^* has become a fixed point Hamiltonian implies that the constants have acquired their fixed point values K_α^* [note that m^2, $\lambda^4/4!$ etc. of our Hamiltonian Eq.(13.2.1) constitute the set $\{K_\alpha\}$].

If we linearize the transformation $(K_\alpha') = R_b(K_\alpha)$ near the fixed point, we get

$$K_\alpha' = K_\alpha^* + \left(\frac{\partial K_\alpha'}{\partial K_\beta}\right)_{\{K^*\}} \delta K_\beta$$

or (with the summation convention being implied for the Greek indices only)

$$\delta K_\alpha' = T_{\alpha\beta} \delta K_\beta \tag{13.3.2}$$

where

$$\delta K_\alpha = K_\alpha - K_\alpha^* \tag{13.3.3}$$

and

$$T_{\alpha\beta} = \left(\frac{\partial K_\alpha'}{\partial K_\beta}\right)_{\{K^*\}}. \tag{13.3.4}$$

If the eigenvalues of matrix $T_{\alpha\beta}$ are $\{\lambda_i\}$ and the eigenvectors $\{\psi_i^{(\beta)}\}$ then

$$T_{\alpha\beta}\psi_i^{(\alpha)} = \lambda_i \psi_i^{(\beta)}. \tag{13.3.5}$$

We now construct the fields u_i out of the set δK_α by using the linear transformation

$$u_i = \psi_i^{(\alpha)} \delta K_\alpha. \tag{13.3.6}$$

The fields $\{u_i\}$ transform as

$$\begin{aligned}
u_i' &= \psi_i^{(\alpha)} \delta K_\alpha' \\
&= \psi_i^{(\alpha)} T_{\alpha\beta} \delta K_\beta \\
&= \delta K_\beta T_{\alpha\beta} \psi_i^{(\alpha)} \\
&= \lambda_i \psi_i^{(\beta)} \delta K_\beta \\
&= \lambda_i u_i. \tag{13.3.7}
\end{aligned}$$

The $u_{(i)}$ transform, under the action of R by a multiplicative factor and are known as scaling fields.

We note that there are three possibilities:

1. $\lambda_i > 1$, the corresponding scaling variables grow under the action of the renormalization group. These are the relevant variables.

2. $\lambda_i < 1$, the corresponding scaling variables shrink and ultimately vanish under the action of the renormalization group. These are the irrelevant variables.

3. $\lambda_i = 1$, the corresponding scaling variables are called marginal.

The correlation length transforms under the action of the group as

$$\xi(R_b H) = \xi(H)/b. \tag{13.3.8}$$

At criticality ξ is infinite and does not transform under the group operation. The fixed point Hamiltonian does not transform under the group operation and in that sense the fixed point Hamiltonian H^* is critical. The general Hamiltonian can be written as

$$H = H^* + \sum_i O_i u_i \tag{13.3.9}$$

where O_i, are the operators corresponding to the variables u_i (which, in turn, are determined by K_α). It is immediately clear that $(R_b)^l H \to H^*$ as $l \to \infty$, if the set $\{i\}$ constitute the irrelevant set and the relevant variables u_i are held equal to zero. Hence at criticality, the relevant fields must be equal to zero. Turning now to ordinary critical phenomena, we note that for criticality the external magnetic field h must be equal to zero and the temperature T must be equal to T_c. If we construct the variable $t = (T - T_c)/T_c$, then $t = 0$ and $h = 0$ at criticality. Hence there must be two relevant variables, t and H, to describe the usual critical point. All other variables must be irrelevant. We anticipate the absence of marginal variables since the existence of a marginal variable u_m would imply that

$$H = H^* + C_m O_m u_m$$

would remain invariant for all values of the constant C_m and this would generate a line of critical points.

The eigenvalues λ_i corresponding to the scaling fields t and h are denoted by λ_t, and λ_h respectively and we denote by λ_1, λ_2 the set of eigenvalues for irrelevant variables.

The free energy per unit volume is determined by the scaling variables t, h, λ_1, λ_2, ... Since $\ln Z$ does not change under the renormalization group transformation, the free energy per unit volume

$$F = -\frac{1}{V} \ln Z$$

has the transformation

$$F(x) = b^{-D} F(x')$$ (13.3.10)

i.e.,

$$
\begin{aligned}
F(t, h, u_1, u_2, \ldots) &= b^{-D} F(\lambda_t t, \lambda_h h, \lambda_1 u_1, \lambda_2 u_2, \ldots) \\
&= b^{-nD} F(\lambda_t^n t, \lambda_h^n h, \lambda_1^n u_1, \lambda_2^n u_2, \ldots).
\end{aligned}
$$ (13.3.11)

For sufficiently large n, we can choose $\lambda_t^n t = 1$, so that $n = \frac{\ln(1/t)}{\ln \lambda_t}$ and this leads to

$$F(t, h, u_1, u_2, \ldots) = t^{D/y_t} F\left(1, \frac{h}{t^{y_h/y_t}}, \frac{u_1}{t^{y_1/y_t}}, \ldots\right)$$ (13.3.12)

where

$$
\begin{aligned}
y_t &= \ln \lambda_t / \ln b > 0, \\
y_h &= \ln \lambda_h / \ln b > 0
\end{aligned}
$$

while,

$$y_1 = \ln \lambda_1 / \ln b, \quad y_2 = \ln \lambda_2 / \ln b, \quad \ldots\ldots \quad \text{all} < 0$$

For sufficiently small t (i.e., close to the critical point) $\frac{u_1}{t^{y_1/y_t}} \to 0$ and we have

$$F(t, h) = t^{D/y_t} F\left(\frac{h}{t^{y_h/y_t}}\right).$$ (13.3.13)

Since the specific heat behaves as $t^{-\alpha}$, the free energy should behave as $t^{2-\alpha}$ for $h = 0$ and that leads to

$$2 - \alpha = \frac{D}{y_t}$$ (13.3.14)

The magnetization $M = \left(\frac{\partial F}{\partial h}\right)_{h=0} \propto t^{\beta}$ and we have from Eq.(13.3.13)

$$\beta = \frac{D - y_h}{y_t}.$$ (13.3.15)

The susceptibility $\chi = \frac{\partial M}{\partial h} \propto t^{-\gamma}$ and differentiating the free energy twice we get

$$\gamma = \frac{2y_h - D}{y_t}.$$ (13.3.16)

From the above relations we readily see that

$$\gamma + 2\beta = \frac{D}{y_t} = 2 - \alpha$$

or

$$\gamma + 2\beta + \alpha = 2$$

which establishes the scaling law given in Eq.(11.6.8).

Differentiating Eq.(13.3.13)

$$M = t^{(D-y_h)/y_t} F'\left(\frac{h}{t^{y_h/y_t}}\right). \tag{13.3.17}$$

As $t \to 0$, $h/t^{y_h/y_t} \to \infty$ and to recover the behaviour $M \sim H^{1/\delta}$, the asymptotic behaviour of $F'(x)$ for large x has to be $x^{1/\delta}$, with δ chosen to remove the t-dependence. This yields,

$$\frac{y_h}{y_t} \cdot \frac{1}{\delta} = \frac{D - y_h}{y_t}$$

$$\Rightarrow \quad \frac{1}{\delta} = \frac{D - y_h}{y_h} \tag{13.3.18}$$

leading to

$$\begin{aligned}
\beta(1 + \delta) &= \beta\delta\left(1 + \frac{1}{\delta}\right) \\
&= \frac{D - y_h}{y_t} \cdot \frac{y_h}{D - y_h} \cdot \frac{D}{y_h} \\
&= \frac{D}{y_t} \\
&= 2 - \alpha
\end{aligned}$$

or

$$\alpha + \beta(1 + \delta) = 2$$

which is another scaling law [see Eq.(11.6.10)].

The correlation length ξ transforms as

$$\xi(t) = \frac{1}{b}\xi(\lambda_t t) \tag{13.3.19}$$

under the renormalization group transformation. Repeated iterations lead to

$$\xi(t) = \frac{1}{b}\xi(\lambda_t t) = \ldots = \frac{1}{b^n}\xi(\lambda_t^n t).$$

Choosing $\lambda_t^n t = 1$, we have

$$\xi(t) = t^{-1/y_t}\xi(1). \tag{13.3.20}$$

Identifying the exponent ν as

$$\nu = \frac{1}{y_t}, \tag{13.3.21}$$

from Eq.(13.3.14) we establish the scaling law

$$D\nu = 2 - \alpha.$$

For the two-dimensional Ising model in zero external field [studied in Sec.13.1(C)], we see that $\lambda_t = \left(\frac{\partial K'}{\partial K}\right)_{K_c}$, which yields the exponent ν stated at the end of that Section.

We now turn to the transformation of the correlation function [Eq.(13.2.8)]

$$G_c^{(2)}(k,t) = b^{2-\eta} G_c^{(2)}(bk, \lambda_t t). \qquad (13.3.22)$$

At $T = T_c$, i.e., $(t = 0)$, it is clear that

$$G_c^{(2)}(k) \propto k^{-2+\eta}$$

and thus η is indeed the anomalous dimension exponent. On the other hand at $k = 0$, repeated application of Eq.(13.3.22) implies $G_c^{(2)} = (b^n)^{2-\eta} G_c^{(2)}(\lambda_t^n t) \propto t^{-(2-\eta)/y_t}$ as we choose $\lambda_t^n.t = 1$. Since $G_c^{(2)}(k = 0)$ is the susceptibility γ, we have

$$\gamma = (2 - \eta)\frac{1}{y_t} = (2 - \eta)\nu$$

which is another scaling law. Thus, using general principle of the renormalization group transformation, we have succeeded in establishing all the scaling laws introduced in Chapter 11.

13.4   Application to the Gaussian Model

In the Hamiltonian of Eq.(13.2.1) we set $\lambda = 0$ and, on carrying out the renormalization group transformation, readily find

$$
\begin{aligned}
R_b H &= \sum_{k < \Lambda/b} (m^2 + k^2)\phi(\mathbf{k})\phi(-\mathbf{k}) \\
&= \sum_{k' < \Lambda} \zeta_b^2 (m^2 + \frac{k'^2}{b^2})\phi(\mathbf{k'})\phi(-\mathbf{k'}). \qquad (13.4.1)
\end{aligned}
$$

The requirement of fixing ζ_b is to keep the coefficient of the k^2 term unchanged (i.e., always equal to 1). Thus from Eq.(13.4.1), $\zeta_b = b$ and the mass transforms as

$$m'^2 = b^2 m^2. \qquad (13.4.2)$$

With $\zeta_b = b^{1-\eta/2}$, we have $\eta = 0$. From Eq.(13.4.2) it is immediately clear that $b^2 = \lambda_t$ since $m^2 \propto t$.

This leads to the exponent ν as

$$\nu = \frac{1}{y_t} = \frac{\ln b}{\ln b^2} = \frac{1}{2}.$$

The specific heat exponent α follows from

$$2 - \alpha = \frac{D}{y_t} = \frac{D}{2}$$

or $\alpha = \epsilon/2$, where $\epsilon = 4 - D$.

The susceptibility exponent γ follows from

$$\gamma = (2 - \eta)\nu = 2\nu = 1$$

The results are in accordance with those of Chapter 12.

13.5 Application to the Landau-Ginzburg Model

In this section, we demonstrate how the renormalization group works in momentum space by working with the scalar field Landau-Ginzburg model. We will set the external magnetic field equal to zero at the outset so that there should be only one relevant (temperature like) variable to describe the critical behaviour. We will be looking for fixed points with one eigenvalue greater than unity and the remaining less than unity. Remembering that the free energy functional is now being termed the Hamiltonian, we have from Eq.(13.2.1) with $n = 1$

$$
\begin{aligned}
H &= \int d^D x \left[\frac{m^2}{2}\phi^2 + (\nabla\phi)^2 + \frac{\lambda}{4!}\phi^4 \right] \\
&= \sum_k^{\Lambda} \left(\frac{m^2}{2} + \frac{k^2}{2} \right) \phi(\mathbf{k})\phi(-\mathbf{k}) \\
&+ \frac{\lambda}{4!} \sum_{\mathbf{p_1},\mathbf{p_2},\mathbf{p_3}} \phi(\mathbf{p_1})\phi(\mathbf{p_2})\phi(\mathbf{p_3})\phi(-\mathbf{p_1} - \mathbf{p_2} - \mathbf{p_3}). \quad (13.5.1)
\end{aligned}
$$

To implement the renormalization group process for the evaluation of the partition function

$$Z = \int D\phi\, e^{-H}$$

we need to

1. integrate over the Fourier components $\phi(k)$ which have $k = |\mathbf{k}|$ lying between Λ/b and Λ where b is a number greater than unity. This reduces the momentum cutoff to Λ/b.

2. rescale the momenta by b so that the cutoff is restored to the previous value Λ.

3. rescale the fields $\phi(x)$, so that the renormalization group correctly focusses on the problem at hand – e.g., gives the correct form for the two point correlation function at $T = T_c$.

We proceed by splitting each field $\phi(x)$ as

$$
\begin{aligned}
\phi(x) &= \frac{1}{(2\pi)^D} \int_0^{\Lambda/b} \phi(\mathbf{k}) e^{i\mathbf{k}\cdot\mathbf{x}} d^D k + \frac{1}{(2\pi)^D} \int_{\Lambda/b}^{\Lambda} \phi(\mathbf{k}) e^{i\mathbf{k}\cdot\mathbf{x}} d^D k \\
&= \frac{1}{(2\pi)^D} \int \phi_<(\mathbf{k}) e^{i\mathbf{k}\cdot\mathbf{x}} d^D k + \frac{1}{(2\pi)^D} \int \phi_>(\mathbf{k}) e^{i\mathbf{k}\cdot\mathbf{x}} d^D k \quad (13.5.2)
\end{aligned}
$$

where $\phi_<$ refers to the Fourier components with wavenumbers less than Λ/b, whereas $\phi_>$ refers to the components lying between Λ/b and Λ. Throughout our calculation, x, k and p_i are to be understood as D-dimensional vectors. The Hamiltonian can now be written as

$$
\begin{aligned}
H &= \frac{1}{2} \sum_{k<\Lambda/b} (m^2+k^2)\phi_<(\mathbf{k})\phi_<(-\mathbf{k}) + \frac{1}{2} \sum_{\Lambda/b<k<\Lambda} (m^2+k^2)\phi_>(\mathbf{k})\phi_>(-\mathbf{k}) \\
&+ \frac{\lambda}{4!} \sum \phi_<(\mathbf{p_1})\phi_<(\mathbf{p_2})\phi_<(\mathbf{p_3})\phi_<(-\mathbf{p_1}-\mathbf{p_2}-\mathbf{p_3}) \\
&+ 4\cdot\frac{\lambda}{4!} \sum \phi_<(\mathbf{p_1})\phi_<(\mathbf{p_2})\phi_<(\mathbf{p_3})\phi_>(-\mathbf{p_1}-\mathbf{p_2}-\mathbf{p_3}) \\
&+ 6\cdot\frac{\lambda}{4!} \sum \phi_<(\mathbf{p_1})\phi_<(\mathbf{p_2})\phi_>(\mathbf{p_3})\phi_>(-\mathbf{p_1}-\mathbf{p_2}-\mathbf{p_3}) \\
&+ 4\cdot\frac{\lambda}{4!} \sum \phi_>(\mathbf{p_1})\phi_>(\mathbf{p_2})\phi_>(\mathbf{p_3})\phi_<(-\mathbf{p_1}-\mathbf{p_2}-\mathbf{p_3}) \\
&+ \frac{\lambda}{4!} \sum \phi_>(\mathbf{p_1})\phi_>(\mathbf{p_2})\phi_>(\mathbf{p_3})\phi_>(-\mathbf{p_1}-\mathbf{p_2}-\mathbf{p_3}) \\
&= H^< + H_2^> + H_4^>
\end{aligned} \tag{13.5.3}
$$

where $H^<$ is that part of the Hamiltonian which involves only the fields $\phi_<$, $H_2^>$ is the quadratic part involving the fields $\phi_>$ and $H_4^>$ is the part of the Hamiltonian involving quartic terms at least one of which is $\phi_>$. We now need to evaluate

$$
\begin{aligned}
Z &= \int D\phi_< D\phi_> e^{-(H^< + H_2^> + H_4^>)} \\
&= \int D\phi_< e^{-H^<} \int D\phi_> e^{-(H_2^> + H_4^>)}.
\end{aligned} \tag{13.5.4}
$$

The evaluation of the functional integration over $\phi_>$ leads to a function of $\phi_<$ which we write as $e^{-\bar{H}}$. Noting that $H_4^>$ is $O(\lambda)$, we can expand the exponential in powers of $H_4^>$ (assuming λ is small) and obtain

$$
\begin{aligned}
e^{-\bar{H}} &= \int D\phi_> e^{-H_2^>} \left[1 - H_4^> + \frac{1}{2}H_4^{>2} + \dots \right] \\
&= \left(\int D\phi_> e^{-H_2^>} \right) \cdot \left[1 - \langle H_4^> \rangle + \frac{1}{2}\langle H_4^{>2}\rangle + \dots \right]
\end{aligned} \tag{13.5.5}
$$

where

$$
\langle \dots \rangle = \frac{\int D\phi_>(\dots)e^{-H_2^>}}{\int D\phi_> e^{-H_2^>}}. \tag{13.5.6}
$$

We can then express $\bar{H}$ as

$$
\begin{aligned}
\bar{H} &= -\ln \left\{ \left[\int D\phi_> e^{-H_2^>} \right] \cdot \left[1 - \langle H_4^> \rangle + \frac{1}{2}\langle H_4^{>2}\rangle + \dots \right] \right\} \\
&= \langle H_4^> \rangle - \frac{1}{2}\left\{ \langle H_4^{>2}\rangle - \langle H_4^> \rangle^2 \right\} + \text{constant} + O(\lambda^3).
\end{aligned} \tag{13.5.7}
$$

We want exactly those parts of $\bar{H}$ which are functions of $\phi_<$ and hence $\ln\left[\int D\phi_> e^{-H_2^>}\right]$, which is a constant, is not important for our purpose (note that it would be important for an exact calculation of the free energy). At $O(\lambda)$, we need to extract the $\phi_<$ dependent part of $\langle H_4^> \rangle$. This is easily seen to be [from Eq.(13.5.3)]

$$\langle H_4^> \rangle = 6 \cdot \frac{\lambda}{4!} \left\langle \sum \phi_<(\mathbf{p_1})\phi_<(\mathbf{p_2})\phi_>(\mathbf{p_3})\phi_>(-\mathbf{p_1}-\mathbf{p_2}-\mathbf{p_3}) \right\rangle$$

$$= 6 \cdot \frac{\lambda}{4!} \sum \phi_<(\mathbf{p_1})\phi_<(-\mathbf{p_1})\langle \phi_>(\mathbf{p_2})\phi_>(-\mathbf{p_2})\rangle$$

$$= 6 \cdot \frac{\lambda}{4!} \left[\int_{\Lambda/b}^{\Lambda} \frac{d^D p_2}{(2\pi)^D} \frac{1}{p_2^2 + m^2} \right] \sum \phi_<(\mathbf{p_1})\phi_<(-\mathbf{p_1}). \qquad (13.5.8)$$

This term then modifies (i.e., renormalizes) the quadratic part of the Hamiltonian H. The quartic part of H is unaffected at this order.

If we now turn to the term $\frac{1}{2}\left[\langle H_4^{>2}\rangle - \langle H_4^> \rangle^2\right] = \frac{1}{2}\langle H_4^{>2}\rangle_c$, we note that it produces contributions to the quadratic and quartic parts of H. Since we are interested in the lowest order calculations, we ignore at present the contributions to the quadratic part and focus on the quartic contribution instead. Thus, the quartic contribution to H from $\frac{1}{2}\langle H_4^{>2}\rangle_c$ is

$$\frac{1}{2}\left(6 \times \frac{\lambda}{4!}\right)^2 \cdot \left\langle \sum \phi_<(\mathbf{p_1})\phi_<(\mathbf{p_2})\phi_>(\mathbf{q_3})\phi_>(-\mathbf{p_1}-\mathbf{p_2}-\mathbf{p_3}) \right.$$

$$\times \left. \sum \phi_<(\mathbf{q_1})\phi_<(\mathbf{q_2})\phi_>(\mathbf{q_3})\phi_>(-\mathbf{q_1}-\mathbf{q_2}-\mathbf{q_3}) \right\rangle_c$$

$$= \frac{1}{2}\left(6 \times \frac{\lambda}{4!}\right)^2 \sum \phi_<(\mathbf{p_1})\phi_<(\mathbf{p_2})\phi_<(\mathbf{q_1})\phi_<(\mathbf{q_2})$$

$$\times \left\langle \sum \phi_>(\mathbf{p_3})\phi_>(-\mathbf{p_1}-\mathbf{p_2}-\mathbf{p_3})\phi_>(\mathbf{q_3})\phi_>(-\mathbf{q_3}-\mathbf{q_1}-\mathbf{q_2}) \right\rangle_c .$$

$$\qquad (13.5.9)$$

The above step shows the first complication in the renormalization group procedure. The "renormalized" coupling constants produced in Eq.(13.5.9) is momentum dependent. The coefficient of $\phi_<(\mathbf{p_1})\phi_<(\mathbf{p_2})\phi_<(\mathbf{q_1})\phi_<(\mathbf{q_2})$ in Eq.(13.5.9) is a function of $\mathbf{p_1}$, $\mathbf{q_1}$, $\mathbf{p_2}$, $\mathbf{q_2}$ with the constraint $\mathbf{p_1} + \mathbf{q_1} + \mathbf{p_2} + \mathbf{q_2} = 0$.

We now expand the momentum-dependent coupling constants about the zero momentum value and recognize the fact that the renormalization group produces entirely new couplings of the form $p^2\phi^4, \ldots$ etc. These would be irrelevant in the long wavelength limit. Consequently the right hand side of

Eq.(13.5.9)

$$
\frac{1}{2}\left(6\times\frac{\lambda}{4!}\right)^2\cdot2\sum\phi_<(\mathbf{p_1})\phi_<(\mathbf{p_2})\phi_<(\mathbf{q_1})\phi_<(\mathbf{q_2})
$$

$$
\times\ \int\frac{d^Dp_3}{(2\pi)^D}\frac{1}{p_3^2+m^2}\int\frac{d^Dq_3}{(2\pi)^D}\frac{1}{q_3^2+m^2}\cdot(2\pi)^D\delta^D(\mathbf{p_3}+\mathbf{q_3}).
$$

Thus the relevant contribution to $\bar{H}$ to $O(\lambda^2)$ is

$$
\begin{aligned}
\bar{H} &= \left(\frac{6\lambda}{4!}\right)\int\frac{d^Dp}{(2\pi)^D}\frac{1}{p^2+m^2}\sum\phi_<(\mathbf{p_1})\phi_<(-\mathbf{p_1}) \\
&\quad - \left(\frac{6\lambda}{4!}\right)^2\int\frac{d^Dp}{(2\pi)^D}\left(\frac{1}{p^2+m^2}\right)^2\sum\phi_<\phi_<\phi_<\phi_<. \quad (13.5.10)
\end{aligned}
$$

And $H^< + \bar{H}$ becomes

$$
\begin{aligned}
H^< + \bar{H} &= \sum\frac{\lambda}{4!}\left(\frac{m^2}{2}+\frac{k^2}{2}+\frac{6\lambda}{4!}\int\frac{d^Dp}{(2\pi)^D}\frac{1}{p^2+m^2}\right)\phi_<\phi_< \\
&\quad + \sum\left[\frac{\lambda}{4!}-\left\{\left(\frac{6\lambda}{4!}\right)^2\int\frac{d^Dp}{(2\pi)^D}\left(\frac{1}{p^2+m^2}\right)^2\right\}\right. \\
&\quad \times\ \left.\phi_<\phi_<\phi_<\phi_<\right].
\end{aligned}
$$

$$(13.5.11)$$

We now rescale k to $k' = bk$, so that the cutoff on k' is restored to Λ. Simultaneously the fields $\phi(x)$ are scaled by ζ which scales $\phi(k)$ by b^D/ζ. Having scaled distances and fields, we find the renormalized Hamiltonian H_R as

$$
\begin{aligned}
H_R &= \sum\frac{b^{2D}}{\zeta^2}\left(\frac{m^2}{2}+\frac{k'^2/b^2}{2}+\frac{6\lambda}{4!}\int\frac{d^Dp}{(2\pi)^D}\frac{1}{p^2+m^2}\right)\phi_<(\mathbf{k'})\phi_<(-\mathbf{k'}) \\
&\quad + \sum\frac{b^{4D}}{\zeta^4}\left\{\frac{\lambda}{4!}-\left(\frac{6\lambda}{4!}\right)^2\int\frac{d^Dp}{(2\pi)^D}\left[\frac{1}{p^2+m^2}\right]^2\right\}\phi_<\phi_<\phi_<\phi_< \\
&= \sum\left(\frac{\bar{m}^2}{2}+\frac{ak'^2}{2}\right)\phi(\mathbf{k'})\phi(-\mathbf{k'}) \\
&\quad + \sum\frac{\bar{\lambda}}{4!}\cdot\phi(\mathbf{k_1'})\phi(\mathbf{k_2'})\phi(\mathbf{k_3'})\phi(-\mathbf{k_1'}-\mathbf{k_2'}-\mathbf{k_3'})
\end{aligned}
$$

$$(13.5.12)$$

where

$$
\bar{m}^2 = \frac{b^D}{\zeta^2}\left(m^2+\frac{12\lambda}{4!}\int\frac{d^Dp}{(2\pi)^D}\frac{1}{p^2+m^2}\right) \quad (13.5.13)
$$

$$
a = \frac{b^D}{\zeta^2}(1+O(u^2)) \quad (13.5.14)
$$

$$
\bar{\lambda} = \frac{b^D\cdot\lambda}{\zeta^4}\left[1-\frac{36\lambda}{4!}\int\frac{d^Dp}{(2\pi)^D}\left(\frac{1}{p^2+m^2}\right)^2\right]. \quad (13.5.15)
$$

The appropriate renormalization group requires that $a = 1$, which forces $\zeta^2 = b^{D-2}$ (we are free to choose an overall scale factor and this is done by requiring $a = 1$). The recursion relation for the coupling constant λ now becomes

$$\bar{\lambda} = b^{4-D}\lambda\left[1 - \frac{3\lambda}{2}\int\frac{d^Dp}{(2\pi)^D}\left(\frac{1}{p^2+m^2}\right)^2\right].$$

(13.5.16)

To the lowest order in λ, we see $\bar{\lambda} = b^\epsilon\lambda$ and $\bar{m}^2 = b^2m^2$ reflecting the canonical dimensions of λ and m^2. The renormalization group to $O(\epsilon)$ can now be set up by expanding Eq.(13.5.16) as

$$
\begin{aligned}
\bar{\lambda} &= \lambda(1+\epsilon\ln b)\left[1 - \frac{3}{2}\lambda\int\frac{d^Dp}{(2\pi)^D}\left(\frac{1}{p^2+m^2}\right)^2\right] \\
&= \lambda\left[1+\epsilon\ln b - \frac{3}{2}\lambda\int\frac{d^Dp}{(2\pi)^D}\left(\frac{1}{p^2+m^2}\right)^2\right]
\end{aligned}
$$

(13.5.17)

where the last line follows from the expectation that the non-zero fixed point in λ is going to be of $O(\epsilon)$. In the critical region m^2 is small and thus evaluating the integral in $D = 4$ at the leading order we have

$$\int\frac{d^Dp}{(2\pi)^D}\left(\frac{1}{p^2+m^2}\right)^2 \simeq \frac{2\pi^2}{(2\pi)^4}\int_{\Lambda/b}^{\Lambda}\frac{p^3dp}{p^4} = \frac{1}{8\pi^2}\ln b$$

leading to

$$\bar{\lambda} = \lambda + \lambda\epsilon\ln b - \frac{3}{2}\frac{\lambda^2}{8\pi^2}\ln b.$$

(13.5.18)

Turning to Eq.(13.5.13), we can evaluate (again at $D = 4$)

$$
\begin{aligned}
\int\frac{d^Dp}{(2\pi)^D}\left(\frac{1}{p^2+m^2}\right) &= \frac{1}{8\pi^2}\int_{\Lambda/b}^{\Lambda}\frac{p^3dp}{p^2+m^2} \\
&= \frac{1}{16\pi^2}\left[\Lambda^2 - \frac{\Lambda^2}{b^2}\right] - \frac{m^2}{8\pi^2}\ln b
\end{aligned}
$$

and write

$$\bar{m}^2 = b^2\left\{m^2 + \frac{\lambda}{2}\left[\frac{\Lambda^2}{16\pi^2}\left(1-\frac{1}{b^2}\right) - \frac{m^2}{8\pi^2}\ln b\right]\right\}.$$

(13.5.19)

It helps to consider b close to unity so that the recursion relations acquire a differential character. We put $b = e^{\delta l} \cong 1 + \delta l$ and to $O(\delta l)$ obtain

$$
\begin{aligned}
\bar{m}^2 &= (1-2\delta l)\left\{m^2 + \frac{\lambda}{2}\left[\frac{\Lambda^2}{16\pi^2}2\delta l - \frac{m^2}{8\pi^2}\delta l\right]\right\} \\
&= m^2 + \delta l\left[2m^2 - \frac{2m^2}{16\pi^2} - \frac{\lambda\Lambda^2}{16\pi^2}\right]
\end{aligned}
$$

or

$$\frac{dm^2}{dl} = \left(2 - \frac{\lambda}{16\pi^2}\right) m^2 - \frac{\lambda\Lambda^2}{16\pi^2} \tag{13.5.20}$$

while

$$\bar{\lambda} = \lambda + \left(\lambda\epsilon - \frac{3}{2}\frac{\lambda^2}{8\pi^2}\right)\delta l$$

$$\frac{d\lambda}{dl} = \lambda\left(\epsilon - \frac{3}{2}\frac{\lambda}{8\pi^2}\right). \tag{13.5.21}$$

Eqs.(13.5.20) and (13.5.21) constitute the renormalization group flow equations to the lowest order.

As explained in the previous section, once the flow has been obtained, one looks for the fixed points and the eigenvalues of the linearized transformation about the fixed point. The fixed points and the associated eigenvalues are:

1. $\lambda = 0$, $m^2 = 0$ – the Gaussian fixed point with eigenvalues b^2 and b^ϵ (note that the eigenvalues refer to the flow equations in the form $\bar{m}^2 = f(m^2, \lambda)$ and $\bar{\lambda} = g(m^2, \lambda)$ .... Clearly for $\epsilon < 0$ (i.e., $D > 4$), one eigenvalue is smaller than unity (the "λ eigenvalue") and hence the fixed point is relevant for describing critical behaviour (one eigenvalue greater than unity and the others smaller). In dimensions greater than four, the quartic term is not involved in the determination of critical behaviour. The Gaussian model results are consequently exact. For $\epsilon > 0$ the Gaussian fixed point has two eigenvalues greater than unity and hence not useful.

2. $\lambda = \frac{16\pi^2}{3}\epsilon$, $m^2 = \frac{\Lambda^2\epsilon/3}{2-\epsilon/3}$ the nontrivial fixed point with eigenvalues $b^{2-\epsilon/3}$ and $b^{-\epsilon}$. For $\epsilon > 0$ (i.e., $D < 4$), the eigenvalue in the λ-direction is now smaller than unity while the other eigenvalue is greater than unity. Consequently, it is this fixed point which now describes the critical behaviour. It is unstable in one direction (corresponding to the temperature like variable) and stable in the λ-direction.

The "relevant" or "thermal" eigenvalue, $b^{2-\epsilon/3}$, determines the exponent ν as [see Eq.(13.3.21)]

$$\nu = \frac{1}{2 - \epsilon/3} = \frac{1}{2} + \frac{\epsilon}{12} + O(\epsilon^2). \tag{13.5.22}$$

At this order the anomalous dimensions exponent $\eta = 0$ and thus

$$\gamma = 2\nu = 1 + \frac{\epsilon}{6} + O(\epsilon^2)$$

Hyperscaling leads to the specific heat index α as

$$\alpha = 2 - D\left(\frac{1}{2} + \frac{\epsilon}{12}\right) = \frac{\epsilon}{6} + O(\epsilon^2).$$

The exponent β follows from the scaling relation $\alpha + 2\beta + \gamma = 2$ as

$$2\beta = 2 - 1 - \frac{\epsilon}{6} - \frac{\epsilon}{6} = 1 - \frac{1}{3}\epsilon$$

or

$$\beta = \frac{1}{2} - \frac{\epsilon}{6} + O(\epsilon^2).$$

It should be noted that η can also be obtained by working to $O(\lambda^2)$. This involves the extraction of the terms quadratic in $\phi_< \phi_<$ from the $O(\lambda^2)$ part in Eq.(13.5.9). This makes the structure of Eq.(13.5.14)

$$a = \frac{b^{D-2}}{\zeta^2}\left[1 + 48\left(\frac{\lambda}{4!}\right)^2 \frac{3}{2}\frac{1}{2}\ln b + \ldots\right]. \qquad (13.5.23)$$

The requirement of setting $a = 1$ now establishes η when we substitute the fixed point value of λ obtained from Eq.(13.5.21). The reader should go through the straightforward algebra of obtaining Eq.(13.5.23) and verify that at this order the anomalous dimension η works out to be $\epsilon^2/54$.

14

Disordered Systems

14.1 Introduction

In our study of critical phenomena so far we have been interested only in the inherent thermal fluctuations in a system. The ordering or disordering seen in Ising models or $X - Y$ or Heisenberg models has been due to the competition between aligning forces of atomic origin and the thermal fluctuations. In this chapter, we explore the presence of other fluctuations. In terms of an Ising model these fluctuations could manifest themselves in some sites on the under-lying lattice being empty at random or the interaction between all neighbouring sites having a random variation or in the existence of a random external field. The essential question will be whether the transition to an ordered state survives in the presence of these random perturbations and if it does what are the new critical exponents and scaling laws. The two terms to be explained before we start our discussion are the ones commonly used in dealing with disorders: quenched disorder and annealed disorder.

Quenched disorder is one where the disordering variables are fixed rigidly in space over the period of observation time. As an example if we consider a "dilute" magnet-a magnet where some of the magnetic atoms are replaced by non-magnetic ones at randomly chosen sites-then at low temperatures the mobility of the atoms is small and the set of sites at which the non-magnetic atoms sit is fixed in time. Now consider the Hamiltonian for such a system. It can be written as

$$H = -J \sum_{\langle ij \rangle} p_i p_j S_i . S_j \qquad (14.1.1)$$

where S_i is the n-component spin vector of unit magnitude ($n = 1$, Ising; $n = 2$, $X - Y$; $n = 3$, Heisenberg), J is the exchange constant, $\langle ij \rangle$ denotes sum over nearest neighbours and p_i is the random variable which takes the value unity if the site i is occupied by a magnetic atom and zero if it is occupied by a non-magnetic one. The partition function is

$$Z(\{p_i\}) = \text{Tr}_{\{S_i\}} e^{-\beta H} \tag{14.1.2}$$

and the free energy is given by

$$F(\{p_i\}) = -\ln Z(\{p_i\})/\beta L^D. \tag{14.1.3}$$

The free energy is dependent on the set $\{p_i\}$ and the correlation functions likewise would be dependent on the set $\{p_i\}$. Thus, there would be no translational invariance and calculations would be extremely difficult.

At this point one simplifies things by trying to imagine what a real free energy would depend upon. For a sufficiently large sample it seems reasonable that the free energy would be determined by the overall impurity concentration p rather than by the set $\{p_i\}$. This implies that the physical free energy will be the one averaged over an ensemble where each member of the ensemble has a different set of sites $\{i\}$ occupied by the non-magnetic impurities. If the distribution $P(\{p_i\})$ for the set $\{p_i\}$ is known, then the observed free energy would be

$$F(P) = \sum_{\{p_i\}} P(\{p_i\}) F(\{p_i\}) \tag{14.1.4}$$

(this ensemble averaging can be visualized by breaking up the physical dimension L into smaller but still macroscopic sizes L_j. In each L_j a set of sites are occupied and the number of subsystems of size L_j is L/L_j so that if we take the limits $L_j \to \infty$ and $L/L_j \to \infty$ the above averaging will be accomplished). Thus for the quenched system the relevant free energy is the impurity averaged free energy. Hence, the prescription for dealing with quenched systems is to average in Z over the impurity distribution. For the above system the $P(\{p_i\})$ takes the simple form

$$P(\{p_i\}) = \prod_i (p\delta_{p_i,1} + (1-p)\delta_{p_i,0}). \tag{14.1.5}$$

To understand what annealed systems are, we need to return to the above system and consider the diffusivity of the atoms to be high so that the non-magnetic ions could diffuse from one site to another over time scales shorter than that of the experiment. Now, stretching one's imagination to say that the Hamiltonian is still given by Eq.(14.1.1), we would obtain for the partition function

$$Z = \text{Tr}_{\{p_i\}} P(\{p_i\}) Z(\{p_i\}) = Tr_{(\{p_i\},\{S_i\})} e^{-\beta H} P(\{p_i\}) \tag{14.1.6}$$

since the averages are measured not only over the possible spin configurations, but also the positions of the non-magnetic ones. The free energy is obtained by simply taking the logarithm. Translational invariance is maintained in the Z of Eq.(14.1.6) and things are relatively straightforward. Henceforth, we will deal with quenched systems only.

14.2 Models with Quenched Disorder

We will consider the model described by the Hamiltonian

$$H = -\sum_{\langle ij \rangle} J_{ij} \vec{S}_i . \vec{S}_j \qquad (14.2.1)$$

where the J_{ij} are taken to be random variables. In this manner the above model includes both the random interaction strength as well as the random site occupation case, when we consider $J_{ij} = J p_i p_j$. To simplify the situation, we consider ferromagnetic models, which is to say that the probability of J_{ij} becoming negative will be taken to be small. The probability distribution of J_{ij} will have to specified and we shall assume that the J_{ij}'s are uncorrelated. Thus the distribution will be of the form

$$P(\{J_{ij}\}) = \prod_{\langle ij \rangle} p(J_{ij}) \qquad (14.2.2)$$

with the mean value of J_{ij} greater than zero.

This sort of disorder is also called homogeneous. We try to retain the ferromagnetic behaviour by requiring the deviation of J_{ij} to be much smaller than its mean.

To see if this model will have a sharp second order transition, we follow an argument due to Harris. Different parts of the sample will, owing to statistical fluctuations, be characterized by different critical temperatures and hence there will be a distribution $p(T_c)$ of the critical temperatures across the sample. The number of parts of the system can be taken to be V/ξ^D, where V is the total sample volume and ξ, the correlation length. We can now invoke the central limit theorem to assert that the fluctuation δT_c would be proportional to the square root of the number of individual parts and hence

$$\delta T_c \sim \xi^{-D/2}. \qquad (14.2.3)$$

For the correlation length to be ξ with a sharp transition temperature T_c, there must be the relation $\xi \propto (T - T_c)^{-\nu} = (\Delta T)^{-\nu}$ or $\Delta T \propto \xi^{-1/\nu}$. If we are to have a sharp transition, then we must have

$$\delta T_c \ll \Delta T \qquad (14.2.4)$$

which implies

$$\xi^{-D/2} \ll \xi^{-1/\nu} \qquad (14.2.5)$$

leading to

$$\frac{D}{2} > \frac{1}{\nu}$$
$$D\nu - 2 > 0$$
$$\alpha < 0 \qquad (14.2.6)$$

where in the last line we have used the hyperscaling relation $D\nu = 2 - \alpha$. Thus the Harris criterion is that for $\alpha < 0$, there will be a sharp transition to the ferromagnetic phase in the random system. The initial exponents should remain the same as that of the pure system. For $\alpha > 0$ on the other hand, the slightest impurity will change the transition in some way. To find out the quantitative answer, we need to set up a renormalization group argument and that requires a trick special to all problems involving randomness.

We first change from a model on a lattice to a Ginzburg-Landau model. This is done by noting that the characteristic of the randomness is to produce local critical temperatures and hence the coefficient 'a' of the quadratic term which depends on the critical temperature will acquire a spatial dependence and consequently the free energy functional will be ($\phi(x)$ is a vector field)

$$F = \int d^D x \left[\frac{1}{2} a(x) \vec{\phi}(x)^2 + \frac{1}{2} (\partial_\alpha \phi_\beta)^2 + \frac{u}{4} (\vec{\phi}^2(x))^2 \right]. \tag{14.2.7}$$

The distribution of the $a(x)$ has to be provided and if the disorder is uncorrelated, we must have

$$P\{a(x)\} = \prod_x P(a(x)) \tag{14.2.8}$$

and a Gaussian choice of $P(a(x))$ would give

$$P(a(x) \propto e^{-(a-a_0)^2/2\Delta} \tag{14.2.9}$$

implying a mean value a_0 for the mass parameter and a fluctuation strength of Δ From Eq.(14.2.9), it should be clear that

$$\langle (a(x) - a_0)(a(0) - a_0) \rangle_{av} = \Delta \delta^D(x) \tag{14.2.10}$$

where $\langle \ldots \rangle_{av}$ denotes an averaging over the impurity distribution.

We shall now introduce the trick which will make it possible to address the problem of evaluating the partition function with the Ginzburg-Landau free energy of Eq.(14.2.7), with the supplementary condition that the averaging over the probability distribution has to be done over the free energy and not the partition function. Thus,

$$\beta F = -\langle \ln\, Z \rangle_{av} = - \int\limits_{-\infty}^{\infty} Da(x) P(\{a(x)\}) \ln\, Z(\{a\}). \tag{14.2.11}$$

Using the identity $\ln\, Z = \lim_{n\to 0} \frac{Z^n - 1}{n}$, we can easily see that

$$\beta F = - \lim_{n\to 0} \frac{1}{n} [\langle Z^n \rangle_{av} - 1]. \tag{14.2.12}$$

Now

$$Z^n = \int D\phi^{(\alpha)}(x)e^{-\beta F_n} \tag{14.2.13}$$

where

$$F_n = \int d^D x \sum_{\alpha=1}^{n} \left[\frac{1}{2}a(x)\left\{\vec{\phi}^{(\alpha)}(x)\right\}^2 + \frac{1}{2}\sum_i \left\{\vec{\nabla}\phi_i^{(\alpha)}(x)\right\}^2 + \frac{u}{4}\left\{\vec{\phi}^{(\alpha)}(x)^2\right\}^2 \right] \tag{14.2.14}$$

Averaging over $a(x)$,

$$\begin{aligned}
\langle Z^n \rangle_{av} &= \int D\phi^{(\alpha)}(x) \int Da(x) P[\{a(x)\}]e^{-\beta F_n} \\
&= \int D\phi^{(\alpha)}(x)e^{-\beta F_n^{\text{eff}}} \tag{14.2.15}
\end{aligned}$$

with

$$\begin{aligned}
F_n^{\text{eff}} &= \int d^D x \left[\frac{a_0}{2}\sum_{\alpha=1}^{n}\left\{\vec{\phi}^{(\alpha)}(x)\right\}^2 + \frac{1}{2}\sum_{\alpha=1}^{n}\left\{\nabla\vec{\phi}^{(\alpha)}(x)\right\}^2 \right. \\
&\quad \left. + \frac{u}{4}\sum_{\alpha=1}^{n}\left\{\left(\vec{\phi}^{(\alpha)}(x)\right)^2\right\}^2 - \Delta \sum_{\alpha,\beta=1}^{n}\left\{\vec{\phi}^{(\alpha)}(x)\right\}^2\left\{\vec{\phi}^{(\beta)}(x)\right\}^2 \right]. \tag{14.2.16}
\end{aligned}$$

The evaluation of $\langle Z^n \rangle$ involves the translationally invariant free energy functional of Eq.(14.2.16) and hence the renormalization group techniques of Chapter 13 can be taken over directly. The only difference is that instead of the two parameters that we had there (a and u), we have three parameters now a_0, u and Δ. Accordingly one can go ahead to find the recursion relations for these three variables. The recursion relations will involve n and in the end the limit $n \to 0$ will have to be taken. This trick of converting the quenched problem into one where an effective translationally invariant Hamiltonian is involved is called the 'replica trick'.

At this point we will not go through the details of the calculation of the recursion relations. We merely discuss the outcome of the calculation. There is a fixed point with $\Delta^r=0$, which corresponds to the pure system. On analyzing the stability of this fixed point, it is found that it is stable if $\alpha < 0$ thus establishing the Harris criterion that if $\alpha < 0$ the behaviour of the system with random impurities is identical to that of the pure system. For $\alpha > 0$ a different fixed point is stable. This implies that there is a sharp transition in this case as well, but the critical exponents are changed from the values corresponding to the pure system. In particular, the new α is negative and thus the Harris criterion for a sharp transition is obeyed. It has to be true that if the deviation Δ is large enough, then the sharp transition to the ferromagnetic state will be lost, but it is not clear from the renormalization group flows as to what qualifies as a sufficiently large Δ. What is true is that for small Δ there is a stable fixed point and a sharp transition and as long as we have a ferromagnetically ordered state at low temperatures we can call the randomness weak.

The next disordered magnetic system that we want to look at is one which is amorphous. This implies there is no global "easy" direction along which the spins would like to align. We can have local easy axes whose orientation varies from point to point. We consider the orientation to be a random function of position and assume that this captures the essence of amorphousness. The model Hamiltonian for an amorphous magnet is thus

$$H = -J \sum S_i.S_j - A \sum_i (\hat{n}_i.S_i)^2 \qquad (14.2.17)$$

where J and A are positive constants and $\hat{n}_i$ is the unit vector at site i and its orientation varies randomly from site to site. The term $(\hat{n}_i.S_i)^2$ maintains the global $S \rightarrow -S$ symmetry of the microscopic interactions. The model of Eq.(14.2.17) has competition built into it. The first term requires the spins to align parallel to each other, while the second term favours the i–th spin to align in the direction $\hat{n}_i$ or $-\hat{n}_i,$. We note in passing that the above model has no meaning if the spins are Ising-like.

The issue that we want to address is whether the above model can show an ordered state. The argument proceeds by assuming that an ordered state exists and we create a domain in which same spins are pointing in a different direction (say at right angles to the original direction). If this new state has a lower energy, then the ordered state cannot exist. The first term of Eq.(14.2.17), as we have seen before, implies that it costs energy to turn the spins and if the domain size is L, then the energy it costs is

$$\Delta E_J \sim JL^{D-2}. \qquad (14.2.18)$$

The second term of course favours the turning of spins. This is because if all spins are pointing in the same direction, a large number of terms $\hat{n}_i.S_i$ will be zero or close to it and by turning some of the spins in a different direction a non-zero contribution to $\hat{n}_i.S_i$ will be achieved and a consequent lowering of the energy. To estimate how big the effect is, we merely note that the energy lowering will be proportional to the excess number of random axes pointing in this new direction and since this number comes from statistical fluctuations, we expect it to be of the order of the square root of the number of sites and thus

$$\Delta E_D \sim -ASL^{D/2}. \qquad (14.2.19)$$

The total energy change is

$$\Delta E = \Delta E_J + \Delta E_D \sim JL^{D-2} - AL^{D/2}. \qquad (14.2.20)$$

It should be noted that we do not need to consider thermal fluctuations we are assuming that there is an ordered state in the presence of thermal fluctuations and are testing its stability against the disorder. Hence these calculations refer to $T = 0$ and the thermodynamic stability can be inferred from an energy argument. From Eq.(14.2.20) it is clear that the energy lowering term dominates if

$$D - 2 < \frac{D}{2} \quad \text{i.e.,} \quad D < 4 \tag{14.2.21}$$

Hence in dimensions below four, there can be no ordered state in the amorphous magnet.

The renormalization group argument proceeds by starting with the free energy functional

$$F = \int d^D x \left[\frac{a}{2} \vec{\phi}^2 + \frac{1}{2} \sum_i (\phi_i)^2 + \frac{u}{4} (\vec{\phi}^2)^2 - A(\hat{n}.\vec{\phi})^2 \right] \tag{14.2.22}$$

where the unit vector $\hat{n}(x)$ is assumed to have an isotropic distribution. The replica trick is now carried out and implementing the ϵ-expansion, it is found that there are no stable fixed points for $D < 4$.

The last model to be considered in this section involves a random external field. We imagine a field h_i at site i and that it varies randomly from site to site. The Hamiltonian for this random field problem has the form

$$H = -J \sum_{\langle ij \rangle} S_i.S_j - \sum_i h_i.S_i. \tag{14.2.23}$$

Clearly the model is defined for Ising spins as well. As before, our first consideration will be to see if there can be a long-range order in this model. Assuming that we have an ordered phase at $T = 0$, we consider turning over some of the spins, namely those in a domain of linear dimension L. The exchange term in Eq.(14.2.23) will oppose this spin flip and it will cost an energy of an amount JL^{D-2} if we are talking about $X-Y$ or Heisenberg models and JL^{D-1} if we are discussing Ising models. The spin flip is favoured by the random magnetic field because of the number of excess sites where the field points in the new spin direction.

Because this excess is due to statistical fluctuations, the number of excess sites is of $O(L^{D/2})$. Hence the energy gained by spin flips is $HL^{D/2}$. For $X-Y$ and Heisenberg models the energy change due to the formation of the spin flipped domain is (H is some constant)

$$\Delta E \sim JL^{D-2} - HL^{D/2} \tag{14.2.24}$$

leading to the favouring of disordering for $D < 4$. For Ising systems, on the other hand, the change in energy is

$$\Delta E \sim JL^{D-1} - HL^{D/2} \tag{14.2.25}$$

and hence the disordering is favoured for $D < 2$. This lower critical dimension for the random field Ising model turns out to be quite interesting. If one uses the renormalization group to confirm the above result for the Ising model in the ϵ-expansion, one comes up with a lower critical dimension of three. The perturbative renormalization group must clearly be in error since the correctness of the lower critical dimension of two obtained from the above heuristic arguments has been rigorously established since. The reason for this failure of the perturbative expansions has been the subject of a number of investigations.

14.3   Electron Localization in Disordered Systems

In this last Section we discuss briefly a problem that has been of considerable interest: the motion of electrons in a random medium. A random medium consists of static distribution of scattering or binding centres placed at random in a D-dimensional space. We will ignore the spin of the electron and all interactions among the electrons themselves. The last assumption will allow us to concentrate on the motion of a single electron in a random medium. The classical version of this model is the classical Lorentz model where one considers the motion of a classical particle through a random array of fixed scatterers (hard spheres in $D = 3$, hard discs in $D = 2$). If the scatterers are placed completely at random without regard to the position of the scatterers, we have the overlapping version of the model and if there is interaction among the scatterers and consequently a constraint on the positions of the scatterers, we have the non-overlapping model. Yet another variant is the wind tree model in $D = 2$, where the scatterers (trees) are placed randomly with their diagonals in the x and y directions. The wind is allowed to travel only in the x and y directions, changing directions if it collides with a tree.

In the classical models, the quantity of interest is the diffusion coefficient D, which is related to the velocity correlation function by the Green-Kubo formula

$$D = \int\limits_0^\infty C(t)dt = \int\limits_0^\infty \langle \nu_x(t)\nu_x(0)\rangle dt \qquad (14.3.1)$$

where v_x is the x-component of the velocity and the average denoted by angular brackets is an average over all possible configurations of the scatterers. For the quantum problem it is more usual to talk about the conductivity σ which measures the amount of current produced in response to a small applied electric field by the relation $J_x = \sigma E_x$. The conductivity σ can be related to the diffusion coefficient D. For the classical Lorentz model, the most interesting result is that the correlation function decays slowly (power law) at large times and the behaviour is $C(t) \sim t^{-(1+D/2)}$.

In the quantum model, the quantum mechanical result is well known if the medium is perfectly ordered: a perfect crystal. The electron eigenstates are Bloch functions and Bloch's theorem implies that the conductivity is infinite for a perfect crystal. The eigenstates themselves are extended over the crystal instead of being localized about any particular point in the solid. The energy levels of the system form bands with characteristic gaps between bands. The fact that the conductivity of a solid is finite is due to the lattice vibrations which disturb the perfect crystal structure and provide the mechanism (phonon-electron scattering) for the finite conductivity. In this Section our concern is with static disorder and we consider circumstances where the atoms of the solid do not form a crystalline solid. The basic results on the electron motion in a disordered system were obtained by Anderson who considered a tight-binding model where atoms are placed at the regular sites of a lattice, but the atoms differ from each other in a random way. For example, the binding energy of an electron at the various sites may be a random variable with some distribution. In addition, an electron can hop from one site to another. The parameters of importance in the Anderson model are W the width of the binding energy distribution and V the measure of the matrix element of the part of the Hamiltonian responsible for hopping. Anderson found that, in $D = 3$, if W/V is large enough then the eigenstates for an electron are all localized.

Localized states fall away exponentially from some central point. To have a non-zero value of the conductivity, the electron's wave packet must have some contribution from extended states since a wave packet constructed entirely from localized states cannot travel across the solid under the influence of a weak electric field. For small values of W/V, however, there are both extended states and localized states separated by a mobility edge. In $D = 1$, however, the smallest non-zero value of W/V causes all states to be localized and it is more than likely that the same situation holds in $D = 2$. Hence the transition from localized to conducting states can occur only in $D = 3$.

The calculational tricks that one employs for such problems are usually some special techniques that generally work in $D = 1$, the scaling theory which is presumably quite accurate for $D > 2$ but whose conclusion for $D = 2$ may just have been built into the theory and finally the evaluation of velocity correlation functions which lead to evaluation of the frequency-dependent diffusivity $D(\omega)$ and conductivity $\sigma(\omega)$. If there is complete localization, then the mean square displacement $\langle \Delta x(t)^2 \rangle$ must remain bounded and hence $\sigma(\omega)$ must approach zero at least as fast as ω as $\omega \to 0$.

We begin with a disordered one-dimensional solid where it is universally agreed that all states are localized and that the conductivity vanishes. We will demonstrate first that the electrical conductance $G(L) = \sigma/L$, where σ is the electrical conductivity of a material of length L, is given by the formula

$$G(L) = e^2 T / 2\pi h R \qquad (14.3.2)$$

where T and R are the transmission and reflection coefficients of a spinless electron of energy E across the sample. In the above formulae e is the electronic charge and h is Planck constant. Instead of the electrical conductivity, we will consider the diffusion of an electron across the sample in response to a density gradient. We imagine a steady state maintained by having a wave incident from the left with probability P_L and from the right with probability P_R. If the transmission probability is T, then the probability of the wave incident on the left ending up at the right is TP_L and that for the one incident from the right ending up at the left is TP_R. The reflection probability R means there will be a fraction PL_R on the left and a fraction $P_R R$ on the right due to reflection. The total densities of particles on the left and right are then

$$
\begin{aligned}
n_L &= P_L + TP_R + RP_L \\
n_R &= P_R + TP_L + RP_R.
\end{aligned}
\tag{14.3.3}
$$

If v is the velocity of the particles in free space then the current moving into the sample from the left is

$$
j_L = v(P_L - TP_R - RP_L)
\tag{14.3.4}
$$

and that moving in from the right is

$$
j_R = v(P_R - TP_L - RP_R)
\tag{14.3.5}
$$

Using $R + T = 1$, we find from Eq.(14.3.3)

$$
n_L - n_R = (P_L - P_R)2R
\tag{14.3.6}
$$

and from Eqs.(14.3.4) and (14.3.5),

$$
j_R = vT(P_R - P_L) = -j_L.
\tag{14.3.7}
$$

Thus the current from the left is

$$
j_L = \frac{vT}{2R}(n_L - n_R) = -\frac{vT}{2R}.L\frac{\partial n}{\partial x}
\tag{14.3.8}
$$

and remembering that $j = -D\frac{\partial n}{\partial x}$, where D is the diffusion coefficient, we have

$$
D = \frac{vT}{2R}.L
\tag{14.3.9}
$$

The connection between the electrical conductivity and the diffusivity is given by the relation

$$
\sigma = e^2 N(E)D(E)
\tag{14.3.10}
$$

where $N(E)$ is the density of states per unit length. For a one-dimensional system

$$
N(E) = \frac{1}{\pi h}\frac{dp}{dE} = \frac{1}{\pi hv}.
\tag{14.3.11}
$$

Here p is the momentum of the particles of energy E and $v = dE/dp$ group velocity. Using Eq.(14.3.9) in Eq.(14.3.10), we get back Eq.(14.3.2).

We now present arguments about why σ should vanish for a random medium. The medium is a set of scatterers (each scatterer is a group of atomic scatterers) placed at random separations along a line. The phase of the electron's wavefunctions at any scatterer is the random variable. Consider two adjacent scatterers with transmission coefficients t_1 and t_2 and reflection coefficients r_1 and r_2. Using straightforward quantum mechanics, it is easy to show that the transmission coefficient for the two scatterers together is

$$t = \frac{t_1 t_2}{1 - r_1 r_2} \tag{14.3.12}$$

The total conductance of the two scatterers is then

$$G = \frac{e^2}{2\pi\hbar} \frac{|t|^2}{1 - |t|^2} = \frac{e^2}{2\pi\hbar} \frac{|t_1 t_2|^2}{1 + |r_1 r_2|^2 - |t_1 t_2|^2 - (r_1 r_2 + \bar{r}_1 \bar{r}_2)}, \tag{14.3.13}$$

where bars denote complex conjugates. To use this formula as a basis for a theory of localization, we consider a dimensionless resistance ρ defined as

$$\rho = e^2 (2\pi\hbar G)^{-1} = R/T. \tag{14.3.14}$$

The total resistance of the two scatterers in series is then given by Eq.(14.3.13) as

$$\rho = \frac{1 + |r_1|^2 |r_2|^2}{|t_1|^2 |t_2|^2} - 1 - \frac{r_1 r_2 + \bar{r}_1 \bar{r}_2}{|t_1|^2 |t_2|^2}. \tag{14.3.15}$$

At this point, we need to average over the randomness, with the random interval between the scatterers and the random interval between the atomic scatterers in each of the two scatterers considered. Since we assume that there is no correlation between the two scatterers, the last term in Eq.(14.3.15), averages to zero and we arrive at

$$\begin{aligned}
\langle \rho \rangle &= \frac{1 + |r_1|^2 |r_2|^2}{|t_1|^2 |t_2|^2} - 1 \\
&= \left(\frac{1}{|t_1|^2} - 1 \right) + \left(\frac{1}{|t_2|^2} - 1 \right) + 2 \left(\frac{1}{|t_2|^2} - 1 \right) \left(\frac{1}{|t_2|^2} - 1 \right) \\
&= \langle \rho_1 \rangle + \langle \rho_2 \rangle + 2 \langle \rho_1 \rangle \langle \rho_2 \rangle. \tag{14.3.16}
\end{aligned}$$

The above equation is the composition law of two resistances in series and the important point is that it is not Ohm's law. We now use Eq.(14.3.16) by assuming that $\langle \rho_1 \rangle$ corresponds to a region of length L and $\langle \rho_2 \rangle$ corresponds to a macroscopically small length δL at some random distance from L. If we write $\langle \rho_2 \rangle = \alpha \delta L$, where α is some undetermined proportionality constant, then we can cast Eq.(14.3.16) as

$$\begin{aligned}
\delta\rho &= \langle\rho\rangle - \langle\rho_1\rangle \\
&= \langle\rho_2\rangle[1 + 2\langle\rho_1\rangle] \\
&= \alpha\delta L[1 + 2\langle\rho_2\rangle] \\
&= \alpha\delta L(1 + 2\langle\rho\rangle)
\end{aligned} \qquad (14.3.17)$$

where, in the last step, we have dropped terms of $O(\delta L)^2$. Integration of Eq.(14.3.17) leads to

$$\langle\rho(L)\rangle = (e^{2\alpha L} - 1)/2. \qquad (14.3.18)$$

Thus we have an exponentially growing resistance or equivalently an exponentially decaying conductance. This exhibits the localization of all states in a disordered one dimensional solid.

The above manipulation which depended crucially on the one-dimensional nature of the problem cannot be repeated in higher dimensions. Instead we will explain a technique due to Abrahams *et al.* We follow a demonstration due to Thouless.

The solid is taken to be a D-dimensional cube of length L on a side. Thouless pointed out that the extended and localized states respond differently to perturbations at the boundaries of the cube. The localized states which are confined to the interior of the cube are insensitive to the perturbations while the extended states are sensitive to it. The main idea is that if two cubes with possibly different energy levels are put into contact, then if the perturbations produced by the new interface can shift the energy levels sufficiently, then electrons will be able to move freely from one cube to the other and diffusion will occur. On the other hand, if motion from cube to cube is unlikely, localization will occur. Thus, what we need to characterize is energy shift produced by a perturbation.

Let δE be the energy shift produced by altered boundary conditions. Using the uncertainty principle

$$\delta E \sim \hbar/t_D \qquad (14.3.19)$$

where t_D is the time taken by the electron to diffuse to the boundary and is estimated as

$$t_D \sim L^2/D(L) \qquad (14.3.20)$$

with $D(L)$ as the diffusion coefficient. Thus

$$\delta E \sim \hbar D(L)/L^2 \qquad (14.3.21)$$

and using the connection between conductivity and diffusivity, we have

$$\delta E \sim (\alpha(L)\hbar/e^2)(L^2 N(E))^{-1} \qquad (14.3.22)$$

where $N(E)$ is the density of states per unit volume. If ΔE is the average spacing between energy levels, then

$$N(E) \sim \frac{1}{\Delta E}\frac{1}{L^D} \qquad (14.3.23)$$

and thus

$$\frac{\delta E}{\Delta E} \sim \frac{\sigma(L)\hbar L^{D-2}}{e^2}. \tag{14.3.24}$$

If $\delta E/\Delta E \ll 1$, then boundary alterations have little effect on the electron and one would have localization. On the other hand, diffusion occurs for $\delta E/\Delta E \gg 1$. The dimensionless quantity on the right hand side of Eq.(14.3.24) is called the scale dependent conductance $g(L)$. Thus

$$g(L) = \frac{\sigma(L)\hbar L^{D-2}}{e^2}. \tag{14.3.25}$$

This quantity determines what happens to the electrical conductivity when cubes of volume L^D are put together. We will now concentrate on the L-dependence of $g(L)$ for large L.

The central question is as follows. Suppose we know $g(L)$ for some large but finite value L_0 of L. What would happen if cubes of side $L_0(1+\epsilon)$ were put together. The scaling assumption is that $g[L_0(1 + \epsilon)]$ is determined by $g(L_0)$ and the scale factor $(1 + \epsilon)$ only, i.e.,

$$g\{L_0(1 + \epsilon)\} = f\{g(L_0), (1 + \epsilon)\}. \tag{14.3.26}$$

If we expand the two sides of the above equation in powers of ϵ,

$$g(L_0)+\epsilon L_0\frac{\partial g}{\partial L_0}+\ldots = f\{g(L_0), 1\}+\epsilon \left.\frac{\partial f\{g(L_0), (1 + \epsilon)\}}{\partial \epsilon}\right|_{\epsilon=0} +\ldots. \tag{14.3.27}$$

Noting that $\left.\frac{\partial f}{\partial \epsilon}\right|_{\epsilon=0}$ is a function of $g(L_0)$ alone, we can write it as $g\beta(g)$ and obtain from the $O(\epsilon)$ part of Eq.(14.3.27)

$$\frac{d \ln g}{d \ln L} = \beta(g). \tag{14.3.28}$$

The sign of $\beta(g)$ is vital. If $\beta(g)$ is positive, then for large L, g scales up to a non-zero value, while for negative $\beta(g)$, g will approach zero as $L \to \infty$.

For large g, the system would be conducting and hence $\sigma(L)$ should be non-zero. Returning to Eq. (14.3.25), we assume that $\sigma(L)$ is independent of L and hence for large g obtain

$$\beta(g) = D - 2 \qquad (g \text{ large}). \tag{14.3.29}$$

For small g, on the other hand, the system will be localized and we can assume a form

$$g = g_0 e^{-L/\xi} \tag{14.3.30}$$

which leads to

$$\beta(g) = \ln \frac{g}{g_0}, \qquad (g << g_0). \tag{14.3.31}$$

A further assumption of monotonic passage from one end [Eq.(14.3.31)] to the other [Eq.(14.3.29)] yields the curves, shown in Fig.14.1 for the β-function.

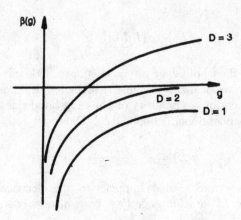

Figure 14.1: β-functions in different dimensions for the electron localization problem.

For $D = 1, 2$, all states are localized, while for $D = 3$, there is a critical g_0 above which there is diffusive behaviour, but below which the system scales down to $g = 0$ and hence all states are localized.

It should be noted that the prediction of the above analysis that all states are localized in $D = 2$ is in a way built into the theory due to the assumption that $\sigma(L)$ is L-independent for large values of g and by the assumption of monotonic behaviour of $\beta(g)$.

15

Transport Equation - I

15.1 Introduction

In this chapter we discuss the theoretical foundations of the Boltzmann transport equation. The macroscopic state of a system in equilibrium is characterized by a few parameters. The state of a gas contained in a box is determined by its pressure, volume and temperature while the temperature, magnetization and volume characterize a piece of magnetic material. In an equilibrium state, these characterizing quantities do not change with time. What is more, if an unforced system happens to be in a non-equilibrium state, the general tendency of the system is to approach an equilibrium state. In the case of a magnetic material at very high temperatures, we know from observation that the equilibrium state is one in which the system shows no net magnetization. The material is composed of individual molecules which carry magnetic moments, all having the same magnitude but capable of pointing in any direction. The zero magnetization at high temperatures is produced by an equal probability of the molecules to orient themselves in all directions. Now suppose that one forcibly produces a state of non-zero magnetization, for example by applying a strong external magnetic field and then removing it. The state with finite magnetization thus produced is a non-equilibrium state. If one watches it over a period of time, the state evolves and eventually the equilibrium state with zero magnetization is achieved. This equilibration is brought about by the magnetic interaction between the individual molecules which causes the reorientation of the magnetic moments.

As another example we may consider one mole of a gas confined to occupy half of a given volume V at a maintained temperature T. If the constraint, e.g., a partition which divides V into two parts, is removed then we have a non-equilibrium situation where the whole volume V is now available to the gas at temperature T, but its pressure it nearly double than that would be obtained in an equilibrium situation. Once again, if one waits for the system to evolve, the collisions between the gas molecules help bring about equilibrium and the final state is characterized by the temperature T, volume V and the pressure appropriate to that obtained from the thermodynamic equation of state. The equilibrium is brought about by interaction. Collisions are caused only because the molecules interact with each other (however weakly) and can be understood only on the microscopic scale.

The above examples make it clear that the equilibrium state is one which does not change with time. This is where the question of time scales comes in. It is virtually impossible to talk about a state of absolute equilibrium-one that would never change with time-it is proper to talk about a time τ_S for a particular state to change and the time τ_0 over which the state is being observed. The observed state can be considered in equilibrium only if $\tau_S \gg \tau_0$. A trivial example is an amount of water in a beaker which for most practical purposes can be taken to be in equilibrium and concepts of equilibrium thermodynamics applied to it (e.g., at any given instant its temperature can be measured). If, however, one observes over a periods of days, the water level in the beaker will fall due to evaporation and hence over this prolonged period of observation the state is not an equilibrium one. The other point to note it that the approach to equilibrium is a consequence of interactions at a microscopic level and any attempt to describe the approach to equilibrium must proceed at the microscopic level. For a gas of N molecules in a volume V, this elaborate description was achieved by Boltzmann and this is what we will describe in the next few chapters.

15.2   Distribution Function

As discussed in the previous section, the approach to equilibrium can be understood at a microscopic level. If we consider N molecules of a gas in a volume V then N is of the order of Avogadro number which is 10^{23} and hence specifying the dynamics of all the N particles is virtually impossible task. One has to opt for a coarser description as has been emphasized in Chapter 1 and this involves specifying the number $dN(\mathbf{r}, \mathbf{p}, t)$ of molecules lying within a momentum range $\mathbf{p}$ to $\mathbf{p} + d\mathbf{p}$ and a spatial range $\mathbf{r}$ to $\mathbf{r} + d\mathbf{r}$. A distribution function $f(\mathbf{r}, \mathbf{p}, t)$ is the density in phase space defined by

$$dN = f d^3r d^3p \tag{15.2.1}$$

with the normalization

$$N = \int dN = \int f(\mathbf{r}, \mathbf{p}, t) d^3r d^3p. \tag{15.2.2}$$

The volume element $d^3r d^3p$ in the coordinate and momentum space is an infinitesimal when compared to macroscopic dimensions, but certainly large compared to atomic dimensions and thus we expect the function f (related to the physical density) to be smooth if we are not concerned about atomic dimensions. We can generally expect the function f to be spatially uniform (especially if external forces and boundary effects are absent) and then the integration over d^3r produces the total volume and we have

$$N = V \int f(\mathbf{p}, t) d^3p. \tag{15.2.3}$$

Thus $\int f(\mathbf{p}, t) d^3p$ is the time-dependent particle density.

Equilibrium corresponds to the $\lim_{t \to \infty}$ if it exists, whence $f(\mathbf{r}, \mathbf{p}, t)$ presumably becomes independent of time. The task of the theory is to construct an equation of motion for the function f. This, in principle, is a much simpler task than solving the $6N$ Hamilton's equation of motion associated with the N particles. This simplifying point of view where the microscopic dynamics is given a coarser description is a sort of bridge between the microscopic and macroscopic description. As was seen in Chapter 1, it is the equilibrium distribution function that, together with a linking hypothesis, allows for the calculation of all thermodynamic properties of all systems. In this chapter we intend to obtain the equation for the time development of $f(\mathbf{r}, \mathbf{p}, t)$.

The expected structure of the time development equation is

$$\frac{\partial f}{\partial t} = Lf \tag{15.2.4}$$

where L is an operator which can be of any sort, in general nonlinear. The equilibrium distribution f_0 corresponds to

$$\frac{\partial f_0}{\partial t} = Lf_0 = 0 \tag{15.2.5}$$

which is a fixed point of the time development of f. The equilibrium is stable only if a small perturbation from the equilibrium state leads to a time development where the perturbation dies away or keeps oscillating about the zero value. This dynamics of the perturbation can be obtained from a linear stability analysis where one sets $f = f_0 + \delta f$ and linearizes the right hand side of Eq.(15.2.4) about f_0 to write

$$\frac{\partial}{\partial t} \delta f = K \delta f \tag{15.2.6}$$

where K is now a linear operator. The solution can lead to oscillatory behaviour which generally corresponds to wave propagation or relaxational behaviour which leads to damping coefficients like viscosity, thermal conductivity, etc. We now proceed to find the equation of motion for f.

15.3 In the Absence of Collisions

In the simplest situation all interaction between the molecules is ignored. An external force F is supposed to act which may be uniform or may depend on position. If the position and momentum of the 'i'-th particle at time t is $\mathbf{r}_i$ and

$\mathbf{p_i}$ respectively, and if the external force at the position of the 'i'-th particle is $\mathbf{F_i}$, then in an interval δt, the position and momentum are altered to

$$\mathbf{r_i'} = \mathbf{r_i} + \frac{\mathbf{p_i}}{m}\delta t \qquad (15.3.1)$$

$$\mathbf{p_i'} = \mathbf{p_i} + \mathbf{F_i}\delta t \qquad (15.3.2)$$

where $\mathbf{r_i'}$, and $\mathbf{p_i'}$ are respectively the position and momentum at the time $t + \delta t$. If we consider a volume element $d^3r d^3p$ about $\mathbf{r}$ and $\mathbf{p}$, then the number of particles in this volume element is

$$dN = f(\mathbf{r}, \mathbf{p}, t)d^3r d^3p. \qquad (15.3.3)$$

In a time δt, the volume element $d^3r d^3p$ changes to the element $d^3r' d^3p'$ where $\mathbf{r}'$ and $\mathbf{p}'$ are given by Eqs.(15.3.1) and (15.3.2). In the absence of collisions, all the dN particles in $d^3r d^3p$ at time t are transferred to the volume $d^3r' d^3p'$ at time $t + \delta t$. The distribution function is now given by

$$dN = f(\mathbf{r}', \mathbf{p}', t + \delta t)d^3r' d^3p'. \qquad (15.3.4)$$

From Eqs.(15.3.3) and (15.3.4), we see

$$f(\mathbf{r}, \mathbf{p}, t)d^3r d^3p = f(\mathbf{r}', \mathbf{p}', t + \delta t)d^3r' d^3p' \qquad (15.3.5)$$

To see how $d^3r d^3p$ is related to $d^3r' d^3p'$, we need only consider the element $dx dp_x$. Here x, y and z are the components of $\mathbf{r}$ and (p_x, p_y, p_z) are the components of $\mathbf{p}$. Since

$$x' = x + \frac{p_x}{m}\delta t$$

$$p_x' = p_x + F_x\delta t$$

implying

$$dx' dp' = \begin{vmatrix} \dfrac{\partial x'}{\partial x} & \dfrac{\partial x'}{\partial p_x} \\ \dfrac{\partial p_x'}{\partial x} & \dfrac{\partial p_x'}{\partial p_x} \end{vmatrix}$$

$$= \begin{vmatrix} 1 & \dfrac{\delta t}{m} \\ \dfrac{\partial F_x}{\partial x}\delta t & 1 \end{vmatrix} dx dp_x$$

$$= \left[1 + \frac{1}{m}\frac{\partial F_x}{\partial x}(\delta t)^2 \right] dx dp_x. \qquad (15.3.6)$$

Dropping all terms of $O((\delta t)^2)$, we arrive at

$$dx' dp_x' = dx dp_x. \qquad (15.3.7)$$

Identical arguments lead to $dy' dp_y' = dy dp_y$ and $dz' dp_z' = dz dp_z$, so that $d^3r' d^3p' = d^3r d^3p$, and consequently Eq.(15.3.5) becomes

$$f(\mathbf{r}, \mathbf{p}, t) = f(\mathbf{r}', \mathbf{p}', t + \delta t) = f(\mathbf{r}, \mathbf{p}, t) + \frac{\partial f}{\partial t}\partial t + \frac{\partial f}{\partial \mathbf{r}}\cdot\frac{\mathbf{p}}{m}\delta t + \frac{\partial f}{\partial \mathbf{p}}.\mathbf{F}\delta t$$

or

$$\frac{\partial f}{\partial t} + \frac{\mathbf{p}}{m}\cdot\frac{\partial f}{\partial \mathbf{r}} + \mathbf{F}\cdot\frac{\partial f}{\partial \mathbf{p}} = 0. \tag{15.3.8}$$

Thus in the absence of collisions, the distribution function satisfies a linear evolution equation. In the next section we consider the effect of collisions.

15.4 In the Presence of Collisions

Interactions between molecules which lead to collisions can cause molecules to leave and enter the volume element $d^3r d^3p$ that we have been considering. The gas is considered to be sufficiently dilute so that only binary interactions need to be considered. At large distances attractive interaction is sufficiently weak so that we can ignore it. At close range, namely at a separation of the order of molecular diameter, the interaction is very strongly repulsive so that the entire effect of the interaction between two molecules manifests itself in the form of elastic collisions. In what follows, we will ignore all quantum effects, i.e., the de Broglie wavelength $\lambda \sim \hbar/\sqrt{mk_BT}$ is much smaller than the typical intermolecular separation $(V/N)^{1/3}$. For a dilute gas this condition is easily satisfied. In the presence of collisions, if molecules enter or leave the volume element at the rate $|\partial f/\partial t|_{\text{coll}}$ then Eq.(15.3.8) becomes

$$\frac{\partial f}{\partial t} + \frac{\mathbf{p}}{m}\cdot\frac{\partial f}{\partial \mathbf{r}} + \mathbf{F}\cdot\frac{\partial f}{\partial \mathbf{p}} = \left.\frac{\partial f}{\partial t}\right|_{\text{coll}} = \left.\frac{\partial f}{\partial t}\right|_{\text{in}} - \left.\frac{\partial f}{\partial t}\right|_{\text{out}} \tag{15.4.1}$$

where $|\partial f/\partial t|_{\text{in}}$ is the contribution from collisions which cause molecules to enter the volume $d^3r d^3p$ and $|\partial f/\partial t|_{\text{out}}$ is the contribution from collisions which cause the molecules to leave the volume $d^3r d^3p$.

We consider molecules with velocity $\mathbf{v}$ to collide elastically with molecules of velocity $\mathbf{v}_1$ and acquire velocities $\mathbf{v}_2$ and $\mathbf{v}_3$ respectively after the collision. This process causes loss of molecules from the volume element $d^3r d^3p$. The reverse process where a molecule of velocity $\mathbf{v}_2$ strike a molecule of velocity $\mathbf{v}_3$ producing velocities $\mathbf{v}$ and $\mathbf{v}_1$ after the collision causes an increase in the number of molecules in the volume $d^3r d^3p$. The first process will be characterized by the cross section $\sigma(\mathbf{v}, \mathbf{v}_1 \to \mathbf{v}_2, \mathbf{v}_3)$ and the second by the cross-section $\sigma(\mathbf{v}_2, \mathbf{v}_3 \to \mathbf{v}, \mathbf{v}_1)$. So far as the spatial part is concerned the whole process occurs in the volume d^3r. The number of molecules with velocity between $\mathbf{v}$ and $\mathbf{v} + d\mathbf{v}$ in the volume d^3r is $f(\mathbf{r}, \mathbf{v}, t)d^3r d^3v$ (instead of the momentum it is simpler to consider the velocity as the variable in what follows). The number of possible collisions per unit time per unit area of these molecules with those with velocity $\mathbf{v}_1$ is determined by their own number and the number of molecules with velocity between $\mathbf{v}_1$ and $\mathbf{v}_1 + d\mathbf{v}_1$, lying in the physical volume with unit cross-sectional area and height $|\mathbf{v} - \mathbf{v}_1|$. The number of such molecules is $f(\mathbf{v}, \mathbf{v}_1, t)|\mathbf{v} - \mathbf{v}_1|d^3v$, and hence the number of possible collisions per unit time (the incident flux) is $f(\mathbf{r}, \mathbf{v}, t)f(\mathbf{r}, \mathbf{v}_1, t)|\mathbf{v} - \mathbf{v}_1|d^3r d^3v d^3v_1$.

At this point, an assumption creeps in almost unnoticed. The two varieties of molecules-those with velocity $\mathbf{v}$ and those with velocity $\mathbf{v_1}$–are uncorrelated in the sense that the joint probability distribution of these molecules has been taken as the product of the individual probability distributions. This lack or correlation is an assumption-generally called the assumption of molecular chaos. This assumption is at the heart of the success of statistical mechanics and as seen is Chapter 1, the molecular dynamics is such that the lack of correlation is a plausible dynamical state.

The elastic collisions change the direction of relative velocity of the two molecules from the incident $\mathbf{v} - \mathbf{v_1}$ to the final $\mathbf{v_2} - \mathbf{v_3}$. In a solid angle $d\Omega$ about the final direction for a given $\mathbf{v}, \mathbf{v_1}, \mathbf{v_2}$ and $\mathbf{v_3}$, the number of particles scattered per unit time is the number of particles leaving the element $d^3r d^3v$ in that direction and is given by

$$dN_{out} = |\mathbf{v} - \mathbf{v_1}| f(\mathbf{r}, \mathbf{v}, t) f(\mathbf{r}, \mathbf{v_1}, t) \sigma(\mathbf{v}, \mathbf{v_1} \to \mathbf{v_2}, \mathbf{v_3}) d\Omega d^3r d^3v d^3v_1. \quad (15.4.2)$$

The total number leaving the element $d^3r d^3v$ per unit time as obtained by integrating over all possible $\mathbf{v_1}$ and solid angle is

$$N_{out} = \int_{\mathbf{v_1}, \Omega} |\mathbf{v} - \mathbf{v_1}| f(\mathbf{r}, \mathbf{v}, t) f(\mathbf{r}, \mathbf{v_1}, t) \sigma(\mathbf{v}, \mathbf{v_1} \to \mathbf{v_2}, \mathbf{v_3}) d\Omega d^3v d^3v_1 d^3r.$$

$$(15.4.3)$$

The reverse collision, i.e., molecules with velocities $\mathbf{v_2}$ and $\mathbf{v_3}$ colliding to give rise to final velocities $\mathbf{v}$ and $\mathbf{v_1}$ is obtained by arguments identical to those given above and lead to

$$N_{in} = \int_{\mathbf{v_2}, \mathbf{v_3}} |\mathbf{v_2} - \mathbf{v_3}| f(\mathbf{r}, \mathbf{v_2}, t) f(\mathbf{r}, \mathbf{v_3}, t) \sigma(\mathbf{v_2}, \mathbf{v_3} \to \mathbf{v}, \mathbf{v_1}) d\Omega d^3v_2 d^3v_3 d^3r.$$

$$(15.4.4)$$

We now note the following equalities:

i) the collision is elastic and hence the magnitude of the relative velocity of approach is equal to the magnitude of the relative velocity of receding, i.e.,

$$|\mathbf{v_{rel}}| = |\mathbf{v} - \mathbf{v_1}| = |\mathbf{v_2} - \mathbf{v_3}| = |\mathbf{v'_{rel}}| \quad (15.4.5)$$

ii) in terms of the centre of mass velocity $\mathbf{v_{cm}} = \frac{1}{2}(\mathbf{v_1} + \mathbf{v})$ and the relative velocity $\mathbf{v_{rel}} = \mathbf{v} - \mathbf{v_1}$, the volume element $d^3v_1 d^3v = d^3v_{cm} d^3v_{rel}$ as the Jacobian of the transformation is unity

$$d^3v_{cm} d^3v_{rel} = \begin{vmatrix} \frac{\partial \mathbf{v_{cm}}}{\partial \mathbf{v}} & \frac{\partial \mathbf{v_{cm}}}{\partial \mathbf{v_1}} \\ \frac{\partial \mathbf{v_{rel}}}{\partial \mathbf{v}} & \frac{\partial \mathbf{v_{rel}}}{\partial \mathbf{v}} \end{vmatrix} d^3v d^3v_1 = \begin{vmatrix} 1 & 0 \\ 0 & 1 \end{vmatrix} d^3v d^3v_1 = d^3v d^3v_1$$

Similarly $d^3v_2 d^3v_3 = d^3v_{cm} d^3v'_{rel}$. The centre of mass velocity is unchanged and hence d^3v_{cm} is unaltered. The volume element $d^3v'_{rel}$ is equal to the element d^3v_{rel} since the elastic collisions simply rotates the direction of $\mathbf{v_{rel}}$ and hence the magnitude of a volume element about $\mathbf{v_{rel}}$ is unchanged.

Thus,

$$d^3v_1 d^3v = d^3v_{\text{cm}}d^3v_{\text{rel}} = d^3v_{\text{cm}}d^3v'_{\text{rel}} = d^3v_2 d^3v_3. \tag{15.4.6}$$

iii) The dynamics is invariant under time reversal, rotation and reflection. The time reversal invariance implies

$$\sigma(\mathbf{v_2}, \mathbf{v_3} \to \mathbf{v}, \mathbf{v_1}) = \sigma(-\mathbf{v}, -\mathbf{v_1} \to -\mathbf{v_2}, -\mathbf{v_3}). \tag{15.4.7}$$

To exploit the invariances under rotation and reflection, we consider a rotation by 180 degrees about the z-axis which sends v_x to $-v_x$ and v_y to $-v_y$. Now a reflection about the $x - y$ plane takes v_z to $-v_z$ and thus under the combined operation $\mathbf{v} \to -\mathbf{v}$. Invariance under rotation and reflection consequently implies

$$\sigma(-\mathbf{v}, -\mathbf{v_1}, \to -\mathbf{v_2}, -\mathbf{v_3}) = \sigma(\mathbf{v}, \mathbf{v_1} \to \mathbf{v_2}, \mathbf{v_3})$$

and hence, using Eq.(15.4.7)

$$\sigma(\mathbf{v_2}, \mathbf{v_3} \to \mathbf{v}, \mathbf{v_1}) = \sigma(\mathbf{v}, \mathbf{v_1} \to \mathbf{v_2}, \mathbf{v_3}). \tag{15.4.8}$$

Using Eqs.(15.4.5), (15.4.6) and (15.4.8), we can rewrite Eq.(15.4.4) as

$$N_{in} = \int_{\mathbf{v_1}, \Omega} |\mathbf{v} - \mathbf{v_1}| f(\mathbf{r}, \mathbf{v_2}, t) f(\mathbf{r}, \mathbf{v_3}, t) \sigma(\mathbf{v}, \mathbf{v_1} \to \mathbf{v_2}, \mathbf{v_3}) d\Omega d^3v d^3v_1 d^3r.$$

$$\tag{15.4.9}$$

Thus the number of extra molecules ending up per unit time in the volume element $d^3r d^3v$ due to collision is

$$
\begin{aligned}
N_{\text{in}} - N_{\text{out}} &= \int_{\mathbf{v_1}, \Omega} |\mathbf{v} - \mathbf{v_1}| \sigma(\mathbf{v}, \mathbf{v_1} \to \mathbf{v_2}, \mathbf{v_3}) \\
&\times [f(\mathbf{r}, \mathbf{v_2}, t) f(\mathbf{r}, \mathbf{v_3}, t) - f(\mathbf{r}, \mathbf{v_1}, t) f(\mathbf{r}, \mathbf{v}, t)] d\Omega d^3v_1 d^3r d^3v \\
&= \left. \frac{\partial f}{\partial t} \right|_{\text{coll}} d^3r d^3v
\end{aligned}
\tag{15.4.10}
$$

and Eq.(15.4.1) becomes

$$
\begin{aligned}
\frac{\partial f}{\partial t} + \mathbf{v} \cdot \frac{\partial f}{\partial \mathbf{r}} + \frac{\mathbf{F}}{m} \cdot \frac{\partial f}{\partial \mathbf{v}} &= \int |\mathbf{v} - \mathbf{v_1}| \sigma(\mathbf{v}, \mathbf{v_1} \to \mathbf{v_2}, \mathbf{v_3}) \\
&\times [f(\mathbf{r}, \mathbf{v_2}, t) f(\mathbf{r}, \mathbf{v_3}, t) - f(\mathbf{r}, \mathbf{v_1}, t) f(\mathbf{r}, \mathbf{v}, t)] d\Omega d^3v_1.
\end{aligned}
\tag{15.4.11}
$$

This is the Boltzmann transport equation, which gives the time evolution of the distribution function f and is of the form $\partial f / \partial t = Lf$ that we had anticipated in Eq.(15.2.4).

We now find the equilibrium distribution f_0 in the absence of any external force $\mathbf{F}$. Since the right hand side of Eq.(15.4.11) cannot produce any spatial dependence in f, in the absence of any external force, the distribution function f cannot depend on $\mathbf{r}$ and thus the equilibrium distribution for which $\partial f_0 / \partial t = 0$, has to satisfy

$$f d^3 v_1 d\Omega \sigma(\mathbf{v}_1, \mathbf{v} \to \mathbf{v}_2, \mathbf{v}_3)\{f_0(\mathbf{v}_2)f_0(\mathbf{v}_3) - f_0(\mathbf{v}_1)f_0(\mathbf{v})\} = 0.$$

A sufficient condition for this equally to hold is

$$f_0(\mathbf{v}_2)f_0(\mathbf{v}_3) = f_0(\mathbf{v})f_0(\mathbf{v}_1). \tag{15.4.12}$$

The product of the distribution function is collision invariant and hence

$$\ln f_0(\mathbf{v}) + \ln f_0(\mathbf{v}_1) = \text{Collision invariant}. \tag{15.4.13}$$

Now, in an elastic collision the additive invariants are momentum and kinetic energy:

$$\mathbf{v} + \mathbf{v}_1 = \mathbf{v}_2 + \mathbf{v}_3 = \text{Collision invariant}$$
$$v_1^2 + v^2 = v_2^2 + v_3^2 = \text{Collision invariant}.$$

Consequently,

$$\begin{aligned} \ln f_0(\mathbf{v}) &= \alpha + \beta.\mathbf{v} + \gamma v^2 \\ &= A_0 - A(\mathbf{v} - \mathbf{v_0})^2 \end{aligned} \tag{15.4.14}$$

where α, β and γ are constants and so are A_0, A and $\mathbf{v_0}$ expressible in terms of α, β and γ. We can write the above as

$$f_0(\mathbf{v}) = Ce^{-A(\mathbf{v}-\mathbf{v_0})^2} \tag{15.4.15}$$

and choosing the origin of $\mathbf{v}$ as $\mathbf{v_0}$ without any loss of generality

$$f_0(\mathbf{v}) = Ce^{-Av^2} \tag{15.4.16}$$

where C is a constant. The constants A and C are related because of the normalization

$$\begin{aligned} N &= \int f_0(v) d^3r d^3v = VC \int_0^\infty e^{-Av^2} d^3v \\ &= 4\pi.VC \int_0^\infty v^2 e^{-Av^2} dv \\ &= (2\pi VC/A^{3/2}) \times \Gamma\left(\frac{3}{2}\right) \\ &= \left(\frac{\pi}{A}\right)^{3/2} VC \end{aligned}$$

or

$$\left(\frac{\pi}{A}\right)^{3/2} C = \frac{N}{V} = n \qquad (15.4.17)$$

where n is the number density.

To make further progress, we can calculate the pressure exerted by the gas enclosed in a cubical volume V. To obtain the pressure we can choose a particular face (say xy) and calculate the number of molecules with the z-component of velocity between v_z and $v_z + dv_z$ incident per unit time per unit area on this face. Clearly, the number of such molecules is

$$dN = C \int\limits_0^\infty dv_x \int\limits_0^\infty dv_y e^{-A(v_x^2+v_y^2)} e^{-Av_z^2} dv_z \times 1 \times v_z. \qquad (15.4.18)$$

Every time such a molecule hits the $x-y$ plane and is reflected, the change in momentum is $2mv_z$ and hence the force per unit area is

$$dF = 2mv_z dN. \qquad (15.4.19)$$

To get the total force per unit area and hence the pressure P, we integrate over v_z from 0 to ∞ and get using Eq.(15.4.18),

$$
\begin{aligned}
P &= 2mC \int\limits_{-\infty}^\infty dv_x dv_y e^{-A(v_x^2+v_y^2)} \int\limits_0^\infty v_z^2 e^{-Av_z^2} dv_z \\
&= \frac{2mC}{A} \left\{ \Gamma\left(\frac{1}{2}\right) \right\}^2 \frac{1}{A^{1/2}} \cdot \frac{1}{2A} \cdot \Gamma\left(\frac{3}{2}\right) \\
&= \frac{m}{2A} \left(\frac{\pi}{A}\right)^{3/2} C \\
&= \frac{m}{2A} n.
\end{aligned} \qquad (15.4.20)
$$

In equilibrium the pressure P of the gas is related to its volume V and temperature T. The only interaction that we have assumed is elastic collision between the gas molecules and that is a fair approximation to an ideal gas. For an ideal gas, the relation between P, V and T is

$$PV = Nk_BT$$

where N is the number of molecules and k_B is a constant known as Boltzmann's constant with the numerical value $1.38 \times 10^{-16} \text{erg/K}$.

Thus

$$P = nk_BT \qquad (15.4.21)$$

for the ideal gas and comparing with Eq.(15.4.20) we see that

$$A = \frac{m}{2k_BT}. \qquad (15.4.22)$$

The equilibrium distribution can then be written as

$$f_0(v) = n \left(\frac{m}{2\pi k_B T} \right)^{3/2} e^{-\frac{mv^2}{2k_B T}} \qquad (15.4.23)$$

which gives the density of molecules under equilibrium condition with velocity between $\mathbf{v}$ and $\mathbf{v} + d\mathbf{v}$. This is the familiar Maxwell-Boltzmann distribution for velocities in a non interacting gas of classical particles.

16

Transport Equation - II

16.1 Introduction

In the previous chapter, we have obtained the evolution equation for the distribution function f and used it to obtain the equilibrium distribution function. **In this chapter our focus will be on setting up a non-equilibrium distribution and obtaining its time evolution.** We will begin with an equilibrium distribution and give it a small perturbation. The resulting perturbation will evolve in time according to the linearized version of Eq.(15.4.11). The full transport equation [Eq.(15.4.11)] is virtually impossible to solve because of the non-linear collision integral, but for small perturbations about the equilibrium state the nonlinear term can be linearized and under such circumstances the approach to equilibrium can be analytically studied. For attaining equilibrium the collision term is essential and hence the decay of the perturbation to the equilibrium state is possible only if the collision term is included in the transport equation. If we want to consider a simple situation without the collision term, then a small perturbation to the equilibrium state can oscillate about the equilibrium distribution if a net unbalanced force is generated on distortion of the equilibrium. If we ignore collisions and also the presence of external forces, there is no possibility of a non-equilibrium distribution ever approaching the equilibrium distribution.. As an example of this, we may consider a non-equilibrium distribution where the particles have their velocities distributed in a Maxwell Boltzmann fashion but are all concentrated at $r = 0$ at time $t = 0$. Thus, the distribution at $t = 0$ is

$$f(\mathbf{r}, \mathbf{v}, 0) = N \left(\frac{m}{2\pi k_B T} \right)^{3/2} \exp \left(-\frac{mv^2}{2k_B T} \right) \delta^3(r). \qquad (16.1.1)$$

The evolution occurs without collision and no interactions between the particles which can produce an external force. The transport equation takes the form

$$\frac{\partial f}{\partial t} + \mathbf{v} . \frac{\partial f}{\partial \mathbf{r}} = 0 \qquad (16.1.2)$$

with the solution
$$f(\mathbf{r}, \mathbf{v}, t) = g(\mathbf{r} - \mathbf{v}t, \mathbf{v}) \tag{16.1.3}$$
where g is an arbitrary function, which can be fixed by the initial condition:
$$f(\mathbf{r}, \mathbf{v}, 0) = g(\mathbf{r}, \mathbf{v}) = N\left(\frac{m}{2\pi k_B T}\right) \exp\left(-\frac{mv^2}{2k_B T}\right) \delta^3(r). \tag{16.1.4}$$

Having thus fixed $g(\mathbf{r}, \mathbf{v})$, we find from Eq.(16.1.3)
$$f(\mathbf{r}, \mathbf{v}, t) = N\left(\frac{m}{2\pi k_B T}\right)^{3/2} \exp\left(-\frac{mv^2}{2k_B T}\right) \delta^3(r - vt). \tag{16.1.5}$$

The number of particles $n(r, t)d^3r$ lying between $\mathbf{r}$ and $\mathbf{r}+\mathbf{dr}$ is obtained by integrating over all velocities and thus

$$
\begin{aligned}
n(\mathbf{r}, t)d^3r &= \int_v d^3 v f(\mathbf{r}, \mathbf{v}, t) d^3 r \\
&= N\left(\frac{m}{2\pi k_B T}\right)^{3/2} d^3 r \int e^{-\frac{mv^2}{2k_B T}} \delta^3(\mathbf{r} - \mathbf{v}t) d^3 v \\
&= N\left(\frac{m}{2\pi k_B T}\right)^{3/2} \frac{1}{t^3} e^{-\frac{mr^2}{2k_B T t^2}} d^3 r
\end{aligned}
$$

using the standard rules of δ-function integration. This leads to
$$n(r, t) = N\left(\frac{m}{2\pi k_B T}\right)^{3/2} \frac{1}{t^3} \cdot \exp\left(-\frac{mr^2}{2k_B T t^2}\right). \tag{16.1.6}$$

As time evolves the particles concentrated at $r = 0$ start spreading out because of the initial velocities and ultimately diffuse out over all space. As expected, there is no equilibrium distribution due to the absence of interactions and boundaries.

16.2   Interaction without Collisions

We consider a collection of electrons spread out on a two-dimensional surface. A sample of such a two-dimensional electron gas is obtained by spraying the surface of liquid helium with electrons. Charge neutrality is maintained by having a positively charged background. The areal density of electrons is low enough for quantum effects to be ignored. In an equilibrium distribution, the external force is zero-on any given electron the Coulomb force due to the other electrons and the charged background cancel out. The equilibrium distribution f_0 is then independent of the position coordinate and equilibrium is achieved by collisions with f_0 given by (in this chapter we will write k_B for Boltzmann constant, reserving $\mathbf{k}$ for the wavevector)
$$f_0 = n.\frac{m}{2\pi k_B T} \cdot \exp\left(-\frac{mv^2}{2k_B T}\right) \tag{16.2.1}$$

where $n = N/A$, N being the total number of electrons, A the area over which they are spread and the distribution, as should be checked, is normalized in two dimensions, i.e., $A \int f_0(v)d^2v = N$. If a small deviation from the equilibrium distribution occurs, then the resultant force on any electron is no longer zero and this unbalanced force can check the perturbation from growing, without the effect of collisions. In what follows we will ignore the effect of collisions on the perturbation and work with the left hand side of the transport equation [Eq.(15.4.11)]. We imagine that the equilibrium distribution f_0 is changed to

$$f = f_0 + \delta f. \tag{16.2.2}$$

The deviation δf leads to local density variations $\delta n(r,t) = \int (\delta f)d^2v$ and due to local density variations there will be an unbalanced force $\delta F(r,t)$ obtained from a potential $\delta \phi(r,t)$ which are related as

$$\delta \mathbf{F} = -e\nabla \delta \phi(\mathbf{r},t) \qquad \text{(e is electronic charge)} \tag{16.2.3}$$

and

$$\delta \phi = -e \int \frac{\delta n(\mathbf{r}',t)d^2\mathbf{r}'}{|\mathbf{r}-\mathbf{r}'|}. \tag{16.2.4}$$

The distribution function f of Eq.(16.2.2) must satisfy the transport equation

$$\frac{\partial}{\partial t}f_0 + \frac{\partial}{\partial t}\delta f + \mathbf{v} \cdot \frac{\partial}{\partial \mathbf{r}}(f_0 + \delta f) + \frac{\delta \mathbf{F}}{m} \cdot \frac{\partial}{\partial \mathbf{v}}(f_0 + \delta f) = \text{Collision term with } (f_0 + \delta f). \tag{16.2.5}$$

As stated above, we ignore the contribution coming from the collision term. Consequently, the right hand size is zero in Eq.(16.2.5). Now $\partial f_0/\partial t = \partial f_0/\partial \mathbf{r} = 0$ and working to order δf, we obtain

$$\frac{\partial}{\partial t}\delta f + \mathbf{v} \cdot \frac{\partial}{\partial \mathbf{r}}\delta f + \frac{\delta \mathbf{F}}{m} \cdot \frac{\delta}{\delta \mathbf{v}}f_0 = 0. \tag{16.2.6}$$

Note that in the third term $\delta \mathbf{F}$ is already $O(\delta f)$ and hence $(\delta \mathbf{F}/m.\partial/\partial v)\delta f$ is of higher order, i.e. $O(\delta f)^2$, and has been dropped.

To proceed further we have to make a statement for the form of δf. If, as explained before, we expect δf to be oscillatory, then the temporal and spatial dependence of δf may be taken to be of the form

$$\delta f = Ae^{i(\mathbf{k}\cdot\mathbf{r}-\omega t)}, \qquad (i = \sqrt{-1} \text{ as usual}) \tag{16.2.7}$$

where the velocity dependence is contained in A. The spatial dependence $e^{i\mathbf{k}\cdot\mathbf{r}}$ is assumed to be such that the wavenumber k (magnitude of the wave vector $\mathbf{k}$) is small which would mean a mild spatial dependence. Similarly we expect the frequency ω to be small as well and the final answer would have to be consistent with these expectations. Consistent with Eq.(16.2.7), the spatial and temporal dependence of $\delta n(\mathbf{r},t)$ would be

$$\delta n = Be^{i(\mathbf{k}\cdot\mathbf{r}-\omega t)} \tag{16.2.8}$$

with B, a constant, obtained from A through the relation

$$B = \int Ad^2v. \qquad (16.2.9)$$

Inserting in Eq.(16.2.6), the forms of Eqs.(16.2.7) and (16.2.8), we get

$$
\begin{aligned}
iA(\mathbf{k.v} - \omega)e^{i(\mathbf{k.r}-\omega t)} &= -\frac{1}{m}\delta\mathbf{F} \cdot \frac{\partial}{\partial \mathbf{v}}\left(n\frac{m}{2\pi k_B T}\right)e^{-\frac{mv^2}{2k_B T}} \\
&= \frac{1}{m}\delta\mathbf{F} \cdot \frac{nm}{2\pi k_B T} \cdot \left(-\frac{m}{k_B T}\right)\mathbf{v}e^{-\frac{mv^2}{2k_B T}} \\
&= \frac{f_0}{(k_B T)}\mathbf{v} \cdot (-e^2)\nabla \int \frac{\delta n(\mathbf{r}',t)d^2r'}{|\mathbf{r}-\mathbf{r}'|} \\
&= -\frac{f_0 e^2 B}{(k_B T)}\mathbf{v} \cdot \nabla \int \frac{e^{i(\mathbf{k.r}'-\omega t)}}{|\mathbf{r}-\mathbf{r}'|}d^2r' \\
&= -\frac{f_0 e^2 B}{(k_B T)}\mathbf{v} \cdot \nabla e^{i(\mathbf{k.r}-\omega t)} \times \int \frac{e^{i\mathbf{k.}(\mathbf{r}'-\mathbf{r})}d^2r'}{|\mathbf{r}'-\mathbf{r}|} \\
&= -\frac{if_0 e^2 B}{(k_B T)}(\mathbf{v.k})\frac{2\pi}{k}e^{i(\mathbf{k.r}-\omega t)}.
\end{aligned}
$$

(The two-dimensional Fourier transform $\int (e^{i\mathbf{k.r}}/r)d^2r = 2\pi/k$) leading to

$$A = B\frac{e^2 f_0}{k_B T} \cdot \frac{2\pi}{k} \cdot \frac{\mathbf{v.k}}{\omega - \mathbf{v.k}}. \qquad (16.2.10)$$

Using Eq.(16.2.9), we obtain

$$
\begin{aligned}
B &= \int Ad^2v \\
&= B\frac{2\pi e^2}{k} \cdot \frac{1}{k_B T} \int \frac{\mathbf{v.k}}{\omega - \mathbf{v.k}}d^2v \\
&= B\frac{2\pi e^2}{k}\frac{1}{k_B T} \int \frac{\mathbf{v.k}}{\omega}f_0(1 + \frac{\mathbf{v.k}}{\omega} + \ldots)d^2v. \qquad (16.2.11)
\end{aligned}
$$

The last step follows on the assumption that though ω and k are both small, ω is large compared to k – for self consistency the final result must bear this out. Since $B \neq 0$, we find from Eq.(16.2.11), working to the lowest order,

$$
\begin{aligned}
1 &= \frac{2\pi e^2}{k} \cdot \frac{k^2}{k_B T}\frac{1}{2\omega^2}2\pi \int\limits_0^\infty \frac{nm}{2\pi k_B T}e^{-\frac{mv^2}{2k_B T}}v^3 dv \\
&= \frac{\pi ne^2 k}{m\omega^2}\int\limits_0^\infty x^3 e^{-x^2}dx \\
&= \frac{2\pi ne^2 k}{m\omega^2}
\end{aligned}
$$

or

$$\omega^2 = \frac{2\pi n e^2}{m} k \qquad (16.2.12)$$

confirming that for small k, $\omega \propto k^{1/2}$ is indeed much larger than k. Thus, we find that our *ansatz* for the form of δf is indeed consistent and there is an oscillation about the equilibrium distribution if the charged particle is disturbed from equilibrium and the frequency of oscillation is related to the wavenumber by Eq.(16.2.12). These oscillations are known as plasma oscillations.

16.3 Collisions with Fixed Scatterers

The nonlinear collision integral of Eq.(15.4.11) shall henceforth be denoted by C, i.e.,

$$\begin{aligned} C &= \int d^3v_1 d\Omega |\mathbf{v} - \mathbf{v_1}| \sigma(\Omega)[f(\mathbf{r}, \mathbf{v_2}, t)f(\mathbf{r}, \mathbf{v_3}, t) - f(\mathbf{r}, \mathbf{v_1}, t)f(\mathbf{r}, \mathbf{v}, t)] \\ &= \int d^3v_1 d\Omega |\mathbf{v} - \mathbf{v_1}| \sigma(\Omega)[f_2 f_3 - f_1 f] \end{aligned} \qquad (16.3.1)$$

where instead of the elaborate $\sigma(\mathbf{v}, \mathbf{v_1} \to \mathbf{v_2}, \mathbf{v_3})$, we use $\sigma(\Omega)$ since the cross-section for the elastic collision will be determined by the solid angle Ω alone, and instead of $f(\mathbf{r}, \mathbf{v_i}, t)$ we use the notation f_i. For the rest of this chapter, we will assume a complete absence of external forces $\mathbf{F}$: an assumption almost always valid for uncharged particles in a force-free environment. Boltzmann transport equation now reads

$$\frac{\partial f}{\partial t} + \mathbf{v} \cdot \frac{\partial f}{\partial \mathbf{r}} = C \qquad (16.3.2)$$

with the equilibrium distribution appropriate to a three-dimensional space setup by binary collisions as

$$f_0 = n \left(\frac{m}{2\pi k_B T} \right)^{3/2} e^{-\frac{mv^2}{2k_B T}}. \qquad (16.3.3)$$

A fixed density of scatterers (n_0) is assumed and particles colliding with the scatterers have the directions of their velocity changed but the magnitude is unchanged. Particles with velocity $\mathbf{v}$ when reflected from the scatterer go off with velocity $\mathbf{v'}$ (with $|\mathbf{v'}| = |\mathbf{v}|$) and thus there is a depletion of particles in the volume element $d^3r d^3v$. Similarly, a velocity $\mathbf{v'}$ when scattered gives rise to a velocity $\mathbf{v}$ and there is addition to the number of particles in $d^3r d^3v$. The collision integral simplifies to

$$C = \int d\Omega n_0 |\mathbf{v}| \sigma(\Omega)(f' - f). \qquad (16.3.4)$$

(Note that in $f f_1$, f_1 refers to scatterers and is simply n_0, while in $f_2 f_3$, if f_2 refers to the particles with velocity $\mathbf{v'}$ and is f', then f_3 refers to the scatterers and is n_0. The relative velocity between scatterer and particle is $|\mathbf{v}| = |\mathbf{v}'|$. If $\sigma(\Omega)$ is taken to be a constant, we have Eq.(16.3.4) further simplified to

$$\begin{aligned} C &= n_0|\mathbf{v}|\sigma \int d\Omega(f' - f) \\ &= \Gamma \int d\Omega(f' - f) \end{aligned} \tag{16.3.5}$$

where $\Gamma = n_0|\mathbf{v}|\sigma$ is the collision frequency. For the equilibrium distribution f_0 set up by the binary collisions, C of Eq.(16.3.5) vanishes and for the time development of perturbations to f_0, we will consider the effect of the fixed scatterers alone in this section. The algebra is much simpler than if we were dealing with equilibration by binary collision and hence the basic mathematical technique will be easier to implement.

If we produce a small deviation from the equilibrium distribution, so that

$$F = F_0 + \delta f \tag{16.3.6}$$

then δf satisfies

$$\frac{\partial}{\partial t}\delta f + \mathbf{v}.\frac{\partial}{\partial \mathbf{r}}\delta f = \Gamma \int d\Omega(\delta f' - \delta f) \tag{16.3.7}$$

For the spatial variation of δf, we assume again a slowly varying oscillatory form (as being the smallest deviation from a uniform state) and thus write

$$\delta f(\mathbf{r}, \mathbf{v}, t) = \psi(\mathbf{v}, t)e^{i\mathbf{k}.\mathbf{r}}. \tag{16.3.8}$$

Inserting in Eq.(16.3.7), we obtain

$$\frac{\partial \psi}{\partial t} = -K\psi = -(K_0 + K_1)\psi \tag{16.3.9}$$

where

$$K_0\psi = \Gamma \int d\Omega(\psi - \psi') \tag{16.3.10}$$

and

$$K_1\psi = i\mathbf{k}.\mathbf{v}\psi. \tag{16.3.11}$$

If we can determine the eigenvalues and eigenfunctions of K (λ_n and ϕ_n respectively, $n = 0, 1, 2, ...$) then the solution ψ can be written as

$$\psi = \sum a_n \phi_n e^{-\lambda_n t} \tag{16.3.12}$$

where a_n are constants which can be determined from the initial condition.

For the equilibrium to make sense, we must have the λ_n such that $\text{Re}(\lambda) > 0$ for all n, as otherwise there will be a runaway solution and the equilibrium will not be stable. Further, of all the λ_n our interest is only in the lowest λ, i.e., λ_0 that determines the persistent time scale. If one waits for a time $T \gg \lambda_0^{-1}$, all other terms become negligible and only the term $a_0\phi_0 e^{-\lambda_0 t}$

in Eq.(16.3.12) is of appreciable magnitude. To find λ_n, one has to employ perturbation theory. The operator K_1 having an explicit factor of k is small and hence can be treated perturbatively once the eigenvalues of K_0 are determined. In general, we will be interested in the lowest eigenvalue of K_0. In the special case of Eq.(16.3.10), however, all the eigenvalues and eigenfunctions of K_0 can be exactly determined.

Inspection of Eq.(16.3.10) shows that $\psi =$ constant is an eigenfunction of K_0 with eigenvalue 0. On the other hand, all functions ϕ having the property $\int \phi d\Omega = 0$ are also eigenfunctions of K_0 with eigenvalue $4\pi\Gamma$

$$K_0\phi = \Gamma \int d\Omega(\phi - \phi') = \Gamma\delta \int d\Omega = 4\pi\Gamma\phi \qquad (16.3.13)$$

(note that ϕ' is the function that is Ω-dependent and is integrated over). A set of functions ϕ having the property $\int \phi d\Omega = 0$ are the spherical harmonics Y_{lm} with $l = 0$, $(m = -l, -(l-1)....l-1, l$ for a given l and $l = 1, 2, ...)$. For $l = 0$, Y_{00} is a constant and hence we have the following complete set of eigenfunctions for the operator K_0.

As is well known, Y_{lm} forms a complete set on the space of all solid angles and thus we have solved for K_0 completely. The relevant eigenvalue is clearly $\lambda = 0$ since the correction to it will be proportional to some power of k and hence small. For small k, correction to the other eigenvalue, $4\pi\Gamma$, is not going to be important.

The next step is the evaluation of the correction to $\lambda_0 = 0$ due to the operator K_1. In first-order perturbation theory, the correction is $\langle K_1 \rangle$ which is clearly zero, i.e. $\lambda_0^{(1)} = 0$. The second-order perturbation theory contribution to λ_0 is

$$\lambda_0^{(2)} = \sum_{ex} \frac{(\int \phi_{ex} K_1 \phi_0)}{0 - \lambda_{ex}} \qquad (16.3.14)$$

where ϕ_{ex} stands for all the excited states and λ_{ex} are their respective eigenvalues. In this case

$$\lambda_0^{(2)} = \sum_{l(\neq 0), m} \frac{(\int d\Omega Y_{lm} K_1 Y_{00})^2}{-4\pi\Gamma} \qquad (16.3.15)$$

Now $K_1 = -ikv\cos\theta$ and hence $\int d\Omega Y_{lm} \cdot \cos\theta \cdot Y_{00}$ is non-zero only if $Y_{lm} = Y_{10} = \sqrt{3/4\pi} \cos\theta$ (since Y_{00} is a constant, the orthogonality of Y_{lm} leads to a nonzero integral only for Y_{10}). Thus, $\int \sqrt{1/4\pi}\sqrt{3/4\pi} \cos^2\theta d\Omega = 1/\sqrt{3}$ and Eq.(16.3.15) becomes

$$\lambda_0^{(2)} = \frac{k^2v^2}{4\pi\Gamma} \cdot \frac{1}{3} = \frac{k^2v^2}{2\Gamma'}. \qquad (16.3.16)$$

Where $\Gamma' = 4\pi\Gamma$. Hence to second order in perturbation theory, $\lambda_0 = k^2v^2/3\Gamma'$ and similarly for the eigenvalue $4\pi\Gamma$, $\lambda_{ex} = 4\pi\Gamma + O(k^2) \approx 4\pi\Gamma$

leading to (from Eq.(16.3.12))

$$\psi = a_0 e^{-\frac{k^2 v^2}{3\Gamma'}} + \sum_{l,m(\neq 0)} a_{lm} Y_{lm} e^{-4\pi\Gamma'} \qquad (16.3.17)$$

The result is now transparent. If one waits long enough, only the first term is going to survive if k is small enough. To estimate how small k must be, it is clear that for the first term to survive

$$\frac{k^2 v^2}{\Gamma} << \Gamma$$

or

$$k^2 << \frac{\Gamma^2}{v^2} = \frac{1}{l_{\text{mfp}}^2} \qquad (16.3.18)$$

which implies that the wavelength must be much greater than the mean free path (l_{mfp}). We thus see that with a fixed set of scatterers, a small deviation from equilibrium distribution slowly disappears as the particles undergo diffusive motion on collision with the scatterers.

The general strategy for attacking all problems involving deviation from equilibrium of the collision integral is now clear.

1. Linearize the collision integral in the deviation δf and assuming the spatial variation of the kind shown in Eq.(16.3.8), set up a time evolution equation of the form of Eq.(16.3.9).

2. Find all the zero eigenvalues of K_0.

3. Treat K_1 perturbatively and obtain the first non-vanishing contribution to the zero eigenvalue of K_0.

4. Interpret the eigenvalues and eigenfunctions if possible.

16.4   Binary Collisions

We are now ready to tackle the binary collision integral. A small deviation from the equilibrium distribution is imagined and for later convenience we write this deviation as

$$f = f_0(1 + \delta f). \qquad (16.4.1)$$

The term $f_2 f_3 - f f_1$, when linearized in terms of δ becomes

$$\begin{aligned} f_2 f_3 - f f_1 &= f_{20} f_{30} - f_0 f_{10} + f_{20} f_{30}(\delta f_2 + \delta f_3) \\ &\quad - f_0 f_{10}(\delta f + \delta f_{10}) + O(\delta f)^2 \\ &= f_0 f_{10}(\delta f_2 + \delta f_3 - \delta f_1 - \delta f) + O((\delta f)^2) \end{aligned}$$

where $f_0 f_{10} = f_{20} f_{30}$ for the equilibrium distribution.

The analogue of Eq.(16.3.7) is

$$\frac{\partial}{\partial t}\delta f + \mathbf{v}\cdot\frac{\partial}{\partial \mathbf{r}}\delta f = \int d^3v_1 d\Omega |\mathbf{v} - \mathbf{v_1}|\sigma(\Omega)f_{10}(\delta f_2 + \delta_3 - \delta f_1 - \delta f) \quad (16.4.2)$$

Now with $\delta f = \psi(\mathbf{v},t)e^{i\mathbf{k}\cdot\mathbf{r}}$

$$\frac{\partial \psi}{\partial t} = -K\psi = -K_0\psi - K_1\psi \quad (16.4.3)$$

where

$$K_0\psi = \int d^3v_1 d\Omega |\mathbf{v} - \mathbf{v_1}|\sigma(\Omega)f_{10}(\psi_2 + \psi_3 - \psi_1 - \psi) \quad (16.4.4)$$

$$K_1\psi = i\mathbf{k}\cdot\mathbf{v}\psi. \quad (16.4.5)$$

Before proceeding any further, we need to define a scalar product of two vectors ψ and ϕ in the velocity space as

$$(\psi,\phi) = \int \psi(\mathbf{v})f_0(v)\phi(\mathbf{v})d^3v. \quad (16.4.6)$$

Having defined the scalar product, we now address the question of zero eigenvalues of K_0, if any. This automatically settles the question of the lowest eigenvalue because K_0 cannot have any negative eigenvalues. We establish this by proving

$$(\psi_0, K_0\psi) \geq 0 \quad (16.4.7)$$

The proof is as follows:

$$(\psi, K_0\psi) = \int d^3v_1 d\Omega |\mathbf{v} - \mathbf{v_1}|\sigma(\Omega)f_0 f_{10}(\psi_2 + \psi_3 - \psi_1 - \psi)\psi$$

$$= \int d^3v_1 d^3v |\mathbf{v} - \mathbf{v_1}|\sigma(\mathbf{v_1}, \mathbf{v} \rightarrow \mathbf{v_2}, \mathbf{v_3})$$

$$\times \quad f_0 f_{10}(\psi_2 + \psi_3 - \psi_1 - \psi)\psi\ldots\ldots\ldots\ldots(a)$$

$$= \int d^3v_1 d^3v |\mathbf{v} - \mathbf{v_1}|]\sigma(\mathbf{v_1}, \mathbf{v}] \rightarrow \mathbf{v_2}, \mathbf{v_3})$$

$$\times \quad f_0 f_{10}(\psi_2 + \psi_3 - \psi_1 - \psi)\psi\ldots\ldots\ldots\ldots(b)$$

(by interchanging $\mathbf{v}$ and $\mathbf{v_1}$)

$$= \int d^3v_2 d^3v_3 |\mathbf{v_2} - \mathbf{v_3}|\sigma(\mathbf{v_2}, \mathbf{v_3} \rightarrow \mathbf{v_1}, \mathbf{v})f_{20}f_{30}(\psi + \psi_1 - \psi_2 - \psi_3)\psi_2$$

(by exchanging $\mathbf{v}$ and $\mathbf{v_2}$, $\mathbf{v_1}$ and $\mathbf{v_3}$) $\ldots\ldots\ldots\ldots(c)$

$$= \int d^3v_2 d^3v_3 |\mathbf{v_2} - \mathbf{v_3}|\sigma(\mathbf{v_2}, \mathbf{v_3} \rightarrow \mathbf{v_1}, \mathbf{v})f_{20}f_{30}(\psi + \psi_1 - \psi_2 - \psi_3)\psi_3$$

(by exchanging $\mathbf{v}$ and $\mathbf{v_3}$, $\mathbf{v_1}$ and $\mathbf{v_2}$) $\ldots\ldots\ldots\ldots(d)$

Noting that $f_{10}f_0 = f_{20}f_{30}$, $|\mathbf{v} - \mathbf{v_1}| = |\mathbf{v_2} - \mathbf{v_3}|$, $d^3v_2 d^3v_3 = d^3v d^3v_1$ and $\sigma(\mathbf{v_1}, \mathbf{v} \to \mathbf{v_2}, \mathbf{v_3}) = \sigma(\mathbf{v_2}, \mathbf{v_3} \to \mathbf{v}, \mathbf{v_1})$ (see Chapter 15) we can add Eqs.(a)-(d) above to write

$$(\psi, K_0\psi) = \frac{1}{4} \int d^3v_1 d^3v |\mathbf{v} - \mathbf{v_1}| \sigma(\mathbf{v}, \mathbf{v_1} \to \mathbf{v_2}, \mathbf{v_3}) f_0 f_{10} (\psi + \psi_1 - \psi_2 - \psi_3)^2 \geq 0.$$

$$(16.4.8)$$

This completes the proof.

We now list the zero eigenvalues of K_0. It is obvious from Eq.(16.4.4) that the following wavefunctions lead to $\lambda_0 = 0$ (each of them is a collision invariant and hence $\psi + \psi_1 = \psi_2 + \psi_3$).

$$(i) \quad \psi \;=\; \text{constant}$$
$$(ii) \quad \psi \;=\; p_x, p_y, p_z$$
$$(iii) \quad \psi \;=\; p^2/2m.$$

As in our previous example, the other eigenvalues are of the order Γ, the collision frequency.

The effect of the perturbation operator K_1 on the zero eigenvalue now needs to be calculated. For this purpose we need to employ first order perturbation theory as appropriate to degenerate states since the zero eigenvalue is five-fold degenerate. The procedure for degenerate state perturbation theory is:

(i) Form normalized orthogonal functions out of the wavefunctions of the degenerate state.

In our case we call these wavefunctions $\phi_1, \phi_2, \phi_3, \phi_4, \phi_5$ and write

$$\phi_1 \;=\; 1$$
$$\phi_2 \;=\; \sqrt{\frac{2}{3}} \frac{1}{k_B T} \left(\frac{1}{2} m v^2 - \frac{3}{2} k_B T \right)$$
$$\phi_3 \;=\; \sqrt{\frac{m}{k_B T}} v_z$$
$$\phi_4 \;=\; \sqrt{\frac{m}{k_B T}} v_y$$
$$\phi_5 \;=\; \sqrt{\frac{m}{k_B T}} v_x.$$

$$(16.4.9)$$

It should be verified that the above ϕ_i do satisfy $(\phi_i, \phi_j) = \delta_{ij}$. While the conserved quantities in ϕ_2, ϕ_3, ϕ_4, and ϕ_5 are clear, ϕ_1 can be taken to represent the number density.

(ii) The second step is to diagonalize the perturbation operator in the space of the above orthonormal set.

In this case, we have to from the matrix elements $(K_1)_{ij} = (\phi_i, K_1\phi_j)$. For this purpose it helps to choose the direction of the $\mathbf{k}$ vector. Without any loss of generality we can consider the wavevector in the z-direction. The operator K_1 is ikv_z and clearly all its diagonal elements are zero. Further

$(\phi_4, K_1\phi_i) = (\phi_5, K_1\phi_i) = 0$ for all i and hence we simply need to consider $(K_1)_{ij}$ where i and j range from 1 to 3. Straightforward calculation of the matrix elements of K_1 leads to the matrix

$$K_1 = ik\frac{k_B T}{m}\begin{pmatrix} 0 & 1 & 0 \\ 1 & 0 & \sqrt{2/3} \\ 0 & \sqrt{2/3} & 0 \end{pmatrix}. \qquad (16.4.10)$$

The eigenvalues of the above matrix are 0 and $\pm ik\sqrt{5k_B T/3m}$. Thus, the first-order perturbation theory does not lift the degeneracy altogether. For a wavevector in the z-direction the momentum wavefunctions in the x- and y-direction and a combination of the number and energy (ϕ_1) and ϕ_3 wavefunctions: a combination, which can be identified with the entropy, continue to be degenerate. Finally, in first-order perturbation theory the perturbation δf has the form

$$\delta f = A_0 e^{ik(z \pm \sqrt{5k_B T/3mt})} + \sum_{m=1}^{\infty} A_m e^{-\lambda_m t} \qquad (16.4.11)$$

where the eigenvalues λ_m are of the order Γ and hence for times greater than Γ^{-1} we have

$$\delta f \approx A_0 e^{ik(z \pm \sqrt{5k_B T/3mt})} \qquad (16.4.12)$$

with three eigenvalues still zero which can acquire nonzero values only from second-order perturbation theory and hence are expected to be of order k^2.

The result at $O(k)$ exhibited in Eq.(16.4.12) shows the propagation of a wave along the z-direction with velocity being $\sqrt{5k_B T/3m}$. It is the familiar sound wave with the velocity appropriate to a monatomic gas correctly obtained. The isentropic condition is also correctly obtained as the eigenvalue corresponding to the entropy variable continues to be zero, indicating the conservation of entropy.

Proceeding to second order in perturbation theory, we note that the three-fold degenerate zero eigenvalue will be split producing eigenvalues of $O(k^2)$. Exactly as in Sec.16.3, the contribution of second-order perturbation theory has the typical form $\alpha v^2 k^2/\Gamma$, where v is an average velocity and α is a number of order unity. The modes ϕ_4 and ϕ_5 produce a coefficient $\propto v^2/\Gamma$ which corresponds to shear viscosity, while the combination of modes ϕ_1 and ϕ_3 that corresponds to entropy, produces a coefficient which is the thermal diffusivity. The modes corresponding to the sound wave which give rise to an imaginary eigenvalue at $O(k)$ also acquire a correction at $O(k^2)$ from second-order perturbation theory: the coefficient now corresponding to a combination of thermal and mass diffusivity. The actual calculations of the constants α are laborious, but some idea of what they could be, can be had by writing the typical velocity as $v \approx l_{\text{mfp}}\Gamma$, whence the dissipative (diffusive) contributions $\alpha v^2/\Gamma$ are of the form $\alpha l_{\text{mfp}}.v$. This should be compared with the elementary kinetic theory results: kinematic shear viscosity = thermal diffusivity = mass diffusivity = $(1/3)l_{mfp}v$.

16.5   The H-theorem

The foregoing discussion shows that the equilibrium distribution obtained by
setting the collision term $C = 0$ in Eq.(16.3.1) is stable against small perturbations due to the collision term itself. It is the collision term which is responsible
for the setting up of equilibrium and for providing the form of the equilibrium
distribution. In this section, we establish that the condition $f_2 f_3 = f f_1$ that
obviously sets $C = 0$, is not only sufficient but also necessary. We proceed by
defining

$$H = \int f \ln f d^3 v \qquad (16.5.1)$$

whence

$$\frac{dH}{dt} = \int \frac{\partial f}{\partial t}(1 + \ln f) d^3 v. \qquad (16.5.2)$$

For $\partial f / \partial t$ to be zero it is necessarily true that dH/dt has got to be zero. In
what follows we will assume an absence of external force and the distribution f
to be independent of $\mathbf{r}$, and hence the evolution will be governed by $\partial f / \partial t = C$.
Explicitly writing out the collision term we obtain

$$\frac{dH}{dt} = \int d^3 v d^3 v_1 \sigma(\mathbf{v}, \mathbf{v_1} \to \mathbf{v_2}, \mathbf{v_3}) d\Omega |\mathbf{v} - \mathbf{v_1}| (f_2 f_3 - f f_1) \times (1 + \ln f).$$
$$(16.5.3)$$

The next few steps are identical to those carried out in arriving at
Eq.(16.4.8) [essentially recognizing that the above integral is unaffected by in-
terchanging $\mathbf{v}$ and $\mathbf{v_1}$, also $\mathbf{v}$ with $\mathbf{v_2}$ and $\mathbf{v_1}$ with $\mathbf{v_3}$ as well as $\mathbf{v}$ with $\mathbf{v_3}$ and
$\mathbf{v_1}$ with $\mathbf{v_2}$]. We easily obtain

$$\frac{dH}{dt} = \frac{1}{4} \int d^3 v d^3 v_1 |\mathbf{v} - \mathbf{v_1}| \sigma(\mathbf{v}, \mathbf{v_1} \to \mathbf{v_2}, \mathbf{v_3}) d\Omega (f_2 f_3 - f f_1)(\ln f f_1 - \ln f_2 f_3).$$
$$(16.5.4)$$

The above shows that $\frac{dH}{dt}$ is never positive. this is known as the H-
theorem. For equilibrium , we need

$$\int d^3 v d^3 v_1 |\vec{v} - \vec{v_1}| \sigma(\vec{v}, \vec{v_1} \to \vec{v_2}, \vec{v_3}) d\Omega (f_2 f_3 - f f_1)(\ln f f_1 - \ln f_2 f_3) = 0.$$

The integral cannot change sign and hence $\frac{dH}{dt} = 0$ leads to $f f_1 = f_2 f_3$.
Thus, proving that the condition for equilibrium is necessary requires the use
of Boltzmann's H - theorem. The equality holds only if $f f_1 = f_2 f_3$ and hence
it is a necessary condition for the equilibrium distribution.

17

Transport Equation - III

17.1 A Theorem

In this chapter we discuss conservation laws and hydrodynamics. In the previous chapter, we have repeatedly seen that quantities which are conserved in the elastic collision play a privileged role in the solution of the Boltzmann transport equation. We can now prove a general result: if χ is a conserved quantity in a binary elastic collision, i.e. $\chi_1 + \chi_2 = \chi_3 + \chi_4$, where subscripts 1 and 2 refer to the quantities before collision while 3 and 4 refer to the same quantities after collision, then

$$\int \frac{\partial f}{\partial t}\bigg|_{\text{coll}} \chi d^3 v = 0. \tag{17.1.1}$$

The proof uses a technique that has been widely used in Chapter 16. We write

$$\int \frac{\partial f}{\partial t}\bigg|_{\text{coll}} \chi d^3 v_1 = \int d^3 v_1 d^3 v_2 \sigma(\vec{v}_1, \vec{v}_2 \to \vec{v}_3, \vec{v}_4) d\Omega |\vec{v}_1 - \vec{v}_2|(f_3 f_4 - f_1 f_2)\chi_1. \tag{17.1.2}$$

Now the following interchanges cannot change the integral on the right hand side:

i) interchange $\vec{v}_1$ and $\vec{v}_2$

ii) interchange $\vec{v}_1$ with $\vec{v}_3$ and $\vec{v}_2$ with $\vec{v}_4$

iii) interchange $\vec{v}_1$ with $\vec{v}_4$ and $\vec{v}_2$ with $\vec{v}_3$.

Each of these interchanges leaves $|\vec{v}_1 - \vec{v}_2|$, $d^3v_1 d^3v_2$ and $\sigma(\vec{v}_1, \vec{v}_2 \rightarrow \vec{v}_3, \vec{v}_4)d\Omega$ unchanged and hence

$$
\begin{aligned}
\int \chi_1 \left.\frac{\partial f}{\partial t}\right|_{\text{coll}} d^3v_1 &= \int d^3v_1 d^3v_2 \sigma(\vec{v}_1, \vec{v}_2 \rightarrow \vec{v}_3, \vec{v}_4)d\Omega|\vec{v}_1 - \vec{v}_2| \\
&\times (f_3 f_4 - f_1 f_2)\chi_1 \\
&= \int d^3v_1 d^3v_2 \sigma(\vec{v}_1, \vec{v}_2 \rightarrow \vec{v}_3, \vec{v}_4)d\Omega|\vec{v}_1 - \vec{v}_2| \\
&\times (f_3 f_4 - f_2 f_1)\chi_2 \\
&\quad (1 \leftrightarrow 2 \text{ interchanged}) \\
&= \int d^3v_1 d^3v_2 \sigma(\vec{v}_1, \vec{v}_2 \rightarrow \vec{v}_3, \vec{v}_4)d\Omega|\vec{v}_1 - \vec{v}_2| \\
&\times (f_1 f_2 - f_3 f_4)\chi_3 \\
&\quad (\text{First } 1 \leftrightarrow 3 \text{ and } 2 \leftrightarrow 4 \text{ interchanged,} \\
&\quad \text{then } 1 \leftrightarrow 4 \text{ and } 2 \leftrightarrow 3 \text{ interchanged}) \\
&= \frac{1}{4} \int d^3v_1 d^3v_2 \sigma(\vec{v}_1, \vec{v}_2 \rightarrow \vec{v}_3, \vec{v}_4)d\Omega|\vec{v}_1 - \vec{v}_2| \\
&\times (f_3 f_4 - f_1 f_2)(\chi_1 + \chi_2 - \chi_3 - \chi_4) \\
&= 0
\end{aligned}
$$

which completes the proof.

Using the transport equation (Eq.(15.4.11)) we now rewrite Eq.(17.1.1) as

$$
\int \chi \left(\frac{\partial f}{\partial t} + \vec{v} \cdot \frac{\partial f}{\partial \vec{r}} + \frac{\vec{F}}{m} \cdot \frac{\partial f}{\partial \vec{v}} \right) d^3v = 0. \tag{17.1.3}
$$

This is the general statement of the conservation law.

17.2   Transport Equation for Conserved Quantities

For any conserved quantity χ, we want to transform the conservation law of Eq.(17.1.3) into an equation for the time dependence of the average $\langle \chi \rangle$, which is defined as

$$
\begin{aligned}
\langle \chi \rangle &= \int \chi f d^3v / \int f d^3v \\
&= \int \chi f d^3v / n(\vec{r}, t) \tag{17.2.1}
\end{aligned}
$$

where

$$
n(\vec{r}, t) = \int d^3v f(\vec{r}, v, t) \tag{17.2.2}
$$

which is a space- and time-dependent number density. From Eq.(17.2.1), we find

$$
\int \chi f d^3v = n\langle \chi \rangle = \langle n\chi \rangle \tag{17.2.3}
$$

where in the last step we have taken n inside the angular bracket since the angular bracket averaging is over velocity, while $n(r, t)$ has already been averaged over velocity. With the above definition we can rewrite Eq.(17.1.3) as

$$\frac{\partial}{\partial t} \int \chi f d^3 v + \int \left[\frac{\partial}{\partial \vec{r}} \cdot (\chi f \vec{v}) - f \vec{\nabla} . (\chi \vec{v}) \right] d^3 v$$

$$+ \frac{1}{m} \int \left[\frac{\partial}{\partial \vec{v}} (\chi f \vec{F}) - f \vec{F} . \frac{\partial \chi}{\partial \vec{v}} - \chi f \frac{\partial}{\partial \vec{v}} . \vec{F} \right] d^3 v$$

$$= 0.$$

$$(17.2.4)$$

Now,

$$\int \frac{\partial}{\partial \vec{v}} (\chi f \vec{F}) d^3 v = \int_S \chi f \vec{F} . d\vec{S}_v = 0 \qquad (17.2.5)$$

under the very reasonable assumption that for very large velocities, the distribution function f tends to zero.

If we ignore velocity-dependent forces, then

$$\frac{\partial \vec{F}}{\partial \vec{v}} = 0 \qquad (17.2.6)$$

and using the definition that, in general, for any quantity A,

$$\langle A \rangle = \int A f d^3 v / n \qquad (17.2.7)$$

we can write Eq.(17.2.4) as

$$\frac{\partial}{\partial t} \langle n \chi \rangle + \vec{\nabla} . \langle n \chi \vec{v} \rangle - \langle n \vec{\nabla} . (\chi \vec{v}) \rangle - \langle n \frac{F}{m} . \frac{\partial \chi}{\partial \vec{v}} \rangle = 0. \qquad (17.2.8)$$

This is the form of the time development for the expectation value of a conserved quantity χ. For the binary collision there are three such conserved qualities χ.

i) a trivial constant which we can take as the mass m of the individual colliding particles

ii) the momentum $\vec{p} = m\vec{v}$ of the individual particles

iii) the thermal kinetic energy $\epsilon = \frac{1}{2} m (\vec{v} - \langle \vec{v} \rangle)^2$
The above form is a linear combination of conserved quantities (namely, v^2, v and constant) and hence is itself a conserved quantity.
We take up the three cases separately.
i) If $\chi = m$, we can write Eq.(17.2.8) as

$$\frac{\partial}{\partial t} \langle mn \rangle + \vec{\nabla} . \langle (mn\vec{v}) \rangle - \langle n \vec{\nabla} . (m\vec{v}) \rangle = 0 \qquad (17.2.9)$$

the last term in Eq.(17.2.8) not contributing for obvious reasons. Noting that $mn = \rho$ the mass density, and that $\vec{\nabla}.\vec{v} = 0$ as $\vec{v}$ and $\vec{r}$ are independent of each other, we have

$$\frac{\partial}{\partial t}\langle\rho\rangle + \vec{\nabla}.\langle(\rho\vec{v})\rangle = 0. \qquad \cdot \qquad (17.2.10)$$

If we write $\langle\vec{v}\rangle = \vec{u}$, which can be a function of $\vec{r}$, and note that $\langle\rho\rangle$ since ρ is a function of $\vec{r}$ and t alone, we can write the above as the continuity equation

$$\frac{\partial\rho}{\partial t} + \vec{\nabla}.(\rho\vec{u}) = 0. \qquad (17.2.11)$$

It should be borne in mind that the distribution function f which we are using to average over $\vec{v}$ is not, in general, an even function of $\vec{v}$ and hence $\langle\vec{v}\rangle \neq 0$. The function f is a solution of the transport equation (discussed in Chapter 16).

ii) If $\vec{\chi} = m\vec{v}$, then for the ith component $\chi_i = mv_i$, we have from Eq.(17.2.8)

$$\frac{\partial}{\partial t}\langle mnv_i\rangle + \frac{\partial}{\partial x_j}(mnv_iv_j) - \langle n\frac{\partial}{\partial x_j}\rangle(mv_iv_j) - \langle nF_j\frac{\partial v_i}{\partial v_j}\rangle = 0$$

or

$$\frac{\partial}{\partial t}\langle\rho v_i\rangle + \frac{\partial}{\partial x_j}\langle\rho v_iv_j\rangle = n\langle F_i\rangle \qquad (17.2.12)$$

where use has been made of the fact that **v** and **r** are independent variables. We now define a tensor

$$T_{ij} = \rho\langle(v_i - u_i)(v_j - u_j)\rangle = \rho\langle v_iv_j\rangle - \rho u_iu_j \qquad (17.2.13)$$

having made use of the fact $\langle v_i\rangle = u_i$. In terms of this tensor, Eq.(17.2.12) becomes

$$\frac{\partial\rho}{\partial t}u_i + \rho\frac{\partial u_i}{\partial t} + \frac{\partial}{\partial x_j}T_{ij} + \frac{\partial}{\partial x_j}(\rho u_iu_j) = \frac{\langle F_i\rho\rangle}{m}$$

leading to

$$\left[\frac{\partial\rho}{\partial t} + \frac{\partial}{\partial x_j}(\rho u_j)\right]u_i + \rho\frac{\partial u_i}{\partial t} + \rho u_j\frac{\partial u_i}{\partial x_j} = -\frac{\partial}{\partial x_j}T_{ij} + \frac{\langle F_i\rho\rangle}{m}$$

or

$$\frac{\partial u_i}{\partial t} + u_j\frac{\partial u_i}{\partial x_j} = -\frac{1}{\rho}\frac{\partial}{\partial x_j}T_{ij} + \frac{1}{m}\langle F_i\rangle \qquad (17.2.14)$$

using Eq.(17.2.11) in arriving at the final form.

iii) If $\chi = m\epsilon = m.\frac{m}{2}|\vec{v} - \vec{u}|^2$, i.e., ϵ is the kinetic energy of motion relative to the mean, then Eq.(17.2.8) becomes

$$\frac{\partial}{\partial t}\langle mn\epsilon\rangle + \vec{\nabla}.\langle mn\vec{v}\epsilon\rangle - \langle mn(\vec{v}.\vec{\nabla})\frac{m}{2}|(\vec{v} - \vec{u})|^2\rangle = \langle mn\vec{F}.(\vec{v} - \vec{u})\rangle = 0$$

$$\rho\frac{\partial\langle\epsilon\rangle}{\partial t} + \langle\epsilon\rangle\frac{\partial\rho}{\partial t} + \vec{\nabla}.\langle\rho\epsilon\vec{v}\rangle - m\langle\rho(v_i - u_i)(v_j - u_j)\rangle\frac{\partial u_j}{\partial x_i} = 0.$$

$$(17.2.15)$$

If we define

$$\vec{\theta} = \rho\epsilon(\vec{v} - \vec{u})$$ (17.2.16)

and note that $\rho\langle(v_i - u_i)(v_j - u_j)\rangle = T_{ij}$, then Eq.(17.2.15) becomes

$$\langle\epsilon\rangle\frac{\partial\rho}{\partial t} + \rho\frac{\partial}{\partial t}\langle\epsilon\rangle + \vec{\nabla}.\langle\vec{\theta}\rangle + \vec{\nabla}.\langle(\rho\epsilon\vec{u})\rangle + mT_{ij}\frac{\partial u_j}{\partial x_i} = 0$$

or

$$\rho\frac{\partial}{\partial t}\langle\epsilon\rangle + \rho(\vec{u}.\vec{\nabla})\langle\epsilon\rangle + \langle\epsilon\rangle\left[\vec{\nabla}.(\rho\vec{u}) + \frac{\partial\rho}{\partial t}\right] = -\vec{\nabla}\langle\vec{\theta}\rangle - mT_{ij}\frac{\partial u_j}{\partial x_i}$$

or

$$\frac{\partial}{\partial t}\langle\epsilon\rangle + (\vec{u}.\vec{\nabla})\langle\epsilon\rangle = -\frac{1}{\rho}\vec{\nabla}.\langle\vec{\theta}\rangle - \frac{m}{2\rho}T_{ij}\left(\frac{\partial u_j}{\partial x_i} + \frac{\partial u_i}{\partial x_j}\right)$$ (17.2.17)

where the symmetry of T_{ij} has been used in the last term.

The five conservation laws are thus expressed through Eqs.(17.2.11), (17.2.14) for $i = 1, 2, 3$ and Eq.(17.2.17). To make them more useful, however, we must be able to calculate T_{ij} and $\langle\theta_i\rangle$.

17.3 Equations of Hydrodynamics

As stated at the end of the last section, the conservation laws become usable only if we know the distribution function f, as otherwise it is not possible to calculate the expectation values. Since solving the transport equation and obtaining f is virtually impossible, we have to make an *ansatz* for it. We take it to be locally Maxwell-Boltzmann with a mean velocity $\vec{u}$, i.e.,

$$\vec{f}(\vec{v}, \vec{r}, t) = n\left(\frac{m}{2\pi kT}\right)^{3/2}\exp\left[-\frac{m(\vec{v} - \vec{u})^2}{2k_BT}\right]$$ (17.3.1)

where n and T (the temperature) are both local variables depending on $\vec{r}$ and t. Clearly

$$\langle v_i\rangle = u_i$$ (17.3.2)

and

$$T_{ij} = \rho\langle(v_i - u_i)(v_j - u_j)\rangle = \frac{k_BT}{m}\rho\delta_{ij}.$$ (17.3.3)

If we consider a local equation of state as

$$P = nk_BT = \rho\frac{k_BT}{m}$$ (17.3.4)

then

$$T_{ij} = P\delta_{ij}$$ (17.3.5)

and Eq.(17.2.14) becomes Euler's equation of hydrodynamics. With the above distribution $\langle\theta_i\rangle = 0$ and hence Eq.(17.2.17) becomes

$$\rho\frac{\partial}{\partial t}\langle\epsilon\rangle + \rho(\vec{u}.\vec{\nabla})\langle\epsilon\rangle = -mP\frac{\partial u_i}{\partial x_i}.$$ (17.3.6)

It should be noted that

$$\langle \epsilon \rangle = \langle \frac{1}{2}m(\vec{v} - \vec{u})^2 \rangle = \frac{3}{2}k_B T \qquad (17.3.7)$$

and substituting in Eq.(17.3.6), we obtain

$$\frac{3}{2}k_B \left(\frac{\partial T}{\partial t} + (\vec{u}.\vec{\nabla})T \right) + \rho k_B T \frac{\partial u_i}{\partial x_i} = 0$$

which gives the time dependence of the local temperature as

$$\frac{\partial T}{\partial t} + (\vec{u}.\vec{\nabla})T = -\frac{2}{3}T(\vec{\nabla}.\vec{u}). \qquad (17.3.8)$$

The quantity $\partial/\partial t + \vec{u}.\vec{\nabla}$ is the convective derivative since it is the time rate of change as seen by an observer moving with the local average velocity $\vec{u}$. Such an observer moves along a streamline. We note that at this order of evaluation,

$$\frac{3}{2T} \left(\frac{\partial}{\partial t} + \vec{u}.\vec{\nabla} \right) T = -(\vec{\nabla}.\vec{u}) \qquad (17.3.9)$$

while from the equation of continuity

$$\frac{1}{\rho} \left(\frac{\partial}{\partial t} + \vec{u}.\vec{\nabla} \right) \rho = -(\vec{\nabla}.\vec{u}). \qquad (17.3.10)$$

Subtracting Eq.(17.3.9) from Eq.(17.3.10), we get

$$\left(\frac{\partial}{\partial t} + \vec{u}.\vec{\nabla} \right) \rho T^{-3/2} = 0$$

or

$$\rho T^{-3/2} = \text{constant along a streamline.} \qquad (17.3.11)$$

If we use $P = \rho k T/m$, then $P\rho^{-5/3}$ is constant along the streamline which is the adiabatic law since for an ideal gas $\gamma = C_p/C_v = 5/3$ and we have found that $PV = $ constant. Thus for the streaming observer only adiabatic processes occur in a dilute gas in this approximation.

Finally, using Eq.(17.3.5) in Eq.(17.2.1), the velocity is found to satisfy

$$\frac{\partial u_i}{\partial t} + u_j \frac{\partial u_i}{\partial x_j} = -\frac{1}{\rho} \frac{\partial P}{\partial x_i} + \frac{F_i}{m}. \qquad (17.3.12)$$

If we have steady state conditions ($\partial/\partial t = 0$) under a conservative force $F_i = -m\vec{\nabla}\phi$, then we can write Eq.(17.3.12) as

$$\vec{\nabla} \left(\frac{1}{2}u^2 + \frac{P}{\rho} + \frac{\phi}{m} \right) = \vec{u} \times (\vec{\nabla} \times \vec{u}) - \frac{k_B T}{m} \frac{\vec{\nabla}\rho}{\rho}$$

where the equation of state and the identity $(\vec{u}.\vec{\nabla})\vec{u} = 1/2\vec{\nabla}(u^2) - \vec{u} \times (\vec{\nabla} \times \vec{u})$ have been used. Two cases can be noted:

i) if the flow is irrotational ($\vec{\nabla} \times \vec{u} = 0$) and the density is uniform ($\rho =$ constant)

$$\frac{u^2}{2} + \frac{P}{\rho} + \frac{\phi}{m} = \text{constant in space.} \tag{17.3.13}$$

ii) if the flow is irrotational and the temperature is constant then

$$\rho = \rho_0 \exp\left[-\frac{1}{k_B T}\left(\frac{1}{2}mu^2 + \phi\right)\right] \tag{17.3.14}$$

What is missing from the above hydrodynamic equations are the dissipative terms. This is due to the particular form of the distribution that we have chosen. With that as the starting *ansatz*, we can build up corrections to it. These corrections can be obtained to the lowest order by solving the transport equation in the relaxation time approximation. This involves writing $f = f + \delta f$, with the collision term replaced by an expression of the form $-\delta f/\tau$, where τ is the relaxation time. In the absence of external forces, δf satisfies the relation $\partial \delta f/\partial t + \vec{v}.\partial f/\partial \vec{r} = -\delta f/\tau$ and re-evaluation of the averages using $f = f + \delta f$, we arrive at Navier-Stokes equation and the heat conduction equations including the shear viscosity and thermal conductivity in terms of the relaxation time τ.

18

Transport Equation - IV

18.1 Form of the Transport Equation for Quantum Particles

In this chapter we discuss the transport equation for quantum particles. We have so far discussed situations where quantum effects were not important since the De Broglie wavelength ($\lambda \sim h\sqrt{mk_BT}$) was much smaller than the mean free path (inversely proportional to the density). At low temperatures or high densities this approximation breaks down and we have to consider the quantum effects. If we consider the transport equation for classical particles [see Chapter 15, Section-4] we find that the left hand side of the equation is simply the total rate of change in the particle number and hence is independent of the classical or quantum nature of the particles. The right hand side, on the other hand, depends on the collision process going on in the system and that is going to be sensitive to the nature of the particles. This is immediately obvious if the particles considered are identical and are fermions. The requirement that there can be no more than one fermion in a given state implies that in the scattering process all final states which are occupied cannot be accessed and this will change the collision integral. The case of the bosons is more subtle but certainly there will be an effect on the collision integral as we will see in what follows. Hence the effect of the quantum nature of the particles will be showing up in the collision term and it will depend on whether the particles are fermions or bosons.

Accordingly, the collision term will be altered to take into account the quantum nature of the particles. However, the Boltzmann equation that we will be writing down is sensible only in the quasi-classical approximation. It is only in this limit that it is permissible to talk about $f(\mathbf{r}, \mathbf{v}, t)$. In a strict quantum theory, uncertainty principle puts constraints on the simultaneous measurements of $\mathbf{r}$ and $\mathbf{v}$. In the semiclassical picture that we will be employing, one can interpret the $\mathbf{r}$ and $\mathbf{v}$ as quantum expectation values with the quantum fluctuations small. It is in this spirit that we keep the left hand side of Eq.(15.4.11) unaltered and proceed to alter the collision term depending on whether the particles are fermions or bosons.

18.1.1   Fermions

As we have already stated the collision term has to account for the fact that all occupied final states are inaccessible by collisions. Thus, if we consider the number of particles with velocity in the vicinity of $\mathbf{v}$, then the depletion rate due to the elastic two-body scattering process for classical particles was found in Eq.(15.4.3). The fermionic nature implies that in the scattering process where a particle with velocity v collides with an identical particle with velocity $\mathbf{v_1}$ producing final states with velocities $\mathbf{v_2}$ and $\mathbf{v_3}$, the scattering cross- section will be zero if there is already a particle with velocity $\mathbf{v_2}$ or $\mathbf{v_3}$ in the physical volume d^3r under consideration. To incorporate this fact, we need to multiply the expression for N_{out} (following the notation of Chapter 15, Section-4) by the factor $(1 - f_2)(1 - f_3)$ which ensures that if either of the possible final states is occupied, the cross-Section vanishes. Hence the depletion rate is

$$\int d^3r d^3v_i d^3v\sigma(\mathbf{v}, \mathbf{v_1} \to \mathbf{v_2}, \mathbf{v_3})|\mathbf{v} - \mathbf{v_1}|$$
$$\times \quad f(\mathbf{r}, \mathbf{v}, t)f(\mathbf{r}, \mathbf{v_1}, t)[(1 - f(\mathbf{r}, \mathbf{v_3}, t)][(1 - f(\mathbf{r}, \mathbf{v_2}, t))].$$

Similarly, if we consider the rate of addition to the number of particles in the volume d^3r with velocities in the neighbourhood of velocity $\mathbf{v}$, the additional factor that will enter the expression for N_{in} (again following the notation of Chapter 15, Section-4) is $(1 - f)(1 - f_1)$ which ensures that the scattering cannot occur if there is already a particle with velocity $\mathbf{v}$ or $\mathbf{v_1}$. Hence the addition rate is

$$\int d^3r d^3r_2 d^3v_3\sigma(\mathbf{v_2}, \mathbf{v_3} \to \mathbf{v}, \mathbf{v_1})|\mathbf{v_2} - \mathbf{v_3}|$$
$$\times \quad f(\mathbf{r}, \mathbf{v_2}, t)f(\mathbf{r}, \mathbf{v}, t)[1 - f(\mathbf{r}, \mathbf{v}, t)][1 - f(\mathbf{r}, \mathbf{v_1}, t)].$$

The Boltzmann equation follows from arguments identical to those given in Chapter 15 and we have the Boltzmann transport equation for fermions as

$$\frac{\partial f}{\partial t} + \mathbf{v} \cdot \frac{\partial f}{\partial \mathbf{r}} + \frac{\mathbf{F}}{m} \cdot \frac{\partial f}{\partial \mathbf{v}} = \int d^3v_1|\mathbf{v} - \mathbf{v_1}|\sigma(\mathbf{v}, \mathbf{v_1} \to \mathbf{v_2}, \mathbf{v_3})d\Omega$$
$$\times \quad [f_2 f_3(1 - f)(1 - f_1) - f f_1(1 - f_2)(1 - f_3)].$$
$$(18.1.1)$$

18.1.2   Bosons

For bosons there is no restriction to the number of particles that can occupy a given state. The consequence of this on a scattering process can be found in textbooks on quantum mechanics (e.g. see Feynman Lectures, Vol. III). The result is that in a scattering process if there are N particles already in a potential final state, then the probability of scattering into that state is enhanced by the factor $(N + 1)$. Hence when we consider the depletion rate due to the two-body scattering, the cross-section is enhanced if there are particles present with velocities v_2 or v_3, the velocities of the final state in the scattering process, and the factor which expresses this is $(1 + f_2)(1 + f_3)$. Similarly, the rate of addition of particles in the phase space volume $d^3r d^3v$ is enhanced by the factor $(1 + f)(1 + f_1)$ since the final state in this two-body process now has particles with velocities v and v_1. It should be obvious that the transport equation for bosons will consequently be

$$\frac{\partial f}{\partial t} + \mathbf{v} \cdot \frac{\partial f}{\partial \mathbf{r}} + \frac{\mathbf{F}}{m} \cdot \frac{\partial f}{\partial \mathbf{v}} = \int d^3 v_1 |\mathbf{v} - \mathbf{v_1}| \sigma(\mathbf{v}, \mathbf{v_1} \to \mathbf{v_2}, \mathbf{v_3}) d\Omega$$
$$\times \quad [f_2 f_3 (1 + f)(1 + f_1) - f f_1 (1 + f_2)(1 + f_3)].$$

$$(18.1.2)$$

18.2 Equilibrium Distribution

As in the case of classical particles, the equilibrium distribution is first obtained for the specific case of no external force, i.e., $\mathbf{F} = 0$. In this case the distribution f_0 is independent of $\mathbf{r}$ and since in equilibrium $\partial f_0 / \partial t = 0$, we must have the collision term vanish when f is taken to be the equilibrium distribution. In the following, we will be using the upper sign for fermions and the lower sign for bosons. In equilibrium conditions, we thus have

$$f_0 f_{10} (1 \pm f_{20})(1 \pm f_{30}) = f_{20} f_{30} (1 \pm f_0)(1 \pm f_{10})$$

or

$$\frac{f_0}{1 \pm f_0} \times \frac{f_{10}}{1 \pm f_{10}} = \frac{f_{20}}{1 \pm f_{30}}$$

or

$$\ln \frac{f_0}{1 \pm f_0} + \ln \frac{f_{10}}{1 \pm f_{10}} = \ln \frac{f_{20}}{1 \pm f_{20}} + \ln \frac{f_{30}}{1 \pm f_{30}}. \qquad (18.2.1)$$

In complete analogy with the classical case, we now argue that in $f_0/(1 \pm f_0)$ must be a linear function of the collision invariant. Hence,

$$\ln \frac{f_0}{1 \pm f_0} = \alpha - \beta \epsilon + \gamma . \mathbf{v_0}.$$

We can choose the reference velocity to be zero without any loss of generality and obtain

$$\ln \frac{f_0}{1 \pm f_0} = \alpha - \beta \epsilon$$

or

$$f_0 = \frac{1}{e^{\epsilon \beta - \alpha} \pm 1}. \qquad (18.2.2)$$

The constants β and α now need to be specified. To do so, we note that the classical distribution obtained in chapter 15 must be a limiting case of the quantum distribution of Eq.(18.2.2) at high temperatures and low densities. The low density condition implies $f_0 \ll 1$, which is achieved if we drop the constant term in the denominator of Eq.(18.2.2). This leads to

$$f_0 \approx e^{\alpha}e^{-\beta\epsilon} \tag{18.2.3}$$

and comparing with the last equation of Chapter 15, we infer that

$$\beta = (k_BT)^{-1} \tag{18.2.4}$$

$$e^{\alpha} = n\left(\frac{m}{2\pi k_BT}\right)^{3/2}. \tag{18.2.5}$$

As can be checked by referring to chapter 6, $\alpha = \mu\beta$, where μ is the chemical potential. Thus for fermions,

$$f_0 = \frac{1}{e^{(\epsilon-\mu)/k_BT}+1} \tag{18.2.6}$$

while for bosons

$$f_0 = \frac{1}{e^{(\epsilon-\mu)/k_BT}-1}. \tag{18.2.7}$$

A brief comment should be made here about the interpretation of $f_0(\mathbf{v})$ in this quantum case. The classical interpretation was in terms of a density in phase space, the number of particles in the volume $d^3r d^3v$ being given by $f_0(\mathbf{v})d^3r d^3v$ in equilibrium. In the quantum case it is more appropriate to interpret $f_0(\mathbf{v})$ as the average number of particles in the state with velocity v (the average being over different realizations) and $d^3r d^3v$ as being the number of states available to the quantum particle in that region of phase space multiplied by the factor h^3/m^3.

For the fermions, we notice that if the limit $T \to 0$ is taken, then the distribution reduces to a theta function, with $f_0(\mathbf{v}) = 1$, if $\epsilon = mv^2/2 < \mu$ and $f_0(\mu) = 0$ if $\epsilon > \mu$. All states below the energy μ, called the fermi energy ϵ_F are thus fully occupied and those above the energy μ are totally empty.

$$f_0(\epsilon) = \theta(\epsilon_F - \epsilon), \quad \text{for } T = 0 \tag{18.2.8}$$

where the theta function is defined by $\theta(x) = 1$ if $x > 0$ and $\theta(x) = 0$ if $x > 0$. For the bosons, it should be noted that the average number of particles with $\epsilon = 0$ is $(e^{-\mu/k_BT} - 1)^{-1}$ and this constrains μ to be less than or equal to zero. If we consider bosons, for which the number conservation need not be valid then we must have $\alpha = 0$ in Eq.(18.2.2) and consequently $\mu = 0$. This is true for bosons which can be viewed as excitations in a system characterized as a collection of oscillators (photons in a cavity, phonons arising from the vibrations of the atoms in a solid about their equilibrium position or oscillations in density in a liquid at very low temperatures, specifically helium which alone remains a liquid at temperatures where quantum effects matter).

18.3 Approach to Equilibrium (F≠0 but no Collisions)

We now see how a system of fermions behaves when a small deviation from equilibrium conditions is made. We ignore the collisions and consider the weak interaction between the particles as responsible for bringing the system towards equilibrium. This is analogous to the case of plasma oscillations dealt with in Chapter 18, where for a system of charged particles, against a neutralizing background we studied deviations from equilibrium. It was the Coulomb forces which came into play on departure from equilibrium in that case that was responsible for the oscillations. We now consider an uncharged system of fermions: this is provided by atoms at extremely low temperatures. There is an attractive force (weak) between atoms but in an equilibrium situation, however, the net force on each atom by the collection of the other atoms is zero. The equilibrium distribution is consequently given by Eq.(18.2.8). We take the weak interaction between atoms to be governed by some potential $u(\mathbf{r})$ and if there is a deviation δf from the equilibrium distribution, then the deviation in local number density is $\delta n(\mathbf{r}, t)$ and the potential at the point r will be

$$\delta\phi = \int u(\mathbf{r} - \mathbf{r}')\delta n(\mathbf{r}', t)d^3r'. \tag{18.3.1}$$

The potential is first order in δn on and hence the force F is also first order in δn. The transport equation in the form $\partial f/\partial t = \mathbf{v}.\partial f/\partial\mathbf{r} + \mathbf{F}/m.\partial f/d\mathbf{v} = 0$ then becomes to first order in δn,

$$\frac{\partial}{\partial t}\delta f + \mathbf{v}.\frac{\partial}{\partial\mathbf{r}}\delta f - \frac{\vec{\nabla}\delta\phi}{m}.\frac{\partial f_0}{\partial\mathbf{v}} = 0. \tag{18.3.2}$$

Assuming that the deviation δf will be oscillatory in nature, we can make the *ansatz*

$$\delta f = Ae^{i(\omega t - \mathbf{k}.\mathbf{r})} \tag{18.3.3}$$

$$\delta n = Be^{i(\omega t - \mathbf{k}.\mathbf{r})} \tag{18.3.4}$$

where B is a constant and A is a function of velocity such that

$$B = \int A(\mathbf{v})d^3v. \tag{18.3.5}$$

We can write the force $\vec{\nabla}\delta\phi$ as

$$\vec{\nabla}\delta\phi = \vec{\nabla}\int u(\mathbf{r} - \mathbf{r}')\delta n(\mathbf{r}', t)d^3r'$$

$$= Be^{i\omega t}\vec{\nabla}\left[e^{-i\mathbf{k}.\mathbf{r}}\int u(\mathbf{r} - \mathbf{r}')e^{i\mathbf{k}.(\mathbf{r}-\mathbf{r}')}d^3r'\right]$$

$$= Be^{i\omega t}u(\mathbf{k})(-i\mathbf{k})e^{-i\mathbf{k}.\mathbf{r}} \tag{18.3.6}$$

where $u(\mathbf{k}) = \int d^3 r u(\mathbf{r}) e^{i\mathbf{k}\cdot\mathbf{r}}$ is the Fourier transform of $u(\mathbf{r})$ and Eq.(18.3.2) becomes

$$A(i\omega - i\mathbf{k}\cdot\mathbf{v})e^{i(\omega t - \mathbf{k}\cdot\mathbf{r})} = -Bu(\mathbf{k})\left(i\mathbf{k}\cdot\frac{\partial f}{\partial \mathbf{v}}\right)e^{i(\omega t - \mathbf{k}\cdot\mathbf{r})}. \tag{18.3.7}$$

Now, $\partial f_0/\partial \mathbf{v} = -m\mathbf{v}\delta(\epsilon_F - \frac{1}{2}m\mathbf{v}^2)$ and hence Eq.(18.3.7) becomes

$$A = Bmu(\mathbf{k})\frac{\mathbf{k}\cdot\mathbf{v}}{\omega - \mathbf{k}\cdot\mathbf{v}}\delta(\epsilon_F - \frac{1}{2}m\mathbf{v}^2). \tag{18.3.8}$$

Insertion of the value of A from Eq.(18.3.8) in Eq.(18.3.5) yields

$$\begin{aligned}
1 &= mu(\mathbf{k})\int \frac{kv\cos\theta}{\omega - kv\cos\theta}\delta\left(\epsilon_F - \frac{1}{2}m\tilde{v}^2\right)d^3v \\
&= 2\pi mu(\mathbf{k})\int_{-1}^{1}\frac{\cos\theta d(\cos\theta)}{\frac{\omega}{k}\left(\frac{m}{2\epsilon_F}\right)^{1/2} - \cos\theta}\frac{2}{m}\frac{1}{2v_F}\cdot v_{\mathbf{F}}^2 \\
&= 2\pi v_{\mathbf{F}} u(\mathbf{k})\left[\ln\frac{\frac{c}{v_{\mathbf{F}}} + 1}{\frac{c}{v_{\mathbf{F}}} - 1} - 2\right] \tag{18.3.9}
\end{aligned}$$

where $\frac{1}{2}m v_{\mathbf{F}}^2 = \epsilon_F$ and $\omega = ck$. Clearly if $u(\mathbf{k}) \to 0$, then $c \to |\mathbf{v}|$ for Eq.(18.3.9) to make sense. For small $u(\mathbf{k})$, we accordingly get

$$\frac{c}{v_F} = 1 + 2\exp\left[-\frac{1}{2\pi u(\mathbf{k})v_F}\right]. \tag{18.3.10}$$

Thus, the interaction between the helium atoms allows a wave to be set up in this system of fermions and the wave velocity is given by Eq.(18.3.10) which is a remarkable result in the sense that it is non-perturbative in the strength of the interaction. The deviation of the propagation speed from the Fermi velocity $v_{\mathbf{F}}$ is small if the interaction is small, but this smallness cannot be expressed in a perturbation series. These oscillations are known as zero sound. They are very different from the usual sound (first sound) wave which, as we have seen, for the classical gas, can be propagated only in the presence of collisions. In the next section, we will investigate the effect of collisions and find the velocity of first sound at very low temperatures. However, it should be noted that as $T \to 0$, the damping of first sound increases very strongly and hence the zero sound oscillations dominate in that temperature range.

18.4   Effect of Collisions in a Fermion Gas

We now consider a situation where the external force $\mathbf{F} = 0$ and a deviation from equilibrium has to relax or propagate by the action of the collision term.

The Boltzmann equation now takes the form

$$\frac{\partial f}{\partial t} + \mathbf{v} \cdot \frac{\partial f}{\partial \mathbf{r}} = \int d^3 v d\Omega \sigma |\mathbf{v} - \mathbf{v}_1| \cdot [f_2 f_3 (1 - f)(1 - f_1) - f f_1 (1 - f_2)(1 - f_3)]$$

$$(18.4.1)$$

with the equilibrium distribution

$$f_0 = (e^{(\epsilon - \mu)/k_B T} + 1)^{-1}.$$

$$(18.4.2)$$

The deviation from f_0 is taken to be oscillatory in space with a long wavelength (exactly as in the classical case in Chapter 16) and for convenience we take the small deviation from equilibrium to have the explicit form:

$$f = f_0 + f_0 (1 - f_0) e^{i \mathbf{k} \cdot \mathbf{r}} \psi(\mathbf{v}, t)$$

$$(18.4.3)$$

where $\psi(\mathbf{v}, t)$ is to be found from the explicit solution of the linearized form Eq.(18.4.1). The f-dependent part of the collision integral of Eq.(18.4.1) when linearized in ψ is found by noting that:

$$
\begin{aligned}
f_2 f_3 (1 - f)(1 - f_1) &= f_{20}[1 + (1 - f_{20})\psi_2] f_{30}[1 + (1 - f_{30})\psi_3] \\
&\times [(1 - f_0)(1 - f_0 \psi)][(1 - f_{10})(1 - f_{10}\psi_1)] \\
&= f_{20} f_{30} (1 - f_0)(1 - f_{10}) \\
&\times [1 + (1 - f_{20})\psi_2 + (1 - f_{30})\psi_3 - f_0 \psi - f_{10} \psi]
\end{aligned}
$$

and in a similar way

$$
\begin{aligned}
f_2 f_3 (1 - f_2)(1 - f_3) &= f_{10} f_0 (1 - f_{20})(1 - f_{30}) \\
&\times [1 + (1 - f_{10})\psi + (1 - f_0)\psi_1 - f_{20}\psi_2 - f_{30}\psi_3]
\end{aligned}
$$

leading to

$$f_2 f_3 (1-f)(1-f_1) - f f_1 (1-f_2)(1-f_3) = f_0 f_{10} (1-f_{20})(1-f_{30})[\psi_2 + \psi_3 - \psi_1 - \psi].$$

$$(18.4.4)$$

The transport equation in terms of ψ, becomes

$$\frac{\partial \psi}{\partial t} + i\mathbf{k} \cdot \mathbf{v} \psi = \int d^3 v d\Omega \sigma |\mathbf{v} - \mathbf{v}_1| \frac{f_{10}(1 - f_{20})(1 - f_{30})}{1 - f_0} (\psi_2 + \psi_3 - \psi_1 - \psi)$$

$$(18.4.5)$$

or

$$\frac{\partial \psi}{\partial t} = -K\psi = -K_0 \psi - K_1 \psi$$

$$(18.4.6)$$

where

$$K_0 \psi = \int d^3 v d\Omega \sigma |\mathbf{v} - \mathbf{v}_1| \frac{f_{10}(1 - f_{20})(1 - f_{30})}{1 - f_0} (\psi_2 + \psi_3 - \psi_1 - \psi)$$

$$(18.4.7)$$

$$K_1 \psi = -i\mathbf{k} \cdot \mathbf{v} \psi.$$

$$(18.4.8)$$

Exactly as in Chapter 16, the lowest eigenvalues of the operator K will determine the time dependence of ψ. Since k is small, this lowest eigenvalue is to be found perturbatively in K_1 and the zero eigenvalues of K_0 are the important ones provided K_0 has no negative eigenvalues. By defining the scalar product as

$$(\phi, \psi) = \int \phi(\mathbf{v}) f_0 (1 - f_0) \psi(\mathbf{v}) d^3 v \qquad (18.4.9)$$

it is clear (the reader should prove this using arguments identical to those given in Chapter 16) that

$$(\psi, K_0 \psi) \geq 0 \qquad (18.4.10)$$

and K_0 has non-negative eigenvalues. Consequently the zero eigenvalue alone of K_0 is important and this eigenvalue is five-fold degenerate corresponding to the five collision invariants: the number of particles, the three components of momentum and the kinetic energy. It should be noted that the weighting factor $f_0(1 - f_0)$ in the definition of scalar product in Eq.(18.4.9) is of significance. The probability of a fermion occupying a state is f_0: consequently $1 - f_0$ is the probability of a state being unoccupied or as is customarily described for $\epsilon < \epsilon_F$ as being occupied by a hole. The product $f_0(1 - f_0)$ has to do with the probability of finding electron-hole pairs and this is natural since the collision term acts by producing such pairs – the collision process leading to the formation of an electron state as well as a hole state.

To simplify the calculation we consider the propagation at very low temperatures where the distribution f_0 is essentially the theta function $\theta(\epsilon_F - \epsilon)$ and hence the factor $f_0(1 - f_0)$ behaves like a slightly smeared delta function $\delta(\epsilon - \epsilon_F)$. For integration purposes it is convenient and accurate to take it as delta function. If we consider this form of the weight function, then there is no difference between the wavefunction which is a constant and the wavefunction which corresponds to the kinetic energy since all energies must be considered at the Fermi surface and that effectively makes the energies a constant. In other words, for $T \to 0$ the only pairs which are formed are at the Fermi surface.

We now take the propagation of the disturbance to be along the z-direction, i.e., $\mathbf{k}$ has a z-component alone. The degenerate zero-eigenvalue is now treated in first order perturbation theory. Because of the energy effectively becoming a constant, there are now only four linearly independent eigenfunctions. We need to diagonalize $K_1 = ikv_z$ in the space of these four functions. Obviously he eigenfunctions corresponding to p_x and p_y do not couple at all and hence K_1 has to be diagonalized in the space of the two functions:

  i) $\phi_1 = $ constant $= A$ (say) and ii) $\phi_2 = Bp_z$.

Clearly $(\phi_1, K_1\phi_1) = (\phi_2, K_1\phi_2) = 0$ and we need to worry about the off diagonal elements alone. First, the states $\phi_{1,2}$ need to be normalized. We take the phase space factor to be $\epsilon^{1/2} d\epsilon d\Omega / a\pi$ where constant factors have been dropped without any loss of generality for the purpose at hand. We have,

$$(\phi_1, \phi_1) = A^2 \int \epsilon^{1/2} \delta(\epsilon - \epsilon_F) d\epsilon \frac{d\Omega}{4\pi} = A^2 \epsilon_F^{1/2}.$$

Hence

$$A^2 = \epsilon_F^{-1/2},$$

and

$$
\begin{aligned}
(\phi_2, \phi_2) &= B^2 \int p_z^2 \epsilon^{1/2} \delta(\epsilon - \epsilon_F) d\epsilon d\Omega / 4\pi \\
&= \frac{B^2}{3} \int p^2 \epsilon^{1/2} \delta(\epsilon - \epsilon_F) d\epsilon \\
&= \frac{2mB^2}{3} \int \epsilon^{3/2} \delta(\epsilon - \epsilon_F) d\epsilon \\
&= \frac{2m}{3} B^2 \epsilon_F^{3/2}
\end{aligned}
$$

leading to $B^2 = (3/2m)\epsilon_F^{-3/2}$. The off-diagonal matrix elements can be calculated as

$$
\begin{aligned}
(\phi_1, K_1 \phi_2) &= AB \frac{ik}{m} \int p_z^2 \epsilon^{1/2} \delta(\epsilon - \epsilon_F) d\epsilon \frac{d\Omega}{4\pi} \\
&= \frac{ik2AB}{3} \int \epsilon^{3/2} \delta(\epsilon - \epsilon_F) d\epsilon \\
&= \frac{2}{3} ik AB \epsilon_F^{3/2}.
\end{aligned}
$$

The matrix to be diagonalized is

$$
M = \begin{pmatrix} 0 & \frac{2}{3}ik \ AB\epsilon_F^{3/2} \\ \frac{2}{3}ik \ AB\epsilon_F^{3/2} & 0 \end{pmatrix}.
$$

If the eigenvalue is λ, then it is clear that $\lambda = \pm i\omega$, where we must have

$$\omega^2 = \frac{4}{9} \epsilon_F^3 A^2 B^2 k^2 = \frac{2}{3} \frac{\epsilon_F}{m} k^2 = \frac{1}{3} \mathbf{v_F}^2 k^2$$

writing $\omega = ck$

$$c = \pm \frac{|\mathbf{v_F}|}{\sqrt{3}} \tag{18.4.11}$$

leading to the result that for $T \to 0$ the first sound propagates at the speed $|\mathbf{v_F}|/\sqrt{3}$ which is very different from the speed of the zero sound (Eq. 18.3.9) under the same conditions.

To obtain the damping, we need to go to second order in perturbation theory and obtain the correction to the zero eigenvalue of K_0. This correction is of the form

$$\sum_{m \neq 0} \frac{\langle 0| ik \cos\theta |m \rangle^2}{E_0 - E_m} \sim k^2 \frac{M}{E_1} \tag{18.4.12}$$

where M is $\langle 0| \cos^2 \theta |0 \rangle$ and E_1 is the first nonzero eigenvalue of K_0. This eigenvalue is proportional to the number of electron-hole pairs which as $T \to 0$

is proportional to T_2 and thus the damping diverges as $T \to 0$ obliterating the sound wave in spite of its finite velocity. Thus as $T \to 0$ only the zero sound propagates.

18.5   Collision Term for Bosons

As an application of the transport equation to the case of bosons, we choose a gas of phonons which is obtained when one has density oscillations in a low temperature liquid (superfluid helium) or atomic vibrations in a solid. The Boltzmann equation is ($\mathbf{F} = 0$, no external force)

$$\frac{\partial f}{\partial t} + \mathbf{v}.\frac{\partial f}{\partial \mathbf{r}} = \int d^3 v_1 d\Omega \sigma |\mathbf{v} - \mathbf{v_1}| [f_2 f_3 (1 + f_1)(1 + f) - f f_1 (1 + f_2)(1 + f_3)]$$

$$(18.5.1)$$

with the equilibrium distribution function for phonons being ($\mu = 0$)

$$f_0 = \frac{1}{e^{\epsilon/k_B T} - 1}. \tag{18.5.2}$$

We consider a deviation from equilibrium and write the general distribution f as

$$f = f_0 + f_0(1 + f_0)e^{i\mathbf{k}.\mathbf{r}}.\psi(\mathbf{v}, t). \tag{18.5.3}$$

If we linearize in ψ, then proceeding exactly as in Sec.18.4, we obtain the dynamics of ψ as being governed by

$$\frac{\partial \psi}{\partial t} = -K\psi = -(K_0\psi + K_1\psi) \tag{18.5.4}$$

where

$$K_0\psi = \int d^3 v_1 d\Omega |\mathbf{v} - \mathbf{v_1}| \frac{f_1}{1 + f_0}(1 + f_3)(\psi_2 + \psi_3 - \psi_1 - \psi) \tag{18.5.5}$$

and

$$K_1\psi = i\mathbf{k}.\mathbf{v}\psi. \tag{18.5.6}$$

The scalar product is defined as

$$(\phi, \psi) \int \phi(\mathbf{v}) f_0 (1 + f_0) \psi(\mathbf{v}) d^3 v \tag{18.5.7}$$

which ensures $(\psi, K_0\psi) \geq 0$. Hence the eigenvalues of K_0 are equal to or greater than zero.

Once again the zero eigenvalue is the one which we are interested in. This is fourfold degenerate since number is no longer a conserved quantity. The four eigenvectors are consequently (p_x, p_y, p_z) and the energy ϵ. Once again, we consider the propagation direction to be along the z-axis, so that the perturbation operator is $K_1 = ikp_z/m$. First-order perturbation theory requires the diagonalization of K_1 in the four-dimensional space. Once again p_x and p_y cannot couple and hence the diagonalization is to be carried out in the $p_z - \epsilon$ subspace. Before proceeding further, we need to clarify what p and ϵ are for a phonon. The phonons are energy packets in a sound field and each packet carries energy $\epsilon = \hbar\omega$, where ω is the frequency of the sound wave. If c is the speed of sound and k the wavenumber, then $\omega = ck$ and the momentum of the packet is given by $p = \hbar k = \hbar\omega/c = \epsilon/c$. Once again, dropping irelevent constant factors, one can take the phase space differential element to be $\epsilon^{1/2}d\epsilon d\Omega/4\pi$ and the two functions ϕ_1 and ϕ_2 in the space of which K_1 is to be diagonalized as $\phi_1 = Ap$ and $\phi_2 = B\epsilon$, with the normalization conditions

$$
\begin{aligned}
1 &= (\phi_1, \phi_1) \\
&= A^2 \int p_z^2 f_0 (1 = f_0) \epsilon^{1/2} d\epsilon \frac{d\Omega}{4\pi} \\
&= \frac{A^2}{3} \int p^2 f_0 (1 + f_0) \epsilon^{1/2} d\epsilon \\
&= \frac{A^2}{3c^2} \int \epsilon^{5/2} f_0 (1 + f_0) d\epsilon
\end{aligned}
\tag{18.5.8}
$$

and

$$
\begin{aligned}
1 &= (\phi_2, \phi_2) \\
&= B^2 \int \epsilon^{5/2} f_0 (1 + f_0) d\epsilon s\Omega/4\pi \\
&= B^2 \cdot \frac{3c^2}{A^2}
\end{aligned}
$$

or

$$
A^2/B^2 = 3c^2
\tag{18.5.9}
$$

where we have made use of Eq.(18.5.8). The diagonal elements of K_1 are zero. The off-diagonal element is (note that v for the phonon is the sound speed)

$$
\begin{aligned}
ik \int c.\cos\theta A p_z B\epsilon \frac{d\Omega}{4\pi}\epsilon &= ikc.AB \int p\epsilon \cos^2\theta f_0 p (1 + f_0) \epsilon^{1/2} d\epsilon \frac{d\Omega}{4\pi} \\
&= ik\frac{AB}{3} \int \epsilon^{5/2} f_0 (1 + f_0) d\epsilon \\
&= ik\frac{AB}{3} \cdot \frac{3c^2}{A^2} \\
&= ikc^2 \frac{B}{A}.
\end{aligned}
$$

The matrix to be diagonalized is

$$
M = \begin{pmatrix} 0 & ikc^2\frac{B}{A} \\ ikc^2\frac{B}{A} & 0 \end{pmatrix}.
$$

The eigenvalues λ of M are of the form $i\omega$, where clearly

$$\omega^2 = k^2 c^2 \frac{B^2}{A^2} = k^2 \frac{c^2}{3}.$$

Writing $\omega = c_2 k$, c_2 being the propagation speed of the wave in the phonon gas,

$$c_2 = \pm \frac{c}{\sqrt{3}} \tag{18.5.10}$$

so that the speed of propagation of a disturbance in the phonon gas is related to the sound speed by the factor of $\sqrt{3}$. Phonons dominate in super fluid Helium, as $T \to 0$, and hence at these temperatures, the propagation of second sound, as these waves are called, occurs at a speed of $c/\sqrt{3}$. The experimental confirmation of this in the early forties was one of the spectacular successes of Landau's theory of superfluidity.

19

Metastable States

19.1 Introduction

Our discussion in Chapter 16 has shown that if a small disturbance from equilibrium conditions is made, the system relaxes back to equilibrium or oscillates about the equilibrium position. The relaxation towards the equilibrium distribution is brought about by collisions or interparticle interactions. In all such cases one managed to make progress because it was permissible to linearize around the equilibrium solution. What we have not yet dealt with is a situation where the system is far from equilibrium and attempts to make a transition to the equilibrium state. Dealing with a situation far from equilibrium is analytically difficult and one needs to make some simplifying assumptions. Often, in studying systems far from equilibrium it may so happen that the system may be trapped in a state of local equilibrium. Eventually the transition to the true equilibrium will occur but the time scales involved will be very large. Such states are called metastable states.

One of the simple situations where a system can be put far from equilibrium is to consider the existence of an external force derivable from a potential. If the potential has multiple minimum, then the equilibrium distribution will be peaked in the global minimum. If, however, one considers a distribution peaked in one of the local minima, then this acts as a pseudo-equilibrium state and it takes a lot of time for the distribution to leak out of this state. This is an example of a metastable state.

In the presence of a conservative external force $F = -\vec{\nabla}\phi$, the equilibrium distribution can be written as $f_0(r,v) = f_0(v)g(r)$, where $f_0(v)$ is obtained as before from the vanishing of the collision term and for classical particles will be the usual Maxwell-Boltzmann distribution, while $g(r)$ will have to be determined from the condition that to satisfy the transport equation [see Chapter 15, Section 4], we must have

$$f_0 \mathbf{v} \cdot \vec{\nabla} g(r) + (-\vec{\nabla}\phi) \cdot \frac{\partial f_0}{\partial \mathbf{v}} g(r) = 0. \qquad (19.1.1)$$

But f_0 being the Boltzmann distribution, we have

$$\frac{\partial f_0}{\partial v} = -\frac{m}{k_B T} v f_0,$$

so that Eq.(19.1.1) becomes

$$\frac{\vec{\nabla} g(r)}{g(r)} - \frac{m}{k_B T} \vec{\nabla} \phi = 0 \qquad (19.1.2)$$

implying

$$g(r) = e^{-m\phi/k_B T}. \qquad (19.1.3)$$

The equilibrium distribution is consequently given by $f_0 = Ae^{-m(v^2/2+\phi)/k_B T}$ and it is clear that if ϕ has a minimum, then there will be concentration of the distribution there and if ϕ has multiple minima, then the local minima are the metastable states.

19.2   A Phenomenological Approach

One of the alternatives to using the transport equation is to use the conservation laws as one learnt in Chapter 16. The conservation law corresponding to the number of particles is the equation of continuity

$$\frac{\partial \rho}{\partial t} + \vec{\nabla}.\mathbf{J} = \frac{\partial \rho}{\partial t} = \vec{\nabla}.(\rho \mathbf{v}) = 0. \qquad (19.2.1)$$

The problem is the evaluation of the current $\mathbf{J} = \rho \mathbf{v}$ since this requires the solution of the transport equation. It is here that a phenomenological approach is effective in the form of a connection between the current $\mathbf{J}$ and the density ρ. If there is a gradient in ρ, then there will be a diffusive flow of particles from higher to lower concentrations and this contributes $-D\vec{\nabla}\rho$ to the current where D is a constant. Because of the external force $\mathbf{F}$, there will be an Ohm's law-like contribution ($\mathbf{J} = \sigma \mathbf{E}$) to the current which can be written as $\gamma \mathbf{F}\rho$; thus, we have

$$\begin{aligned} J &= -D\vec{\nabla}\rho + \gamma \mathbf{F}\rho \\ &= -D\vec{\nabla}\rho - \gamma\rho\vec{\nabla}\phi. \end{aligned} \qquad (19.2.2)$$

The equilibrium condition obviously implies no time dependence and zero current and hence the equilibrium distribution ρ_0 is given by

$$\rho_0 = Ae^{-(\gamma/D)\phi} \qquad (19.2.3)$$

where A is a constant. Comparison of Eqs.(19.2.3) and (19.1.2) shows that

$$\frac{\gamma}{D} = \frac{m}{k_B T} \qquad (19.2.4)$$

which is a form of the fluctuation-dissipation theorem.

As an example of the calculation of the current **J** we take the special case of a one-dimensional problem involving charged particles . The particles are in a constant force field **F** which exists up to $x = b$ (this is not an electric field).

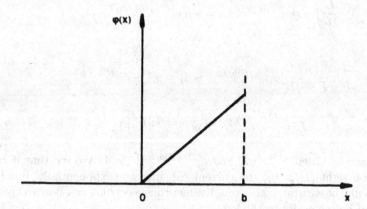

Figure 19.1: A one-dimensional barrier potential.

The potential energy $\phi(x)$ will be (see Fig.19.1)

$$\phi(x) \;=\; Fx \quad \text{for } 0 < x < b,$$
$$\;=\; 0 \quad \text{otherwise.}$$

Particles incident from the left with energy less than Fb will be reflected at $x = b$. Under equilibrium conditions there is no current in the setup. We now impose an electric field qE to the right and find the current. The new potential is

$$\phi'(x) \;=\; Fx - qEx \quad \text{for } 0 < x < b,$$
$$\;=\; -qEx \quad \text{otherwise.}$$

and the current can be written as

$$
\begin{aligned}
J &= -D\frac{\partial\rho}{\partial x} - \frac{D}{k_B T}\frac{\partial\phi'}{\partial x}\rho \\
&= -D\left[\frac{\partial\rho}{\partial x} + \frac{1}{k_B T}\frac{\partial\phi'}{\partial x}\rho\right] \\
&= -De^{-\phi'/k_B T}\frac{\partial}{\partial x}\left(\rho e^{\phi'/k_B T}\right)
\end{aligned}
$$

or

$$\frac{J}{D}e^{\phi'/k_B T} = -\frac{\partial}{\partial x}\left(\rho e^{\phi'/k_B T}\right). \tag{19.2.5}$$

We can now integrate from $x = 0$ to $x = b$, keeping in mind that J is independent of x in the steady state situation and hence $\partial\rho/\partial t = 0$. Also ρ

is constant outside the region $0 < x < b$ under the same conditions. From Eq.(19.2.5), we then obtain

$$\frac{J}{D}\int_0^b dx e^{(F-qE)x/k_BT} = -\rho[e^{(F-qE)b/k_BT} - 1]$$

or

$$\frac{J}{D}\cdot\frac{k_BT}{F-qE}[e^{(F-qE)b/k_BT} - 1] = \rho[1 - e^{-qEb/k_BT}]$$

or

$$J = \frac{DF}{k_BT}e^{-Fb/k_BT}\rho(e^{qEb/k_BT} - 1) \qquad (19.2.6)$$

where we have assumed $F \gg qE$ and $e^{(F-qE)b/k_BT} \gg 1$. We see that if the field is to the right ($E > 0$), the current can increase exponentially, but for $E < 0$, the current saturates at a small value: thus electrons can flow easily to the right and not so easily to the left.

19.3   Decay of Metastable States

We now see how a metastable state will evolve towards equilibrium and over what time scales. We consider a potential per unit mass $\phi(x)$ with two minima – one the true minimum and the other a metastable state. The potential is shown in Fig.19.2.

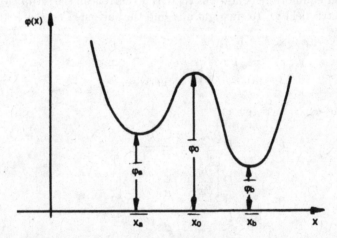

Figure 19.2: An asymmetric double-well potential. The local minimum at x_a corresponds to a metastable state.

The minima at $x = x_a$ and x_b have the values ϕ_a and ϕ_b respectively. The intervening maximum at $x = x_0$ has the value $\phi = \phi_0$. The equilibrium distribution is going to be proportional to $e^{-m\phi/k_BT}$ and hence the maximum population is going to be at $x = x_b$ where ϕ has a true minimum.

We assume that a far-from-equilibrium distribution is formed where the population is put entirely in the left hand well centred about $x = x_a$ which is a local minimum of the potential. This state will decay because of a current that will be set up in this out-of-equilibrium condition.

The current as in Eq.(19.2.5) will be given by

$$J = -D\frac{\partial \rho}{\partial x} - m\frac{D}{k_BT}\rho\frac{\partial \phi}{\partial x}$$

$$= -De^{-m\phi/k_BT}\frac{\partial}{\partial x}(\rho e^{m\phi/k_BT})$$

or

$$\frac{J}{D}e^{m\phi/k_BT} = -\frac{\partial}{\partial x}(\rho e^{m\phi/k_BT}). \tag{19.3.1}$$

We now make the assumption that the deviation of ρ its initial value is small; this implies that the spatial derivative of J small and to a first approximation, we can take J to be a constant. With this approximation we can integrate Eq.(19.3.1) from $x = x_a$ to $x = x_b$ and write

$$\frac{J}{D}\int_{x_a}^{x_b} dx e^{m\phi/k_BT} = -\int_{x_a}^{x_b}\frac{\partial}{\partial x}(\rho e^{m\phi/k_BT})dx$$

$$= \rho(x_a)e^{m\phi_a/k_BT} - \rho(x_b)e^{m\phi_b/k_BT}$$

$$\cong \rho(x_a)e^{m\phi_a/k_BT}$$

$$\cong \rho_0 e^{m\phi_a/k_BT} \tag{19.3.2}$$

where the assumption is that not much of the distribution has leaked out yet and hence $\rho(x_b) \cong 0$ and $\rho(x_a) \cong \rho_0$, the initial density.

In attempting to integrate $e^{m\phi/k_BT}$ between x_a and x_b, we approximate $\phi(x)$ by a quadratic approximation about the maximum ϕ_0 at $x = x_0$. Consequently, we assume

$$\phi(x) = \phi_0 - \frac{1}{2}\phi''(x_0)(x - x_0)^2, \quad \text{for } x_a \leq x \leq x_b. \tag{19.3.3}$$

Now, $e^{-1/2\phi''(x_a)(x_{ab}-x_0)^2/k_BT} \ll 1$ (unless the temperature is very high) and hence we can take the integration limits of the variable $(x - x_0)$ as being from $-\infty$ to ∞ with this approximation Eq.(19.3.2) becomes

$$\frac{J}{D}\int_{-\infty}^{\infty}\exp\left[\frac{m\phi_0}{k_BT} - \frac{m}{2}\frac{\phi''(x_0)}{k_BT}(x - x_0)^2\right]d(x - x_0) = \rho(0)e^{m\phi_a/k_BT}$$

or

$$J = D\rho(0)\sqrt{\frac{m\phi''(x_0)}{2k_BT\pi}}e^{-(m/k_BT)(\phi_0-\phi_a)}. \tag{19.3.4}$$

The important factor is the exponential in Eq.(19.3.4). The term $m(\phi_0 - \phi_a)$ is the potential barrier that the particle at $x = x_a$ has to overcome in order to get over to the other side. At a temperature T, the typical energy available to the system to climb over this barrier is $k_B T$. This gives rise to the factor $e^{-m(\phi_0 - \phi_a)/k_B T}$ in the current.

Having found the current, we can estimate the time it takes the system to come to equilibrium. Since the current is like a velocity, the relevant time scale is proportional to the inverse of J and thus the equilibration time τ_s is of the order of

$$\tau_s = \frac{1}{D\rho(0)} \sqrt{\frac{2\pi k_B T}{m\phi''(x_0)}} e^{(m/k_B T)(\phi_0 - \phi_a)}. \qquad (19.3.5)$$

The time scale is exponentially large and hence the long-lived character of metastable states. It is on this long time scale that the state far-from-equilibrium comes to equilibrium – but the Boltzmann transport equation is a cumbersome apparatus for the treatment of this equilibration. In the next two chapters, we introduce two other techniques for studying systems which are out of equilibrium.

19.4   Beyond Metastability: Glasses

If a liquid is cooled in such a manner that it does not crystallize but rather reaches a supercooled state (far-from-equilibrium situation since at this temperature, the equilibrium state will be a solid) then on keeping lowering the temperature, the viscosity can become enormous. In fact it can be so large that it would be impossible to call it a liquid anymore (no flow) and usual to call it a glass. The glass transition temperature T_g is defined (arbitrarily) as the temperature at which the viscosity becomes 10^{12} Pa.sec (the viscosity of water at room temperatures is 10^{-2} Pa.sec). In this glassy phase, the relaxation times exceed the common observation or numerical time scales and the system never reaches an equilibrium or even a metastable state. It exhibits an aging behaviour where the properties keep evolving with time. Understanding clearly the physical mechanisms responsible for the viscosity increase of liquids approaching the glass phase and the aging phenomena in the glass phase is still an open challenge.

How does one avoid crystallization on cooling a liquid? In general a fast cooling through the freezing point temperature T_f takes a liquid to a metastable "supercooled" state. Near T_f, the typical time scale for relaxation of density fluctuations is $\sqrt{\frac{ma^2}{k_B T}}$ where a is a typical distance between molecules and m is the molecular mass. This is typically a few picoseconds. At T_g which is often $2T_f/3$, the time scale becomes of the order of 100 seconds which is 14 orders of magnitude larger – quite similar to the large increase of viscosity. This increase in the time scales associated with a supercooled liquid is remarkable also due to its temperature dependence. One plots the logarithm of the viscosity as a function of T_g/T (Angell plot) and typical plots are shown in Fig.(19.3).

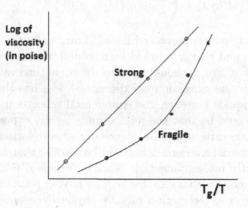

Figure 19.3: A typical Angell plot.

The straight line plot correspond to strong glass formation, which exhibits Arrhenius behaviour. One can extract from the slope an affective activation energy which suggests for a relaxation mechanism – the local breaking of a chemical bond. The energy barrier to activate this breaking sets the slope of the straight line in Figure 3. For the "fragile" glasses if one looks at the local slope (the effective activation energy at a given temperature), then the energy scale increases as the temperature decreases. The increase of the energy barriers suggests that for fragile supercooled liquids, glass transition is a collective phenomenon. This is supported by the fact that the viscosity and hence the relaxation time can be fitted to the form

$$\tau = \tau_0 \exp\left(\frac{DT_0}{T - T_0}\right) \tag{19.4.1}$$

indicating a divergent timescale at a critical temperature T_0.

The divergent timescale raises the question whether the glass transition can really be viewed as a standard second order phase transition where the correlation length diverges at the critical temperature. For a N-particle system the structure factor is defined as

$$S(q) = \left\langle \frac{1}{N} \delta\rho(\vec{q})\delta\rho(-\vec{q}) \right\rangle \tag{19.4.2}$$

with

$$\delta\rho(\vec{q}) = \sum_{i=1}^{N} e^{i\vec{q}\cdot\vec{r}_i} - \frac{N}{V}\delta_{q,0} \tag{19.4.3}$$

where V is the volume and $\vec{r}_i$ is the radius vector of the i-th particle. Measurements of the structure factor did not reveal any interesting low $\vec{q}$ behaviour. A more useful quantity to study is the dynamic structure factor defined as

$$F(q,t) = \left\langle \frac{1}{N} \delta\rho(\vec{q},t)\delta\rho(-\vec{q},0) \right\rangle. \qquad (19.4.4)$$

The mean $F(q,t)$ in a supercooled liquid show, at first, a rather fast relaxation to a plateau and this is followed by a second, much slower relaxation.

It is expected that $F(q,t)$ should relax to the equal time value $S(q)$ as $t \to \infty$, and in most cases the relaxation is exponential, i.e., has the form $e^{-\Gamma(q)t}$. In the supercooled liquids however, the exponential relaxation does not hold. A possible scenario could be that the relaxation is locally exponential but the typical relaxation scale varies spatially. Hence global correlations become non-exponential after a spatial averaging. It could be also be that even locally the relaxation is inherently non-exponential. What is certain is the the relaxation is spatially heterogeneous with coexisting regions having relaxation rates faster or slower than the average relaxation rate. A physical characterization of the dynamic heterogeneity is possible through the typical lifetime and length scales of the heterogeneities.

It would be natural to ask what would be the convenient way to talk about these heterogeneities. We have already defined the dynamic two-point function $F(q,t)$ in Eq.(19.4.4). In real space we define the correlation $C(\vec{r},t)$ as

$$C(\vec{r},t) = \langle \delta\rho(\vec{x},0)\delta\rho(\vec{x}+\vec{r},t) \rangle = \frac{1}{V} \int d^3x \, \delta\rho(\vec{x},0)\delta\rho(\vec{x}+\vec{r},t) \qquad (19.4.5)$$

with $\delta\rho(\vec{r},t) = \rho(\vec{r},t) - \rho_0$, ρ_0 being an average density. This correlation function measures, how on an average, the dynamics causes $\delta\rho(\vec{r},t)$ to be decorrelated. We need to know whether the decorrelation is homogeneous. For that one needs to study the variance of $C(\vec{r},t)$, which can be defined as

$$\chi_4(t) = \int \frac{d^3x \, d^3x'}{V \, V} \delta\rho(\vec{x},0)\delta\rho(\vec{x}+\vec{r},t)\delta\rho(\vec{x}',0)\delta\rho(\vec{x}'+\vec{r},t) - [C(\vec{r},t)]^2. \quad (19.4.6)$$

Using translational invariance this can then be written as

$$\chi_4(t) = \int \frac{d^3x \, d^3y}{V \, V} \delta\rho(\vec{x},0)\delta\rho(\vec{x}+\vec{y},t)\delta\rho(\vec{x}+\vec{r},t)\delta\rho(\vec{x}+\vec{y}+\vec{r},t) - [C(\vec{r},t)]^2$$

$$= \int \frac{d^3x}{V} S_4(\vec{x},t) \qquad (19.4.7)$$

where

$$S_4(\vec{x},t) = \frac{1}{V} \int d^3y \left[\{\delta\rho(\vec{x},0)\delta\rho(\vec{x}+\vec{y},0)\delta\rho(\vec{x}+\vec{r},t)\delta\rho(\vec{x}+\vec{y}+\vec{r},t)\} \right.$$

$$\left. - \{\delta\rho(\vec{y},0)\delta\rho(\vec{y}+\vec{r},t)\}^2 \right]. \qquad (19.4.8)$$

The function $S_4(\vec{x},t)$ is a measure of the spatial correlation between dynamical events between times 0 and t at different spatial points of the system, i.e., it is the spatial extent of the dynamically heterogeneous regions over a time span t. The function $\chi_4(t)$ can be numerically found and a typical behaviour at a given temperature shows a peak at a particular time delay t. This peak becomes more pronounced and moves to larger values of t as the temperature T is decreased (see Fig.(19.4)).

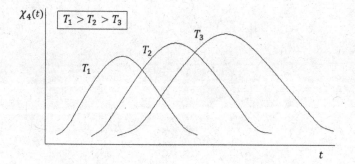

Figure 19.4: A typical Angell plot.

The increase in the height of the peak as the temperature is decreased indicates the existence of a growing correlation length with decreasing temperature. Thus there are growing length scales and growing time scales if one looks at the fourth order correlation functions. Over the last decade a lot of effort has been devoted to the study of these growing scales. While discussing the techniques and results of these studies is beyond the scope of this book, we refer to a few relevant articles at the end of the chapter which will give some idea of what is involved.

20

Langevin Equations

20.1 Introduction

From equations for distribution functions, we are now going to focus on equation for a single particle. However, since it is the same kind of questions that we intend to study, equilibrium distributions and approaches to equilibrium, there must be a probabilistic element to this differential equation for single particle motion. One can easily picture how such a situation may arise in practice. Imagine a particular molecule, in a gas of randomly moving molecules, being dragged with a velocity v. As the different molecules strike this chosen molecule at random, there will be forces acting in all directions and of all possible magnitudes, so that the result of averaging this force (over a period of time say) will be zero. Because of the velocity v of the chosen molecules, there will, however, be a viscous drag corresponding to the larger number of collisions per unit time with molecules travelling in the opposite direction than with molecules travelling in the same direction. For a spherical molecule of radius r, this resistive force is $6\pi\eta rv$ (Stokes' law) opposite in direction to v, where η is the coefficient of viscosity which contains information about the various collision processes. Thus the required form of the equation of motion will be

$$\frac{dv}{dt} = -f(v) + g(t) \qquad (20.1.1)$$

where $f(v)$ is the resistive force due to the surrounding and $g(t)$ is the random force due to the surrounding the average value $\langle g \rangle = 0$. An equation of the above form is called a Langevin equation. It is a stochastic, as opposed to a deterministic, equation because of the random force (i.e., a force that is not definitely known) $g(t)$ and arises whenever the effect of a particular particle's coupling to its environment is represented in some averaged manner. It reflects the fact that one lacks precise knowledge of the environment and hence is required to perform an averaging process over the environmental degrees of freedom. A particularly obvious example of this is provided by the case of a particle interacting with a collection of oscillators. This is presented in the next section.

20.2   An Example with Oscillators

We consider a particle with momentum p and coordinate x (for simplicity, we take the motion to be one-dimensional; additional dimensions bring no complications) interacting with a set of harmonic oscillators having momenta p_i and coordinates x_i. All masses are taken to be unity for convenience. The Hamiltonian describing the system is taken to be

$$H = \frac{p^2}{2} + \sum \frac{p_j^2}{2} + \sum \frac{1}{2}\omega_j^2 \left(q_j - \frac{\gamma_j}{\omega_j^2}x \right)^2 \tag{20.2.1}$$

The key point is that the coupling of the particle to the oscillators is linear. This makes the problem exactly solvable. The equations of motion for the different variables are obtained as follows:

$$\begin{aligned}
\dot{p} &= -\frac{\partial H}{\partial x} \\[2mm]
&= -\sum_j \omega_j^2 \left(q_j - \frac{\gamma_j}{\omega_j^2}x \right)\left(-\frac{\gamma_j}{\omega_j^2} \right) \\[2mm]
&= \sum_j \gamma_j \left(q_j - \frac{\gamma_j}{\omega_j^2}x \right) \tag{20.2.2}
\end{aligned}$$

$$\dot{q}_j = \frac{\partial H}{\partial p_j} = p_j \tag{20.2.3}$$

and

$$\dot{p}_j = -\frac{\partial H}{\partial q_j} = -\omega_j^2 \left(q_j - \frac{\gamma_j}{\omega_j^2}x \right). \tag{20.2.4}$$

From Eqs.(20.2.3) and (20.2.4), we get

$$\ddot{q}_j + \omega_j^2 q_j = \gamma_j x. \tag{20.2.5}$$

The solution of Eq.(20.2.5) can be written down as

$$\begin{aligned}
q_j(t) &= q_j(0)\cos\omega_j t + \frac{p_j(0)}{\omega_j}\sin\omega_j t + \int_0^t \gamma_j \frac{\sin\omega_j(t-s)}{\omega_j}x(s)ds \\[2mm]
&= q_j(0)\cos\omega_j t + \frac{p_j(0)}{\omega_j}\sin\omega_j t + \gamma_j \left. \frac{\cos\omega_j(t-s)}{\omega_j^2}x(s) \right|_0^t \\[2mm]
&\quad - \int_0^t \gamma_j \frac{\cos\omega_j(t-s)}{\omega_j^2}v(s)ds
\end{aligned}$$

$$= \left[q_j(0) - \frac{\gamma_j}{\omega_j^2} x(0) \right] \cos \omega_j t + \frac{p_j(0)}{\omega_j} \sin \omega_j t + \frac{\gamma_j}{\omega_j^2} x(t)$$

$$- \int_0^t \gamma_j \frac{\cos \omega_j (t - s)}{\omega_j^2} v(s) ds$$

Hence,

$$q_j(t) - \frac{\gamma_j}{\omega_j^2} x(t) = \left[q_j(0) - \frac{\gamma_j}{\omega_j^2} x(0) \right] \cos \omega_j t + \frac{p_j(0)}{\omega_j} \sin \omega_j t$$

$$- \int_0^t \gamma_j \frac{\cos \omega_j (t - s)}{\omega_j^2} v(s) ds. \tag{20.2.6}$$

In the above $v = \frac{dx}{dt}$ and is the same as p since the mass has been taken to be unity. Inserting in Eq.(20.2.2), the value of $q_j - \frac{\gamma_j}{\omega_j^2}$ obtained from Eq.(20.2.6), we find

$$\dot{v} = -\sum_j \left[\frac{\gamma_j}{\omega_j^2} \int_0^t \cos \omega_j (t - s) v(s) ds + \gamma_j (q_j(0) \right.$$

$$\left. - \frac{\gamma_j}{\omega_j^2} x(0)) \cos \omega_j t + \frac{\gamma_j}{\omega_j} p_j(0) \sin \omega_j t \right]$$

which can be written as,

$$\dot{v} = -\int K(t - s) v(s) ds + f(t) \tag{20.2.7}$$

where,

$$K(t - s) = \sum_j \frac{\gamma_j^2}{\omega_j^2} \cos \omega_j (t - s) \tag{20.2.8}$$

$$f(t) = \sum_j \gamma_j [q_j(0) - \frac{\gamma_j}{\omega_j^2} x(0)] \cos \omega_j t$$

$$+ \sum_j \frac{\gamma_j p_j(0)}{\omega_j} \sin \omega_j t. \tag{20.2.9}$$

At this point, Eq.(20.2.7) is completely deterministic as expected. It ceases to be deterministic when it is not possible to specify all the $p_j(0)$ and $q_j(0)$ exactly. Now suppose that the distribution of $p_j(0)$ and $[q_j(0) - \frac{\gamma_j}{\omega_j^2} x(0)]$ is specified and that this distribution is a Maxwell-Boltzman distribution at a temperature T (i.e., quadratic in the variables and each j is independent of the other: it should be pointed out that such distributions are also called Gaussian).

If the distribution alone is known, then only averages over this distribution make sense and the assumed nature of the distribution implies

$$\langle p_j(0) \rangle = \langle q_j(0) - \frac{\gamma_j}{\omega_j^2} x(0) \rangle = 0 \qquad (20.2.10)$$

$$\langle p_i(0) p_j(0) \rangle = k_B T \delta_{ij} \qquad (20.2.11)$$

$$\langle [q_i(0) - \frac{\gamma_i}{\omega_i^2} x(0)].[q_j(0) - \frac{\gamma_j}{\omega_j^2} x(0)] \rangle = \frac{k_B T}{\omega_i^2} \delta_{ij}. \qquad (20.2.12)$$

Now, the force $f(t)$ of Eq.(20.2.9) makes sense only in terms of expectation values and Eq.(20.2.10) yields

$$\langle f(t) \rangle = 0. \qquad (20.2.13)$$

For the second moment, on the other hand, we have

$$
\begin{aligned}
\langle f(t) f(s) \rangle &= \left\langle \sum_j \left[\gamma_j [q_j(0) - \frac{\gamma_j}{\omega_j^2} x(0)] \cos \omega_j t + \frac{\gamma_j}{\omega_j} p_j(0) \sin \omega_j t \right] \right. \\
&\quad \times \left. \sum_i \left[\gamma_i [q_i(0) - \frac{\gamma_i}{\omega_j^2} x(0)] \cos \omega_i s + \frac{\gamma_j}{\omega_i}.p_i(0) \sin \omega_i s \right] \right\rangle \\
&= \sum_{i,j} \gamma_i \gamma_j \langle [q_j(0) - \frac{\gamma_j}{\omega_j^2} x(0)][q_j(0) - \frac{\gamma_j}{\omega_j^2} x(0)] \rangle \cos \omega_j t \cos \omega_j s \\
&\quad + \sum_{i,j} \frac{\gamma_i \gamma_j}{\omega_i \omega_j} \langle p_i(0) p_j(0) \rangle \sin \omega_j t \sin \omega_i s \\
&= \sum_i \frac{\gamma_i^2}{\omega_i^2} k_B T \cos \omega_i (t - s) \\
&= k_B T K(t - s). \qquad (20.2.14)
\end{aligned}
$$

In arriving at the above result we have made use of the fact that the cross correlation of the momenta and positions $\left\langle p_i(0) \left(q_j(0) - \frac{\gamma_j x(0)}{\omega_j^2} \right) \right\rangle = 0$ for all (i, j). It should be noted that due to the averaging over the initial values $q_i(0)$ and $p_i(0)$, Eq.(20.2.7) is no longer a deterministic equation. The force $f(t)$ makes sense only when its moments are specified. Before proceeding further, we make a digression on Gaussian distributions.

20.3   Gaussian Distribution and the Central Limit Theorem

A variable x is said to have a Gaussian distribution if the probability $p(x)$ of finding a value x for it is given as

$$p(x) = \frac{1}{\sqrt{2\pi\sigma^2}} e^{-x^2/2\sigma^2} \qquad (20.3.1)$$

where we have arranged the pre-factor to ensure that the distribution is normalized, i.e. $\int_{-\infty}^{\infty} p(x)dx = 1$. The constant '$\sigma$' is called the width of the distribution. Clearly the average value $\langle x \rangle$ of x is zero and the mean square average

$$\langle x^2 \rangle = \int_{-\infty}^{\infty} x^2 p(x)dx = \frac{1}{(2\pi\sigma^2)^{1/2}} \int_{-\infty}^{\infty} x^2 e^{-x^2/2\sigma^2} dx = \sigma^2. \qquad (20.3.2)$$

To find the average value of any power of x, we can perform the integration very easily, but it is instructive to follow a different route. This involves calculating

$$
\begin{aligned}
\langle e^{-ikx} \rangle &= \int_{-\infty}^{\infty} e^{-ikx}.p(x)dx \\
&= \int_{-\infty}^{\infty} \frac{1}{(2\pi\sigma^2)^{1/2}} e^{-x^2/2\sigma^2 - ikx} dx \\
&= \int_{-\infty}^{\infty} \frac{1}{(2\pi\sigma^2)^{1/2}} \exp\left[-\frac{x^2}{2\sigma^2} - 2\frac{ik\sigma}{\sqrt{2}}\frac{x}{\sqrt{2}\sigma} - \left(\frac{ik\sigma}{\sqrt{2}}\right)^2 - \frac{k^2\sigma^2}{2} \right] dx \\
&= \frac{e^{-k^2\sigma^2/2}}{(2\pi\sigma^2)^{1/2}} \int_{-\infty}^{\infty} \exp\left[-\frac{1}{2\sigma^2}\left(x + \frac{ik\sigma}{\sqrt{2}} \right)^2 \right] dx. \qquad (20.3.3)
\end{aligned}
$$

Expanding the two sides of Eq.(20.3.3),

$$\left\langle \sum_{n=0}^{\infty} \frac{(-ik)^n}{n!} x^n \right\rangle = \sum_{m=0}^{\infty} \left(-\frac{k^2\sigma^2}{2} \right)^m \frac{1}{m!}$$

or

$$\sum_{n=0}^{\infty} \frac{(-i)^n k^n}{n!} \langle x^n \rangle = \sum (-1)^m \frac{k^{2m}\sigma^{2m}}{2^m} \frac{1}{m!}. \qquad (20.3.4)$$

Clearly

$$\langle x^{2m+1} \rangle = 0 \qquad (20.3.5)$$

and

$$\langle x^{2m} \rangle = \frac{(2m)!\sigma^{2m}}{2^m m!}. \qquad (20.3.6)$$

The importance of the above derivation of $\langle x^n \rangle$ is understood from the observation that, if we find for an unknown distribution the moments are given by Eqs.(20.3.5) and (20.3.6), then it follows that under this distribution $\langle e^{-ikx} \rangle = e^{-k^2\sigma^2/2}$ and now an inverse Fourier transform shows that the distribution is Gaussian. Thus the above derivation of the moments clearly shows that given the moments, we can arrive at the distribution.

If we have N independent variables $x_1, x_2, \ldots x_N$ and the joint probability distribution is Gaussian, then it implies that

$$P(x_1, x_2, \ldots x_N) = \frac{1}{(2\pi\sigma^2)^{1/2}} \exp\left(-\frac{x_1^2 + x_2^2 + \ldots + x_N^2}{2\sigma^2}\right). \qquad (20.3.7)$$

It will be shown that for N independent random variables $x_1, x_2, \ldots, x_N$ if a variable $Y = \frac{1}{\sqrt{N}}(x_1 + x_2 + \ldots + x_N)$ is formed then the distribution of Y is also Gaussian in the limit $N \to \infty$. The variables $\{x_i\}$ are characterized by the moments $\langle x_i \rangle = 0$ and $\langle x_i^2 \rangle = \sigma^2$ (the distribution of $\{x_i\}$ is not taken to be Gaussian; if it were, then the result would be trivial for all N).

Consider $\langle \exp^{-ikY} \rangle$ and write it as

$$\langle e^{-ikY} \rangle = 1 - ik\langle Y \rangle + \frac{(ik)^2}{2!}\langle Y^2 \rangle - \frac{(ik)^3}{3!}\langle Y^3 \rangle + \frac{k^4}{4!}\langle Y^4 \rangle + \ldots. \qquad (20.3.8)$$

Now $\langle Y \rangle = \langle Y^3 \rangle = 0$ since odd terms give vanishing moments for $\{x_i\}$, while

$$\langle Y^2 \rangle = \frac{1}{N}\langle x_1^2 + x_2^2 + \ldots + x_n^2 + 2x_1x_2 + 2x_2x_3 + \ldots \rangle = \frac{1}{N}.N\sigma^2 = \sigma^2 \quad (20.3.9)$$

and

$$\langle Y^4 \rangle = \frac{1}{N^2}\left\langle \sum_i x_i^4 + 6\sum_{i \rangle j} x_i^2 x_j^2 + 4\sum_{j \rangle k} x_i^2 x_j x_k \right\rangle. \qquad (20.3.10)$$

For large N, we note $\langle \sum x_i^4 \rangle$ is $O(N)$ while $\langle \sum x_i^2 x_j^2 \rangle$ is $O(N^2)$ and hence working to $O(1)$ accuracy,

$$\langle Y^4 \rangle = \frac{1}{N^2}.6.\frac{N^2}{2}.\sigma^4 = 3\sigma^4$$

Thus the pattern being formed is

$$\begin{aligned} \langle e^{-ikp} \rangle &= 1 - \frac{k^2\sigma^2}{2} + \frac{1}{2}.\frac{k^4\sigma^4}{4} + \ldots \\ &= e^{-k^2\sigma^2/2}. \end{aligned} \qquad (20.3.11)$$

Hence from our previous considerations, the distribution of Y is Gaussian as $N \to \infty$ (note that this limit is vital in the reduction of $\langle Y^4 \rangle$ to $3\sigma^4$ and will be so for all higher moments) and has the width σ^2. The variable $\sum x_i$ would then have the width $N\sigma^2$. The above results constitute the central limit theorem.

To end this section, we will consider yet another Gaussian distribution-that is appropriate to a random function $f(t)$. We now talk about the probability of f being a definite function $f(t)$ and express this probability as

$$P(f) = Ae^{-\int_{-\infty}^{\infty} dt f(t)^2/\sigma^2} \qquad (20.3.12)$$

where the constant A is given by

$$A^{-1} = \int D[f] e^{-\int_{-\infty}^{\infty} dt f(t)^2/\sigma^2}. \qquad (20.3.13)$$

The integration $D[f]$ is over all possible functions $f(t)$ and the definition of A assures the normalization $\int D[f]P(f) = 1$. Operationally the functional integral is best treated in the Fourier space by introducing the transform

$$f(t) = \frac{1}{\sqrt{2\pi}} \int_{-\infty}^{\infty} f(\omega)e^{i\omega t} d\omega \qquad (20.3.14)$$

the requirement that $f(t)$ is real implying $f(-\omega) = f(\omega)$. A given set of $f(\omega)$ leads to a given function $f(t)$. If we scan every possible value of $f(\omega)$ for every value of ω, then the space of all functions of $f(t)$ will be taken into account . Accordingly,

$$D[f] = \prod_{\omega} df(\omega),$$

$$\int_{-\infty}^{\infty} f(t)^2 dt = \frac{1}{(2\pi)^2} \int_{-\infty}^{\infty} f(\omega_1)e^{i\omega_1 t} d\omega_1 f(\omega_2)e^{i\omega_2 t} d\omega_2 dt$$

$$= \frac{1}{2\pi} \int_{-\infty}^{\infty} f(\omega_1)f(\omega_2)d\omega_1 d\omega_2 \int_{-\infty}^{\infty} \frac{dt}{2\pi} e^{i(\omega_1+\omega_2)t}$$

$$= \frac{1}{2\pi} \int_{-\infty}^{\infty} f(\omega_1)f(\omega_2)d\omega_1 d\omega_2 \delta(\omega_1 + \omega_2)$$

$$= \int_{-\infty}^{\infty} f(\omega_1)f(-\omega_1)\frac{d\omega_1}{2\pi}$$

$$= \int_{-\infty}^{\infty} |f(\omega)|^2 \frac{d\omega}{2\pi}$$

$$= \sum_{\omega} |f(\omega)|^2.$$

The distribution can now be thought of as

$$p(f) = A \exp \left[-\sum_\omega \frac{|f(\omega)|^2}{\sigma^2} \right] \qquad (20.3.15)$$

with

$$A^{-1} = \int \prod_\omega df(\omega) \exp \left[-\sum_\omega \frac{|f(\omega)|^2}{\sigma^2} \right]. \qquad (20.3.16)$$

One can now express the various odd moments as

$$\langle f(\omega) \rangle = \langle f(\omega_1) f(\omega_2) f(\omega_3) \rangle = \ldots = \langle f(\omega_1) f(\omega_2) \ldots f(\omega_{2m+1}) \rangle = 0. \qquad (20.3.17)$$

The even moments are expressed as

$$\langle f(\omega_1) f(\omega_2) \rangle = 0 \quad \text{if} \quad \omega_1 \neq -\omega_2$$

while for $\omega_1 = -\omega_2$, from the properties of the Gaussian, we have $\langle f(\omega_1) f(\omega_2) \rangle = \sigma^2$, leading to the result

$$\langle f(\omega_1) f(\omega_2) \rangle = \sigma^2 \delta_{\omega_1, -\omega_2}. \qquad (20.3.18)$$

It should be clear that if we consider the fourth moment, then,

$$\langle f(\omega_1) f(\omega_2) f(\omega_3) f(\omega_4) \rangle = \sigma^4 [\delta_{\omega_1, -\omega_2} \delta_{\omega_3, -\omega_4} + \delta_{\omega_1, -\omega_3} \delta_{\omega_2, -\omega_4}$$
$$+ \delta_{\omega_1, -\omega_4} \delta_{\omega_2, -\omega_3}] \qquad (20.3.19)$$

and obviously the fourth moment is zero if $\omega_1 + \omega_2 + \omega_3 + \omega_4 \neq 0$. For an arbitrary even order moment

$$\langle f(\omega_1) f(\omega_2) \ldots f(\omega_{2n}) \rangle$$

$$= \sigma^{2n} \left[\sum_{\substack{\text{(all pairwise} \\ \text{breakups)}}} \delta_{\omega_1, -\omega_2} \delta_{\omega_3, -\omega_4} \ldots \delta_{\omega_{2n-1}, -\omega_{2n}} \right]. \qquad (20.3.20)$$

The number of pairs is $\frac{(2n)!}{n!2n}$.

Finally, we need to discuss the correlations of the original function $f(t)$. Clearly

$$\langle f(t) \rangle = 0 \qquad (20.3.21)$$

and

$$
\begin{aligned}
\langle f(t_1)f(t_2)\rangle &= \left\langle \frac{1}{2\pi} \int d\omega_1 f(\omega_1) e^{i\omega_1 t_1} d\omega_2 f(\omega_2) e^{i\omega_2 t_2} \right\rangle \\
&= \frac{1}{2\pi} \int d\omega_1 d\omega_2 e^{i(\omega_1 t_1 + \omega_2 t_2)} \langle f(\omega_1) f(\omega_2) \rangle \\
&= \frac{\sigma^2}{2\pi} \int d\omega_1 d\omega_2 e^{i(\omega_1 t_1 + \omega_2 t_2)} \delta_{\omega_1, -\omega_2} \\
&= \frac{\sigma^2}{2\pi} \int d\omega_1 e^{i\omega_1(t_1 - t_2)} \\
&= \sigma^2 \delta(t_1 - t_2) \qquad\qquad (20.3.22)
\end{aligned}
$$

where we have made use of Eq.(20.3.18). It should now be obvious that

$$
\langle f(t_1)f(t_2)\ldots f(t_{2n+1})\rangle = 0 \qquad (20.3.23)
$$

and

$$
\begin{aligned}
\langle f(t_1)f(t_2)\ldots f(t_{2n})\rangle &= \sigma^{2n} \left[\delta(t_1 - t_2)\delta(t_3 - t_4)\delta(t_{2n-1} - t_{2n}) \right. \\
&+ \left. \text{all such pairwise breakups} \right]. \qquad (20.3.24)
\end{aligned}
$$

We note that if a random function $f(t)$ is characterized by moments of the form shown in Eqs.(20.3.23) and (20.3.24), then the distribution will be given by Eq.(20.3.12).

20.4 Langevin Equation

We now return to Eq.(20.2.7) with the $f(t)$ stochastic and the second moment given by Eq.(20.2.14). It is clear that all odd moments vanish because of the Gaussian nature of the distribution of $p_i(0)$ and $\bar{q}_i(0) = q_i(0) - \frac{\gamma_i}{\omega_i^2} x(0)$. Thus

$$
\langle f(t_1)f(t_2)\ldots f(t_{2n+1})\rangle = 0. \qquad (20.4.1)
$$

To find the even moments, we write

$$
\begin{aligned}
\langle f(t_1)f(t_2)\ldots f(t_{2n})\rangle &= \left\langle \sum_{j_1} \left[\gamma_{j_1} \bar{q}_{j_1}(0) \cos \omega_{j_1} t_1 + \frac{\gamma_{j_1}}{\omega_{j_1}} p_{j_1}(0) \sin \omega_{j_1} t_1 \right] \right. \\
&\times \sum_{j_2} \left[\bar{q}_{j_2}(0) \cos \omega_{j_2} t_2 + \frac{\gamma_{j_2}}{\omega_{j_2}} p_{j_2}(0) \sin \omega_{j_2} t_2 \right] \times \ldots \\
&\times \left. \sum_{j_{2n}} \left[\gamma_{j_{2n}} \bar{q}_{j_{2n}}(0) \cos \omega_{j_{2n}} t_{2n} + \frac{\gamma_{j_{2n}}}{\omega_{j_{2n}}}(0) \sin \omega_{j_{2n}} t_{2n} \right] \right\rangle
\end{aligned}
$$

and now using the property of Gaussian distribution for an even string of $\{q_j(0)\}, \{p_j(0)\}$ or the combination of $(2n - 2m)$ number of $\{\bar{q}_j(0)\}$ and $2m$

number of $\{p_j(0)\}$, we have

$$
\begin{aligned}
\langle f(t_1)f(t_2)\ldots f(t_1)\rangle &= (k_BT)^n\left[K(t_2-t_1).K(t_4-t_3).K(t_6-t_5)\ldots\right. \\
&\times \left. K(t_{2n}-t_{2n-1})\right] \\
&+ \quad \text{all such pairwise breakups.} \tag{20.4.2}
\end{aligned}
$$

Recalling the definition of $K(t-s)$, we note that it is crucially dependent on what form of frequency weighting we employ. Thus

$$
\begin{aligned}
K(t-s) &= \sum_j \frac{\gamma_j^2}{\omega_j^2}\cos\omega_j(t-s) \\
&= \int_{-\infty}^{\infty} \frac{\gamma^2}{\omega^2}\cos\omega(t-s)P(\omega)d\omega \quad [P(\omega)\text{ is the weighting factor}] \\
&= \Gamma\int_0^{\infty}\cos\omega(t-s)d\omega \quad \left[\text{choosing }\frac{\gamma^2}{\omega^2}P(\omega)=\Gamma\right] \\
&= \Gamma\lim_{\omega\to\infty}\frac{\sin\omega(t-s)}{t-s} \\
&= \Gamma\delta(t-s). \tag{20.4.3}
\end{aligned}
$$

The equation of motion for v, that we had obtained in Eq.(20.2.7), now becomes

$$
\dot{v} = -\Gamma v + f(t) \tag{20.4.4}
$$

where $f(t)$ is a random function with the moments,

$$
\begin{aligned}
\langle f(t)\rangle &= 0 \tag{20.4.5} \\
\langle f(t_1)f(t_2)\rangle &= 2D\delta(t_2-t_1) \tag{20.4.6}
\end{aligned}
$$

where $D/\Gamma = k_BT$. For an arbitrary number of factors, f, we have from Eqs.(20.4.2) and (20.4.3),

$$
\begin{aligned}
\langle f(t_1)f(t_2)\ldots f(t_{2n+1})\rangle &= 0 \tag{20.4.7} \\
\langle f(t_1)f(t_2)\ldots f(t_{2n})\rangle &= (2D)^n\left[\delta(t_2-t_1)\delta(t_4-t_3)\delta(t_6-t_5)\ldots\right. \\
&\times \left. \delta(t_{2n}-t_{2n-1})\right] \\
&+ (2D)^n[\text{all possible pairwise breakups}] \tag{20.4.8}
\end{aligned}
$$

The equation of motion given in Eq.(20.4.4) is a linear Langevin equation with white noise. The linearity comes from the fact that the dissipative term is linear in v (i.e. Γv): a nonlinear function of v at this stage would make it a nonlinear Langevin equation. The characteristic of the stochastic force $f(t)$ given in Eqs.(20.4.7) and (20.4.8) makes it a white noise.

The steps leading to Eq.(20.4.3) show that all frequencies contribute equally when one is dealing with white noise. A physical picture of Langevin equation will appear in Sec.20.6 when we consider the question of Brownian motion. For a different frequency weighting the kernel $K(t-s)$ has "memory" i.e. is non-zero fort $t \neq s$ and we have a 'coloured noise'. The relation between the strength of the correlation of f in Eq.(20.4.6) and the strength of the dissipation (Γ) in Eq.(20.4.4) [namely $D = \Gamma(k_B T)$ is called the fluctuation dissipation theorem. It should be clear when $f(t)$ is a white noise term with the moments given by Eqs.(20.4.7) and (20.4.8), the probability distribution associated with f is

$$P[f] = A \exp\left[-\int_{-\infty}^{\infty} \frac{f(t)^2}{2D} dt\right] \tag{20.4.9}$$

where

$$A^{-1} = \int D[f] \exp\left[-\int_{-\infty}^{\infty} \frac{f(t)^2}{2D} dt\right]$$

20.5 Langevin Equation and Equilibrium

We are going to deal almost exclusively with the linear Langevin equation with a white noise force term, i.e., the equation

$$\dot{v} = -\Gamma v + f(t) \tag{20.5.1}$$

with the statistics of $f(t)$ as specified. The advantage of the linear equation is that it can be easily solved. We start with an initial velocity $v(0)$ and see what the Langevin equation tells us. The solution of Eq.(20.5.1) can be written as

$$v(t) = e^{-\Gamma t} \int_0^t f(t') e^{\Gamma t'} dt' + v(0) e^{-\Gamma t}. \tag{20.5.2}$$

Since $f(t)$ is stochastic, so will the $v(t)$ in Eq.(20.5.2) be stochastic and only the moments of $v(t)$ will make sense.

Clearly $\langle v(t) \rangle = v(0) e^{-\Gamma t}$ and for the second moment, we have

$$
\begin{aligned}
\langle v(t_1) v(t_2) \rangle &= \left\langle \left[e^{-\Gamma t_1} \int_0^{t_1} f(t') e^{\Gamma t'} dt' + v(0) e^{-\Gamma t_1} \right] \right. \\
&\quad \left. \times \left[e^{-\Gamma t_2} \int_0^{t_2} f(t'') e^{-\Gamma t''} dt'' + v(0) e^{-\Gamma t_2} \right] \right\rangle \\
&= e^{-\Gamma(t_1+t_2)} \int_0^{t_1} dt' \int_0^{t_2} dt'' e^{\Gamma(t'+t'')} \langle f(t') f(t'') \rangle + v(0)^2 e^{-\Gamma(t_1+t_2)} \\
&= 2D e^{-\Gamma(t_1+t_2)} \int_0^{t_1} dt' \int_0^{t_2} dt'' e^{\Gamma(t'+t'')} \delta(t' - t'') \\
&\quad + v(0)^2 e^{-\Gamma(t_1+t_2)}. \tag{20.5.3}
\end{aligned}
$$

In the integral on the right hand side of Eq.(20.5.3) if $t_1 > t_2$, then the integration over t' needs to carried out first and the integral becomes

$$e^{-\Gamma(t_1+t_2)} \int_0^{t_2} dt'' e^{2\Gamma t''} = \frac{1}{2\Gamma}(e^{-\Gamma(t_1-t_2)} - e^{-\Gamma(t_1+t_2)}).$$

On the other hand, if $t_2 > t_1$, then the integral over t'' needs to be done first and we have the result as $(e^{-\Gamma(t_2-t_1)} - e^{-\Gamma(t_1+t_2)})/2\Gamma$.

Clearly, then

$$\langle v(t_1)v(t_2) \rangle = \frac{D}{\Gamma}e^{-\Gamma(t_1-t_2)} + \left[v(0)^2 - \frac{D}{\Gamma}\right]e^{-\Gamma(t_1+t_2)}. \qquad (20.5.4)$$

For the equal time correlation function $\langle v(t)^2 \rangle$, we have

$$\lim_{t\to\infty} \langle v(t)^2 \rangle = \frac{D}{\Gamma}. \qquad (20.5.5)$$

If we want to determine $\lim t \to \infty \langle v(t)^n \rangle$ it is not necessary to consider $v(0)$, since the term involving $v(0)$ will not survive in that limit. From the statistics of $f(t)$ it is obvious that $\lim t \to \infty \langle v(t)^{2m+1} \rangle \to 0$ while the relevant part of the even order correlation is

$$
\begin{aligned}
\langle v(t)^{2n} \rangle &= \left\langle \left[e^{-\Gamma t}\int_0^t f(t_1)e^{\Gamma t_1}dt_1\right]\left[e^{-\Gamma t}\int_0^t f(t_2)e^{+\Gamma t_2}dt_2\right] \cdots \right. \\
&\quad \times \left. \left[e^{-\Gamma t}\int_0^t f(t_{2n})e^{\Gamma t_{2n}}dt_{2n}\right]\right\rangle \\
&= e^{-2n\Gamma t}\int_0^t dt_1 \int_0^t dt_2 \cdots \int_0^t dt_{2n} \\
&\quad \times \langle f(t_1)f(t_2)\cdots f(t_{2n})\rangle . e^{\Gamma(t_1+t_2\ldots t_{2n})}. \qquad (20.5.6)
\end{aligned}
$$

At this point we use the statistics of $f(t)$ and note that it breaks up into a chain of pairs in $(2n)!/n!2^n$ ways and thus the time-independent part of the right hand side of Eq.(20.5.6) is extracted as

$$\langle v^{2n}(t)\rangle = \frac{D^n}{(\Gamma)^n}\frac{(2n)!}{n!2^n} + \text{terms which involve powers of } e^{-2\Gamma t},$$

leading to

$$\lim_{t\to\infty} \langle v^{2n}(t)\rangle = \left(\frac{D}{\Gamma}\right)^n \frac{(2n)!}{2^n n!}. \qquad (20.5.7)$$

Since $v(t)$ is a random variable, there is a probability distribution $P\{v(t)\}$ associated with it and as $t \to \infty$, this probability distribution tends to the equilibrium value with the moments specified as in Eq.(20.5.7). The odd moments are zero. The equilibrium probability distribution is then given by

$$P_{eq}(v) = \left(\frac{2\Gamma}{4\pi D}\right)^{1/2} e^{-\frac{v^2}{4D} \cdot 2\Gamma}$$

$$= \left(\frac{1}{\sqrt{2\pi k_B T}}\right) e^{-(v^2/2k_B T)} \tag{20.5.8}$$

which is a Gaussian distribution. The important result is that the Langevin dynamics is another way of describing the approach to equilibrium. If we begin with an initial distribution of v which is not according to equilibrium, then as $t \to \infty$ the equilibrium distribution is attained. If, on the other hand, we begin with an equilibrium distribution, then in Eq.(20.5.4), $v(0)^2 = \frac{D}{2\Gamma}$ and the dynamics governed by the Langevin equation corresponds to decay of correlations.

The above results about the establishment of an equilibrium distribution are not restricted to the linear Langevin equation. If we have a Langevin equation with the structure

$$v = -\Gamma \frac{\partial S}{\partial v} + f(t) \tag{20.5.9}$$

where F is an arbitrary function of v and $f(t)$ is the random noise with $\langle f(t) \rangle = 0$ and $\langle f(t_1)f(t_2) \rangle = 2D\Gamma\delta(t_1 - t_2)$, then the equilibrium distribution is proportional to $e^{-S(v)/D}$. This is most conveniently demonstrated by setting up an equation for the probability distribution $P\{v(t)\}$. This will be done in the next chapter.

20.6 Brownian Motion

One of the celebrated papers of Einstein in 1905 deals with a quantitative study of the random motion. A couple of years later Einstein wrote a simplified account of the seminal work.

A direct test of the molecular kinetic theory was his obsession during his student days. The analysis of motion of suspended particles in a liquid was an attempt to make observable predictions that when tested would establish, beyond doubt, the correctness of the molecular point of view.

The analysis begins by considering a non-uniform solution in a container with a semi-permeable membrane of thickness Δx in the centre, as shown in Fig.20.1.

We need to consider the force balance in the region of width Δx. The pressure will be different on either side of the membrane because of the differing concentration . This will cause a force $-\frac{\partial P}{\partial x} A \Delta x$ to the left where A is the cross-sectional area. Clearly, if the pressure decreases as x increases, then the direction of the force will be to the right. If the local density gradient is negative, that is the concentration is higher to the left, then there will be a drift of the solute molecules to the right. If the drift speed is v and the shear viscosity of the solvent is η, then there will be a resistive force $6\pi\eta r v$ (Stokes Law) on the

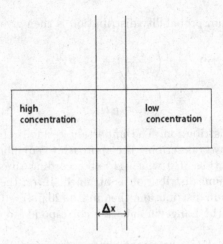

Figure 20.1: A horizontal concentration gradient.

solute molecule, if it is taken to be a sphere of radius r. These are the only forces in the horizontal direction and a steady state would require that the net external force be zero.

If $n(x)$ is the local density of solute molecules, then the total number in the volume $A\Delta x$ is $n(x)A\Delta x$ and the total viscous force will be $6\pi\eta rvn(x)A\Delta x$. The force balance then yields

$$6\pi\eta rvn(x)A\Delta x = -\frac{\partial P}{\partial x}A\Delta x$$

or

$$v = -\frac{1}{6\pi\eta rn}\frac{\partial P}{\partial x}. \tag{20.6.1}$$

The solution pressure can be written as

$$P = nk_BT \tag{20.6.2}$$

at temperature T and thus

$$v = -\frac{k_BT}{6\pi\eta rn}\frac{\partial n}{\partial x}. \tag{20.6.3}$$

The current of the solute molecules is the number of them crossing an unit area per unit time and hence can be written as

$$j = nv \tag{20.6.4}$$

From Eq.(20.6.3) we find

$$j = -\frac{k_BT}{6\pi\eta r}\frac{\partial n}{\partial x}. \tag{20.6.5}$$

The diffusion coefficient D is defined by the relation

$$j = -D\frac{\partial n}{\partial x} \tag{20.6.6}$$

and thus from Eq.(20.6.5) and Eq.(20.6.6) we have

$$D = \frac{k_B T}{6\pi\eta r}. \tag{20.6.7}$$

We now look at the statistical property of the current. In the container shown in Fig.20.2, we consider any cross section (the solid line in Fig.20.2) AB and would like to calculate the current across the section due to the random hopping of the solute molecules.

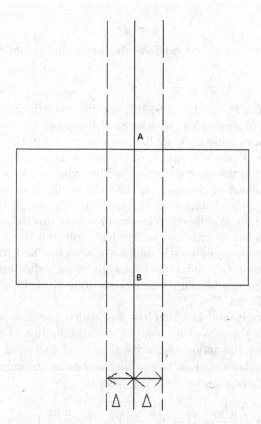

Figure 20.2: Solute molecules within the dashed line hop across AB.

This random motion is produced by the random collisions of the solute molecules with the molecules of the solvent. In a time τ, we assume the root mean square displacement of a solute particle to be Δ. Consequently, a current across the section AB will be set up by those solute particles on the left of it that are moving to the right and solute particles to the right of AB that are moving to the left. Since the probabilities of a left or right jump are equal, half the particles within a distance Δ to the left of AB will cross AB in time τ and half the particles within a distance Δ to the right of AB will cross it in time τ. If n_l is the density of the particles to the left and n_r the density of the particles to the right of AB, then the number n of particles crossing an unit area of AB in time τ is given by

$$n = \frac{1}{2}(n_l \Delta - n_r \Delta) = \frac{1}{2}\Delta \left(-\frac{dn}{dx}\right)\Delta. \tag{20.6.8}$$

Hence, the number crossing unit area per unit time (which is the current) is

$$j = -\frac{1}{2}\frac{\Delta^2}{\tau}\frac{dn}{dx}. \tag{20.6.9}$$

Comparing with Eq.(20.6.6), we have the second important conclusion of Einstein

$$\Delta^2 = 2D\tau. \tag{20.6.10}$$

Between Eq.(20.6.7) and Eq.(20.6.10), one has enough predictive power to find the diameter of suspended particles or Boltzmann's constant k_B.

The verification of Einstein's predictions was primarily the work of Perrin and his students. Early indications of the correctness of Eq.(20.6.7) and Eq.(20.6.10) came from the observations of Seddig, who took two photographs of an aqueous suspension of cinnabar on the same plate at an interval of 0.1 second and measured the distance of corresponding images on the plate. He found that the distances at different temperatures were inversely proportional to viscosity as predicted. Perrin and his students followed the movements of single particles of gamboge or mastic under a microscope and recorded their positions at equidistant time intervals by means of an indicating apparatus. Since the particle size was known, these observations yielded k_B and since R was known, one could get $N = \frac{R}{k_B}$.

Perrin explicitly established that the suspended particles were in thermal equilibrium with the solvent by studying the distribution of particles in a vertical column under the action of gravity. The total energy of a suspended particle at a height z above the base is $\frac{p^2}{2m} + mgz$ and hence the number particles between z and $z + dz$ is given by

$$n(z) = CA \int \frac{d^3p}{h^3}\exp\left(-\frac{p^2}{2mk_BT}\right)\exp\left(-\frac{mgz}{k_BT}\right) \tag{20.6.11}$$

where C is a constant and A is the area of the base of the container. The total number of particles is found from

$$N = CA \int \frac{d^3p}{h^3}\exp\left(-\frac{p^2}{2mk_BT}\right)\int \exp\left(-\frac{mgz}{k_BT}\right)dz. \tag{20.6.12}$$

From Eq.(20.6.11) and Eq.(20.6.12) we have

$$n(z) = N\frac{\exp\left(-\frac{mgz}{k_BT}\right)}{\int_0^h \exp\left(-\frac{mgz}{k_BT}\right)dz} = N\frac{mgz}{k_BT}\frac{\exp\left(-\frac{mgz}{k_BT}\right)}{\left[-\exp\left(-\frac{mgz}{k_BT}\right)\right]}. \tag{20.6.13}$$

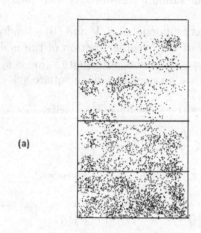

(a)

(b)

Figure 20.3: (a) The distribution as seen by Perrin. (b) The fractal curve

This distribution was confirmed in Perrin's experiment (Fig.20.3a). The displacement of individual particles had the typical from shown in Fig.(20.3b) The path of the particle is an example of a fractal, a curve for which any small section resembles the curve as a whole.

A different way of looking at the problem was derived by Paul Langevin, who was a friend of Einstein. Consider a molecule of mass M colliding elastically with another molecule of mass m. If the velocities of M before and after the collision are $\vec{V}$ and $\vec{V}$ respectively and those of m are $\vec{v}$ and $\vec{v'}$ then

$$M\vec{V} + m\vec{v} = M\vec{V'} + m\vec{v'} \tag{20.6.14}$$

and

$$\frac{1}{2}MV^2 + \frac{1}{2}mv^2 = \frac{1}{2}MV'^2 + \frac{1}{2}mv'^2 \tag{20.6.15}$$

From the above equations we immediately find that

$$\vec{V'} = \frac{M-m}{M+m}\vec{V} + \frac{2m\vec{v}}{M+m}. \tag{20.6.16}$$

We now imagine that the molecule M is being hit randomly by the molecules in from all directions. In each collision, the change in momentum is

$$\triangle \vec{P} = M(\vec{V} - \vec{V'}) = -\frac{2mM}{M+m}\vec{V} + \frac{2mM}{M+m}\vec{v}. \tag{20.6.17}$$

The force on M due to the random collisions, is then clearly composed of two parts:

(i) a resistive part which is proportional to $\vec{V}$ and (ii) a random part which is proportional to the velocity of the irregular motion of the molecule in. The average value of the force is zero and the mean square value is proportional to $k_B T$, where T is the temperature because the mean square velocity of the gas molecule is proportional to $k_B T$.

The equation of motion for the molecule can be written as (setting $M = 1$)

$$\frac{d\vec{V}}{dt} = -\Gamma\vec{V} + \vec{f} \tag{20.6.18}$$

where Γ is the relaxation rate and $\vec{f}$ is a random force with

$$\langle f_i \rangle = 0 \tag{20.6.19}$$
$$\langle f_i(t)f_j(t') \rangle = 2D\delta(t - t')\delta_{ij}. \tag{20.6.20}$$

This is the equation with which we started the chapter (Eq.(20.1.1)).

The above arguments tell us that $\sigma \propto k_B T$ and if M is a spherical solute molecule moving through the fluid, then $\Gamma = \frac{6\pi\eta r}{M}$ where r is the radius. The relation between σ and diffusion constant D is important for the establishment of thermal equilibrium.

The solution of Langevin equation is as written down in Eq.(20.5.2). The displacement Δx_i in time t can them be written as

$$\Delta x_i = \int_0^t V_i(t')dt' = \int_0^t dt' e^{\Gamma t'} \int_0^{t'} dt_1 e^{\Gamma t_1} f_i(t_1) + \frac{V_i(0)}{\Gamma}\left(1 - e^{-\Gamma t}\right). \tag{20.6.21}$$

This leads to

$$\langle(\Delta x_j)^2\rangle = \langle \int_0^t dt' e^{-\Gamma t'} \int_0^{t'} e^{\Gamma t} f_i(t_1)dt_1 \int_0^t dt'' e^{-\Gamma t''} \int_0^{t''} e^{\Gamma t_2} f_i(t_2)dt_2 \rangle$$
$$+ \langle \frac{V_i^2(0)}{\Gamma^2}\left(1 - e^{\Gamma t}\right)^2 + 2\frac{V_i(0)}{\Gamma}\left(1 - e^{-\Gamma t}\right) \int_0^t dt' e^{-\Gamma t'} \int_0^{t'} f_i(t_1)dt_1 \rangle. \tag{20.6.22}$$

Using the moments of $f(t)$

$$\langle(\Delta x_i)^2\rangle = 2\sigma \int_0^t dt' \int_0^t dt'' e^{-\Gamma(t'+t'')}\left[\int_0^{t'} dt_1 \exp(2\Gamma t_1)\theta(t'' - t')\right.$$
$$\left. + \int_0^{t''} dt_1 \exp(2\Gamma t_1)\theta(t' - t'')\right] + \frac{V_i^2(0)}{\Gamma^2}[1 - \exp(-\Gamma t)]^2.$$

Straightforward algebra leads to

$$\langle (\Delta x_i)^2 \rangle = \frac{2\sigma}{\Gamma^2} t + \frac{2\sigma}{\Gamma^3} \left(1 - e^{-\Gamma t} \right) + \frac{V_i^2(0)}{\Gamma^2} \left(1 - e^{-\Gamma t} \right)^2. \tag{20.6.23}$$

For extremely long times, the first term dominates and

$$\langle (\Delta x_i)^2 \rangle = \frac{2\sigma}{\Gamma^2} t = 2 \frac{k_B T}{M} \frac{t}{\Gamma} = 2 \frac{k_B T t}{6\pi \eta r}. \tag{20.6.24}$$

20.7 Quantum Langevin Equation

We end this chapter by showing how a Langevin equation can be obtained for the situation of Sec.20.2, when the dynamics is quantum. Our system S_1 is a single particle (a Brownian particle) and the system S_0 is a system of heat bath particles, all of them connected to the Brownian particle by a spring, thus bringing in the interaction part of the Hamiltonian in Eq.(20.2.1)

$$H = \frac{p^2}{2m} + V(x) + \sum_j \left[\frac{p_j^2}{2m_j} + \frac{1}{2} k_j (q_j - x)^2 \right]. \tag{20.7.1}$$

Here x and p are the coordinate and momentum operators of the Brownian particle and the set $\{q_j, p_j\}$ is the set of coordinates and momentum operators for the heat bath particles. The mass of the jth bath particle is m_j and k_j is the spring constant of the spring connecting. it to the Brownian particle. The potential $V(x)$ is the external potential for the Brownian particle. To this Hamiltonian must be added the commutation relations

$$[x, p] = i\hbar, [q_j, p_k] = i\hbar \delta_{jk}. \tag{20.7.2}$$

The equation of motion for any Heisenberg operators O is

$$\dot{O} = \frac{1}{i\hbar} [O, H] \tag{20.7.3}$$

and hence

$$\dot{x} = \frac{p}{m} \tag{20.7.4}$$

$$\dot{p} = -V'(x) + \sum_j k_j (q_j - x) \tag{20.7.5}$$

$$\dot{q}_j = p_j / m_j \tag{20.7.6}$$

$$\dot{p}_j = -k_j (q_j - x). \tag{20.7.7}$$

The solution of Eqs.(20.7.6) and (20.7.7d) gives

$$q_i(t) = q_i(0) \cos \omega_j t + \frac{p_j(0)}{m_j \omega_j} \sin \omega_j t + x(t) - x(0) \cos \omega_j t$$

$$- \int_0^t dt' \cos \omega_j (t - t') x(t') \tag{20.7.8}$$

where $\omega_j = (k_j/m_j)^{1/2}$. Inserting the value of q_j from Eq.(20.7.8) in Eq.(20.7.5) we easily obtain

$$m\ddot{x} + \int_0^t dt' B(t - t')\dot{x}(t') + V(x) - B(t)x(0) = F(t) \qquad (20.7.9)$$

where

$$F(t) = \sum_j [q_j(0)k_j \cos \omega_j t + p_j(0)\omega_j \sin \omega_j t] \qquad (20.7.10)$$

and

$$B(t) = \sum_j k_j \cos \omega_j t. \qquad (20.7.11)$$

Note that the initial variables of the heat bath occur in the force term in Eq.(20.7.9). Apart from that it is an equation in $x(t)$.

Now, we introduce the statistical average over the initial variables of the heat bath. To do so we assume that at $t = 0$, the oscillators are canonically distributed with respect to the free oscillator Hamiltonian

$$H_0 = \sum \left(\frac{p_j^2}{2m_j} + \frac{1}{2}k_j q_j^2 \right). \qquad (20.7.12)$$

For an arbitrary operator O, we introduce the expectation value,

$$\langle O \rangle = \frac{\mathrm{Tr} O e^{-H_0/k_B T}}{\mathrm{Tr} e^{-H_0/k_B T}} \qquad (20.7.13)$$

where the trace is with respect to the bath particles. It is straightforward to check:

$$\langle q_j(0)q_k(0) \rangle = \frac{\langle p_j(0)p_k(0) \rangle}{\omega_j^2 m_j^2} = \delta_{jk} \frac{\hbar \coth \left(\frac{\hbar \omega_j}{2k_B T} \right)}{2m_j \omega_j}$$

$$\langle q_j(0)p_k(0) \rangle = -\langle p_k(0)q_j(0) \rangle = \frac{1}{2}i\hbar \delta_{jk}. \qquad (20.7.14)$$

In addition, there is the Gaussian property that the expectation value of an odd number of factors of q_j and p_j vanishes and the expectation value of an even number of factors is the sum of the products of the pairwise expectations with the order of the factors preserved. These properties lead to the results:

$$\frac{1}{2}\langle F(t)F(t') + F(t')F(t) \rangle = \sum \frac{1}{2}k_j g\omega_j \left(\coth \frac{\hbar \omega_j}{2k_B T} \right) \cos \omega_j(t - t') \qquad (20.7.15)$$

and

$$[F(t), F(t')] = -i\hbar \sum_j k_j \omega_j \sin \omega_j(t - t')$$

We now turn to $B(t)$ in Eq.(20.7.11) and assuming an infinite number of oscillators in the bath with a continuous distribution of frequency, obtain

$$B(t) = \int_0^\infty N(\omega)k(\omega)\cos\omega t. \qquad (20.7.16)$$

Choosing $N(\omega)k(\omega) \approx$ constant, $B(t) \propto \delta(t)$ and hence at all finite times, $B(t)$ can be dropped, so that Eq.(20.7.9) becomes

$$m\ddot{x} + \int_0^t dt' B(t-t')\dot{x}(t') + V'(x) = F(t) \qquad (20.7.17)$$

with $F(t)$ a random operator prescribed through the correlation in Eq.(20.7.15). We have thus obtained the quantum Langevin equation [Eq.(20.7.17)].

The quantum Langevin equation has been obtained in a very special situation. Can the structure of Eq.(20.7.17) be universal? The best one can assert is that if there is a universal description, then it must be of the form of Eq.(20.7.17).

The last question is: Does one have the approach to the correct equilibrium state in this Langevin equation? This means that we would have to demonstrate for the stationary solution $x(t)$ of the Langevin equations, that

$$\lim_{t\to\infty} \langle x(t)^N \rangle = \frac{\sum_m e^{-E_m/k_B T} \int (\psi_m x^N \psi_n)dx}{\sum_n e^{-E_n/k_B T}}.$$

where ψ_n and E_n are the eigenfunctions and eigenvalues of the Schrodinger equation

$$\left[\frac{-\hbar^2}{2m}\frac{d^2}{dx^2} + V(x)\right]\psi_n = E_n\psi_n. \qquad (20.7.18)$$

This has not been achieved so far.

21

Fokker-Planck Equations

21.1 Introduction

This chapter will consider distributions, equilibrium distributions and the approach to equilibrium from yet another point of view. The dynamics that will be studied is the dynamics of the probability distribution that is associated with Langevin equation. As we have found in the last chapter, the Langevin equation refers to the stochastic dynamics of a variable v (v can actually be anything)

$$\frac{dv}{dt} = -\frac{\partial S}{\partial v} + f(t) \tag{21.1.1}$$

where S is an arbitrary function of v and $f(t)$ is the noise term. The variable $v(t)$ is characterized by its distribution $P\{v(t)\}$ and the Fokker-Planck equation is an equation of motion for the distribution P. A stationary solution for the equation of motion will yield the equilibrium distribution P_{eq}. Finding the equilibrium distribution and studying the approach to equilibrium is one of the major problems associated with Fokker-Planck equations which we will take up in detail. For an arbitrary initial condition, the solution of Fokker-Planck equation can be written down in general, which can be very useful in tackling several out-of-equilibrium situations. Before providing one of the many standard derivations of Fokker-Planck equations in the next section we wish to point out that the Boltzmann transport equation is also an equation for a probability distribution and it stands to reason that we should be able to construct the Fokker-Planck equation from the transport equation. A good clear cut connection, however, is lacking although some literature does exist.

21.2   Fokker-Planck Equation: Derivation

Our starting point will be the Langevin equation for a random variable as written down in Eq.(21.1.1). The statistics of the random force $f(t)$ is specified

$$\langle f(t) \rangle = 0 \qquad\qquad (21.2.1)$$

$$\langle f(t)f(t') \rangle = 2\epsilon\delta(t - t') \qquad\qquad (21.2.2)$$

and all the higher moments are appropriate to a Gaussian distribution for $f(t)$. The probability $P\{f(t)\}$ of getting a particular $f(t)$ is given by

$$P\{f(t)\} = Ae^{-\int f(t)^2 dt/2\epsilon} \qquad\qquad (21.2.3)$$

where

$$A^{-1} = \int D[f]e^{-\int f(t)^2 dt/2\epsilon}. \qquad\qquad (21.2.4)$$

All averaging that we will have occasion to introduce would mean an averaging over the distribution $P\{f(t)\}$, i.e.,

$$\langle \ldots \rangle = \int D[f]P\{f(t)\}(\ldots). \qquad\qquad (21.2.5)$$

It is easiest to proceed by considering time as a discrete variable and taking a sequence $t_0, t_1, t_2, \ldots t_{n+1}$, such that $t_{n+1} - t_n = \delta t$ is an infinitesimal. Writing $v = v(t_n)$, we can write the discretized Langevin equation as

$$v_{n+1} = v_n - g \qquad\qquad (21.2.6)$$

where

$$g = \frac{\partial S}{\partial v}\delta t - \bar{f}(t)\sqrt{\delta t} \qquad\qquad (21.2.7)$$

with the definition

$$\bar{f}(t) = f(t)\sqrt{\delta t} \qquad\qquad (21.2.8)$$

so that the correlation of $\bar{f}(t)$ is

$$\langle \bar{f}(t)\bar{f}(t') \rangle = 2\epsilon\delta_{tt'}, \qquad\qquad (21.2.9)$$

$\delta_{tt'}$ being the Kroenecker delta.

Our aim is to describe the dynamics that takes one from $P\{v(t)\}$ to $P\{v(t + dt)\}$, the probability distribution at time $t + dt$. In terms of the discretized variable we want to go from $P(v_n, t_n)$ to $P(v_{n+1}, t_{n+1})$. The connection is

$$P(v_{n+1}t_{n+1}) = \left\langle \int P(v_n, t_n)\delta(v_{n+1} - v_n + g)dv_n \right\rangle. \qquad\qquad (21.2.10)$$

Note that the integration over the δ-function is not enough although that does take one from step n to step $n + 1$. This step is dependent on g which involves the noise term and hence to get the final answer we must average over the different realizations of the random force. But g is small as shown in Eq.(21.2.7) and we can expand in it as

$$
\begin{aligned}
\delta(v_{n+1} - v_n + g) &= \delta(v_{n+1} - v_n) + g\delta'(v_{n+1} - v_n) \\
&+ \frac{g^2}{2!}\delta''(v_{n+1} - v_n) + O(dt^{3/2}). \quad (21.2.11)
\end{aligned}
$$

Inserting this expansion in Eq.(21.2.10) we find, correct to $O(\delta t)$,

$$
\begin{aligned}
P(v_{n+1}, t_{n+1}) &= \left\langle \int P(v_n, t_n)\delta(v_{n+1} - v_n)dv_n \right. \\
&+ \int P(v_n, t_n)g\delta'(v_{n+1} - v_n)dv_n \\
&+ \left. \int P(v_n, t_n)\frac{g^2}{2}\delta''(v_{n+1} - v_n)dv_n \right\rangle. \quad (21.2.12)
\end{aligned}
$$

Now, the three terms on the RHS of Eq.(21.2.12) can be written as follows:

$$
\int P(v_n, t_n)\delta(v_{n+1} - v_n)dv_n = P(v_{n+1}, t_n), \quad (21.2.13)
$$

$$
\begin{aligned}
\int P(v_n, t_n)g\delta'(v_{n+1} - v_n)dv_n &= -P(v_n, t_n)g\delta(v_{n+1} - v_n) \\
&+ \int \frac{d}{dv_n}\{gP(v_n, t_n)\}\delta(v_{n+1} - v_n)dv_n \\
&= \frac{d}{dv_{n+1}} \cdot (gP(v_{n+1}, t_n)) \quad (21.2.14)
\end{aligned}
$$

(we have assumed that $P \to 0$ at the end points of the integration.)

$$
\begin{aligned}
\int P(v_n, t_n)\frac{g^2}{2}\delta''(v_{n+1} - v_n)dv_n &= -P(v_n, t_n)\frac{g^2}{2}\delta'(v_{n+1} - v_n) \\
&+ \int \frac{d}{dv_n}\left\{\frac{g^2}{2}P(v_n, t_n)\right\}\delta'(v_{n+1} - v_n)dv_n \\
&= \int \frac{d}{dv_n}\left\{\frac{g^2}{2}P(v_n, t_n)\right\}\delta'(v_{n+1} - v_n)dv_n \\
&= -\frac{d}{dv_n}\left\{\frac{g^2}{2}P(v_n, t_n)\right\}\delta(v_{n+1} - v_n) \\
&+ \int \frac{d^2}{dv_n^2}\left\{\frac{g^2}{2}P(v_n, t_n)\right\}\delta(v_{n+1} - v_n)dv_n \\
&= \frac{d^2}{dv_n^2}\left\{\frac{g^2}{2}P(v_{n+1}, t_n)\right\} \quad (21.2.15)
\end{aligned}
$$

(P as well as its derivative have been assumed to vanish at the end points.)

Using Eqs.(21.2.13)-(21.2.15) in Eq.(21.2.12), we arrive at

$$P(v_{n+1}, t_{n+1}) = P(v_{n+1}, t_n) + \left\langle \frac{d}{dv_{n+1}} (gP(v_{n+1}, t_n)) \right\rangle$$

$$+ \left\langle \frac{d^2}{dv_{n+1}^2} \left(\frac{g^2}{2} P(v_{n+1}, t_n) \right) \right\rangle. \qquad (21.2.16)$$

For the averages, we have to $O(\delta t)$

$$\langle g \rangle = \left\langle \frac{\partial S}{\partial v} \right\rangle \delta t - \langle \bar{f} \rangle \sqrt{\delta t} = \frac{\partial S}{\partial v}.\delta t$$

while

$$\langle g^2 \rangle = \langle \bar{f}^2 \rangle \delta t = 2\epsilon \delta t.$$

Using the above averages in Eq.(21.2.16)

$$P(v_{n+1}, t_{n+1}) - P(v_{n+1}, t_n) = \frac{d}{dv_{n+1}} (P\frac{\partial S}{\partial v})\delta t + \epsilon \frac{d^2 P}{dv_{n+1}^2} \delta t. \qquad (21.2.17)$$

Reverting to the continuum variables by taking the limit of $\delta t \to 0$, we have

$$\frac{\partial P}{\partial t} = \frac{\partial}{\partial v} \left(P\frac{\partial S}{\partial v} \right) + \epsilon \frac{\partial^2 P}{\partial v^2}. \qquad (21.2.18)$$

This is the well known Fokker-Planck equation for the probability distribution $P\{v(t)\}$ for a random variable $v(t)$.

What if we start with N variables $\{v_\alpha\}$, $(\alpha = 1, 2, \ldots N)$, which satisfy Langevin equations of the form

$$\frac{dv_\alpha}{dt} = -\frac{\partial S}{\partial v_\alpha} + f_\alpha \qquad (21.2.19)$$

where $S = S(\{v_\alpha\})$ and the random forces f_α have the correlation

$$\langle f_\alpha(t_2) f_\beta(t_1) \rangle = 2\epsilon_\alpha \delta(t_2 - t_1)\delta_{\alpha\beta}.$$

If $P\{v_\alpha(t)\}$ is the probability of the set $\{v_\alpha\}$ acquiring the value $\{v_\alpha(t)\}$ at time t, then the Fokker-Planck equation for P is

$$\frac{\partial P}{\partial t} = \frac{\partial}{\partial v_\alpha} \left\{ P\frac{\partial S}{\partial v_\alpha} \right\} + \epsilon_\alpha \frac{\partial^2 P}{\partial v_\alpha \partial v_\alpha} \qquad (21.2.20)$$

where a repeated index means that it has to be summed over.

To end this subsection, we return to Eq.(21.2.18) and find the equilibrium distribution by setting $\frac{\partial P_{eq}}{\partial t} = 0$. This leads to

$$\frac{\partial}{\partial v} \left(P_{eq}\frac{\partial S}{\partial v} \right) + \epsilon \frac{\partial}{\partial v} \left\{ \frac{\partial P_{eq}}{\partial v} \right\} = 0$$

or

$$\frac{\partial P_{eq}}{\partial v} + \frac{P_{eq}}{\epsilon}\frac{\partial S}{\partial v} = \text{constant.}$$

The constant can be set equal to zero by evaluating the expression for very large values of v, where both P and $\frac{\partial P}{\partial v}$ have to vanish. Thus,

$$\frac{1}{P_{eq}}\frac{\partial P_{eq}}{\partial v} = -\frac{\partial}{\partial v}\left(\frac{S}{\epsilon}\right)$$

or

$$P_{eq} = P_0 e^{-S/\epsilon}. \tag{21.2.21}$$

The constant P_0 can be found by the normalization condition $\int P_{eq}(v)dv = 1$. We have thus the equilibrium distribution to be proportional to $e^{-S/\epsilon}$, as we had said would happen for an arbitrary $S(v)$ in the previous chapter. It now remains to be shown that an arbitrary distribution must reach the equilibrium distribution as $t \to \infty$. This will be dealt with in the next section.

21.3 The General Solution

The standard technique of writing down a formal solution to the Fokker-Planck equation [Eq.(21.2.18)] is to write it in the form of a Schrodinger equation by carrying out the transformation

$$P = \sqrt{P_{eq}}.\psi = e^{-\frac{S}{2\epsilon}}\psi. \tag{21.3.1}$$

where $\sqrt{P_0}$ of Eq.(21.2.21) has been absorbed in ψ. Differentiation yields

$$\frac{\partial}{\partial v}\left(P\frac{\partial S}{\partial v}\right) = \frac{\partial}{\partial v}(e^{-\frac{S}{2\epsilon}}\psi S') = \left(-\frac{S'^2}{2\epsilon}\psi + \psi'S' + \psi S''\right)e^{-\frac{S}{2\epsilon}} \tag{21.3.2}$$

where the prime denotes derivative with respect to v. Again

$$\frac{\partial P}{\partial v} = \frac{\partial}{\partial v}\left(\psi e^{-S/2\epsilon}\right) = \left(\psi' - \frac{S'\psi}{2\epsilon}\right)e^{-S/2\epsilon}$$

and

$$\begin{aligned}
\frac{\partial^2 P}{\partial v^2} &= \frac{\partial}{\partial v}\left(\psi' - \frac{S'\psi}{2\epsilon}\right)e^{-S/2\epsilon}\\
&= \left(\psi'' - \frac{S''\psi}{2\epsilon} - \frac{\psi'S'}{\epsilon} + \frac{S'^2\psi}{4\epsilon^2}\right)e^{-\frac{S}{2\epsilon}}.
\end{aligned} \tag{21.3.3}$$

Inserting Eqs.(21.3.2) and (21.3.3) in Eq.(21.2.18), we get

$$\frac{\partial\psi}{\partial t} = \epsilon\psi'' + \left(\frac{S''}{2} - \frac{S'^2}{4\epsilon}\right)\psi. \tag{21.3.4}$$

Writing the stationary state solutions of this equations as $\psi = e^{-\lambda t}\phi(v)$, we find the eigenvalue equation satisfied by $\phi(v)$ as

$$-\epsilon\phi'' + \left(\frac{S'^2}{4\epsilon} - \frac{S''}{2}\right) = \lambda\phi. \tag{21.3.5}$$

If we define the operator

$$A = \frac{\partial}{\partial v} + \frac{S'}{2\epsilon} \tag{21.3.6}$$

then the adjoint operator is

$$A^\dagger = -\frac{\partial}{\partial v} + \frac{S'}{2\epsilon} \tag{21.3.7}$$

and

$$
\begin{aligned}
A^\dagger A\phi &= \left(-\frac{\partial}{\partial v} + \frac{S'}{2\epsilon}\right)\left(\frac{\partial}{\partial v} + \frac{S'}{2\epsilon}\right)\phi \\
&= -\frac{\partial^2\phi}{\partial v^2} - \frac{S''}{2\epsilon}\phi - \frac{S'\phi'}{2\epsilon} + \frac{S'\phi'}{2\epsilon} + \frac{S'^2\phi}{4\epsilon^2} \\
&= \left(-\frac{\partial^2}{\partial v^2} + \frac{S'^2}{4\epsilon^2} - \frac{S''}{2\epsilon}\right)\phi.
\end{aligned}
\tag{21.3.8}
$$

Consequently, we can write Eq.(21.3.5) as

$$\epsilon A^\dagger A\phi = \lambda\phi. \tag{21.3.9}$$

The spectrum of $\epsilon A^\dagger A$ is non-negative.

The zero eigenvalue of $\epsilon A^\dagger A$ is clearly obtained for $A\phi_0 = 0$ which means that the eigenfunction ϕ_0 corresponding to the zero eigenvalue is obtained from

$$\frac{\partial\phi_0}{\partial v} + \frac{S}{2\epsilon}\phi_0 = 0$$

or

$$\phi_0 = A_0 e^{-S/2\epsilon} \tag{21.3.10}$$

where A_0 is a constant.

The remaining eigenvalues of $\epsilon A^\dagger A$ are written in order of increasing magnitude as $\lambda, \lambda_1, \lambda_3, \ldots$ with eigenfunctions $\phi_1, \phi_2, \phi_3 \ldots$. Since the operator $\epsilon A^\dagger A$ is hermitian, the eigenfunctions of it form a complete set and hence an arbitrary solution of Eq.(21.3.4) can be written as

$$\psi(v,t) = a_0 e^{-S/2\epsilon} + \sum_{n=1}^{\infty} a_n e^{-\lambda_n t}\phi_n \tag{21.3.11}$$

and the probability distribution $P(v, t)$ is obtained from Eq.(21.3.1) as

$$P(v,t) = A_0 e^{-S/\epsilon} + \sum_{n=1}^{\infty} a_n e^{-\lambda_n t} \phi_n e^{-S/2\epsilon}. \tag{21.3.12}$$

Since all λ_n $(n = 1, 2 \ldots) > 0$, as $t \to \infty$, we have

$$\lim_{t \to \infty} P(v, t) = A_0 e^{-S/\epsilon} = P_{\text{eq}}. \tag{21.3.13}$$

Thus we have demonstrated that as $t \to \infty$ the equilibrium distribution is indeed obtained. If the initial distribution is prescribed as $P(v, t = 0) = \psi_0 e^{-S/2\epsilon}$ then from Eq.(21.3.11)

$$\psi_0 = \sum_{n=0}^{\infty} a_n \phi_n$$

so that the a_n are prescribed as,

$$a_n = (\phi_n, \psi_0) \tag{21.3.14}$$

and Eq.(21.3.12) is indeed a complete solution to the problem.

As the final stage of approach to equilibrium is reached, the time scale for reaching equilibrium is set by the smallest eigenvalue λ_1 in the expansion of Eq.(21.3.12). For a variety of problems, this eigenvalue is thus of great importance. Being an excited state eigenvalue it is difficult to obtain it variationally. So we describe a transformation which for the one-dimensional problem casts the eigenvalue λ_1 as the ground state eigenvalue of a Hamiltonian that we shall construct and thence it can be variationally obtained.

To proceed we define the operator

$$\varrho = \begin{pmatrix} 0 & A \\ 0 & 0 \end{pmatrix}. \tag{21.3.15}$$

The adjoint is

$$\varrho^\dagger = \begin{pmatrix} 0 & 0 \\ A^\dagger & 0 \end{pmatrix}. \tag{21.3.16}$$

The operator ϱ is nilpotent with $\varrho^2 = 0 = \varrho^{\dagger 2}$. Now, consider the operator

$$\begin{aligned} H_S &= \varrho\varrho^\dagger + \varrho^\dagger\varrho \\ &= \begin{pmatrix} A^\dagger A & 0 \\ 0 & A A^\dagger \end{pmatrix} \\ &= \begin{pmatrix} H_+ & 0 \\ 0 & H_- \end{pmatrix}. \end{aligned} \tag{21.3.17}$$

Our concern is with the operator H_+, the eigenvalues and the eigenfunctions of which can be written as

$$H_+ \phi_n = \frac{\lambda_n}{\epsilon} \phi_n. \tag{21.3.18}$$

We note that the column vector $\begin{pmatrix} \phi_n \\ 0 \end{pmatrix}$ is an eigenvector of H_S with eigenvalue λ_n/ϵ since

$$H_S \begin{pmatrix} \phi_n \\ 0 \end{pmatrix} = \begin{pmatrix} H_+ & 0 \\ 0 & H_- \end{pmatrix} \begin{pmatrix} \phi_n \\ 0 \end{pmatrix} = \begin{pmatrix} H_+\phi_n \\ 0 \end{pmatrix} = \frac{\lambda_n}{\epsilon} \begin{pmatrix} \phi_n \\ 0 \end{pmatrix}.$$

$$(21.3.19)$$

Further, the operator ϱ commutes with H_S,

$$
\begin{aligned}
[\varrho, \varrho\varrho^\dagger + \varrho^\dagger\varrho] &= \varrho[\varrho,\varrho^\dagger] + [\varrho,\varrho^\dagger]\varrho \\
&= \varrho \begin{pmatrix} -A^\dagger A & 0 \\ 0 & AA^\dagger \end{pmatrix} + \begin{pmatrix} -A^\dagger & 0 \\ 0 & AA^\dagger \end{pmatrix} \varrho \\
&= \begin{pmatrix} 0 & 0 \\ -A & 0 \end{pmatrix} \begin{pmatrix} -A^\dagger A & 0 \\ 0 & AA^\dagger \end{pmatrix} \\
&\quad + \begin{pmatrix} -A^\dagger A & 0 \\ 0 & AA^\dagger \end{pmatrix} \begin{pmatrix} 0 & A \\ A & 0 \end{pmatrix} \\
&= \begin{pmatrix} 0 & -AA^\dagger A \\ 0 & 0 \end{pmatrix} + \begin{pmatrix} 0 & AA^\dagger A \\ 0 & 0 \end{pmatrix}.
\end{aligned}
$$

$$(21.3.20)$$

Hence $\varrho \begin{pmatrix} \phi_n \\ 0 \end{pmatrix}$ is also an eigenvector of H_S with eigenvalue λ_n/ϵ, except when $\lambda_n = \lambda_0 = 0$, since in that case $\varrho\phi_0 = 0$. To each λ_n of H_+, there can be only one ϕ_n in this one-dimensional problem and hence for each λ_n of H_S, we have a double degeneracy: an eigenvector $\begin{pmatrix} \phi_n \\ 0 \end{pmatrix}$, and another $\varrho \begin{pmatrix} \phi_n \\ 0 \end{pmatrix}$.

The structure of $\varrho \begin{pmatrix} \phi_n \\ 0 \end{pmatrix}$ is

$$\begin{pmatrix} 0 & 0 \\ A & 0 \end{pmatrix} \begin{pmatrix} \phi_n \\ 0 \end{pmatrix} = \begin{pmatrix} 0 \\ A\phi_n \end{pmatrix} = \begin{pmatrix} 0 \\ \psi_n \end{pmatrix}.$$

But $\begin{pmatrix} 0 \\ \psi_n \end{pmatrix}$ is an eigenvector of H_S with eigenvalue λ_n/ϵ and consequently

$$\lambda_n \begin{pmatrix} 0 \\ \psi_n \end{pmatrix} = H_S \begin{pmatrix} 0 \\ \psi_n \end{pmatrix} = \begin{pmatrix} H_+ & 0 \\ 0 & H_- \end{pmatrix} \begin{pmatrix} 0 \\ \psi_n \end{pmatrix} = \begin{pmatrix} 0 \\ H_-\psi_n \end{pmatrix}.$$

$$(21.3.21)$$

Thus

$$H_-\psi_n = \frac{\lambda_n}{\epsilon}\psi_n. \qquad (21.3.22)$$

The functions ψ_n are the eigenfunctions of the operator H_- with eigenvalues λ_n/ϵ. The lowest eigenvalue is λ_1/ϵ (the above arguments do not hold for $\lambda_0 = 0$ as explained earlier). A variational calculation on the Hamiltonian H_- for the ground state yields λ_1. Thus, we have the following result:

For the one-dimensional Hamiltonian $\epsilon A^\dagger A$, with the nondegenerate eigenvalues in increasing order $0, \lambda_1, \lambda_2, \ldots$, the eigenvalues $\lambda_1, \lambda_2, \lambda_3, \ldots$ are the nondegenerate eigenvalues of the partner Hamiltonian $\epsilon A A^\dagger$. The eigenvalue λ_1 being the ground state energy of $\epsilon A A^\dagger$, can be obtained variationally.

21.4 Metastable State and the Lowest Eigenvalue

We consider the function $S(v)$ to be of the form shown in Fig.21.1. Instead of using the notation v for the variable, we will shift to x and thus consider $S(x)$ to be a function with two minima at $x = x_1$ and $x = x_2$ and a maximum at $x = x_m$ ($x_1 < x_m < x_2$). The values of $S(x)$ at $x = x_1$, x_m and x_2 are S_1, S_m and S_2 respectively. The function is shown by the solid curve in Fig.21.1.

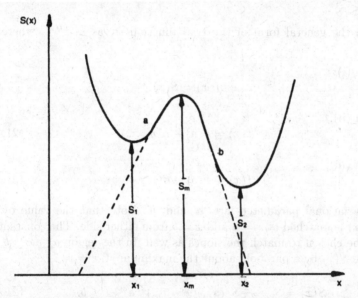

Figure 21.1: $S(x)$ is the potential exhibiting the metastable state. The points $x = a$, $x = b$ and the dashed lines are relevant to the variational trial functions of Eqs.(21.4.2-eq21.4.4).

This is the kind of potential where a metastable state is formed – an initial distribution peaked around $x = x_1$ is such a metastable state and as has been generally demonstrated for the Fokker-Planck equation, eventually this state must reach the equilibrium distribution $P_0 e^{-S(x)/\epsilon}$, where P_0 is a normalizing constant. The approach to equilibrium will be governed by the first excited state eigenvalue λ_1 of the operator $-\epsilon \frac{d^2}{dx^2} + \frac{S'^2}{4\epsilon} - \frac{S''}{2}$ (we note that λ_1 will be exponentially small ϵ since qualitative reasoning shows that the potential $\frac{S'^2}{4\epsilon} - \frac{S''}{2}$ has three minima and the tunneling probabilities between the different minima brings the ground state and the first excited state exponentially close. Since $\lambda_0 = 0$, we have λ_1 exponentially small). As explained in the previous section λ_1 can be found as the ground state energy of the partner Hamiltonian H_-, given by

$$H_- = -\epsilon \frac{d^2}{dx^2} + \frac{S'^2}{4\epsilon} + \frac{S''}{2}. \qquad (21.4.1)$$

We now have to find a good trial wavefunction that will guarantee a small eigenvalue. First note that if $\psi = e^{S/2\epsilon}$, then $H_-\psi$ is identically zero, but clearly ψ is not an admissible eigenfunction since it cannot be normalized. However, it can serve as a guide in choosing the trial function.

We break up the region $-\infty \leq x \leq \infty$ into three distinct regions

1. $-\infty \leq x \leq a$

2. $a \leq x \leq b$

3. $b \leq x \leq \infty$

and assume the general form of the trial function ψ_g as $e^{f(x)/2\epsilon}$, where we choose

1. Region (i):

$$f(x) = S(x) \qquad (21.4.2)$$

2. Region (ii):

$$f(x) = S(a) - \alpha(a - x) \qquad (21.4.3)$$

3. Region (iii):

$$f(x) = S(b) - \beta(x - b). \qquad (21.4.4)$$

The variational parameters are 'a' and 'b'. Note that the value of the function $f(x)$ is matched at $x = a$ and $x = b$ from either side. The constants α and β will be chosen to match the slopes as well. In the region $a \leq x \leq b$, we approximate $S(x)$ by a parabola about the maximum at $x = x_m$, i.e.,

$$S(x) = S_m - \frac{1}{2}S''(x_m)(x - x_m)^2; a \leq x \leq b. \qquad (21.4.5)$$

Consequently the slopes at $x = a$, b, of $S(x)$ are $-S''(x_m)(a - x_m)$ and $-S''(x_m)(b - x_m)$.

Hence, we identify

$$\alpha = -S''(x_m)(a - x_m) \qquad (21.4.6)$$
$$\beta = S''(x_m)(b - x_m). \qquad (21.4.7)$$

The values $S(a)$ and $S(b)$ can be obtained by expanding $S(x)$ about x_1 and x_2 respectively in a parabolic approximation as

$$S(a) = S_1 + \frac{1}{2}S''(x_1)(a - x_1)^2 \qquad (21.4.8)$$
$$S(b) = S_2 + \frac{1}{2}S''(x_2)(b - x_2)^2. \qquad (21.4.9)$$

We thus obtain the following approximation for $f(x)$.

Region(i): $\quad f(x) = S_1 + \dfrac{1}{2}S''(x_1)(a - x_1)^2 + S''(x_m)(a - x_m)(a - x),$

Region(ii): $\quad f(x) = S_m - \dfrac{1}{2}S''(x_m)(x - x_m)^2,$

Region(iii): $\quad f(x) = S_2 + \dfrac{1}{2}S''(x_2)(b - x_2)^2 - S''(x_m)(b - x_m)(x - b).$

The variational estimate for λ_1 is now

$$\lambda_1 = \int\limits_{-\infty}^{\infty} \frac{e^{f(x)/2\epsilon} H_- e^{f(x)/2\epsilon} dx}{\int\limits_{-\infty}^{\infty} e^{f(x)/\epsilon} dx}. \tag{21.4.10}$$

The integral can be evaluated for $\epsilon \ll 1$, by assuming that the range $a \le x \le b$ for integration purposes can be extended from $-\infty$ to ∞. The resulting integral λ_1 is a function of 'a' and 'b' and these variational parameters are found by setting

$$\frac{\partial \lambda_1}{\partial a} = \frac{\partial \lambda_1}{\partial b} = 0 \tag{21.4.11}$$

leading to

$$a = \frac{\sqrt{3\epsilon}}{|u''(a)|}, \qquad b = \frac{\sqrt{3\epsilon}}{|u''(b)|}. \tag{21.4.12}$$

The eigenvalue λ_1 is found to be

$$\lambda_1 = \frac{e^{3/2}}{6\sqrt{6\pi}}[e^{-[S_m - S(a)]/\epsilon} + e^{-[S_m - S(b)]/\epsilon}]. \tag{21.4.13}$$

The characteristic time $\tau_S = \lambda_1^{-1}$ and is known as Kramers' time. It can be obtained from a WKB calculation as well. In that case the pre-factor of the exponential terms of λ_1 is $1/2\pi$. The advantage of Eq.(21.4.13) is that, it being variational, this λ_1 is known to be a strict upper limit on the actual eigenvalue.

21.5 Passage to Equilibrium of a Non-equilibrium State

We consider the sudden switching on of a laser, which can be considered as a far from equilibrium phenomenon. The laser switches on as a pump parameter μ is varied and for $\mu > \mu_0$ the zero intensity state in the cavity is unstable and the intensity begins to grow and saturates at a value determined by μ. The dynamics can be described by the phenomenological equation

$$\frac{\partial n}{\partial t} = (\mu - \mu_0)n - \beta n^3 + f(t). \tag{21.5.1}$$

The noise term $f(t)$ comes from the fact that the atomic levels involved in the lasing action are not sharp but broadened due to various effects and the averaging over the broadening gives noise. It should also be noted that n is not strictly a scalar. The above equation, if derived from a microscopic starting point would involve the electric field $E = \mathrm{Re} E(t)e^{i\omega t}$. The vector $E(t)$ would involve a magnitude and phase for each of its components. Each component (E_x, E_y, or E_z) would thus be characterized by two functions E_1 and E_2 (the real and imaginary parts) and we would actually have two equations (the noise $f = f_1 + if_2$)

$$\frac{\partial E_1}{\partial t} = (\mu - \mu_o)E_1 - \beta(E_1^2 + E_2^2)E_1 + f_1(t) \qquad (21.5.2)$$

$$\frac{\partial E_2}{\partial t} = (\mu - \mu_o)E_2 - \beta(E_1^2 + E_2^2)E_2 + f_2(t) \qquad (21.5.3)$$

with

$$\langle f_1 f_2 \rangle = 0 \qquad (21.5.4)$$

and

$$\langle f_1(t_1)f_1(t_2) \rangle = \langle f_2(t_1)f_2(t_2) \rangle = 2\epsilon\delta(t_1 - t_2). \qquad (21.5.5)$$

We now recall the Fokker-Planck equation for N variables. The Langevin equations are taken to be

$$\dot{q}_\alpha = -\frac{\partial S}{\partial q_\alpha} + f_\alpha \qquad (21.5.6)$$

where

$$S = S(q_\alpha) \qquad (21.5.7)$$

and

$$\langle f_\alpha(t_2)f_\beta(t_1) \rangle = 2\epsilon(t_2 - t_1)\delta_{\alpha\beta}. \qquad (21.5.8)$$

The corresponding Fokker-Planck equation for $P(\{q_\alpha(t)\})$, the probability of the set $\{q_\alpha(t)\}$ acquiring the value $\{q_\alpha(t)\}$ at time t is given by

$$\frac{\partial P}{\partial t} = \frac{\partial}{\partial q_\alpha}\left(P\frac{\partial S}{\partial q_\alpha}\right) + \epsilon\frac{\partial^2 P}{\partial q_\alpha \partial q_\alpha}. \qquad (21.5.9)$$

Hence the Fokker-Planck equation for the problem of Eq.(21.2.2) with $E_1 = q_1$ and $E_2 = q_2$ is

$$\frac{\partial P}{\partial t} = \vec{\nabla}.[P\mathbf{q}(\gamma - \beta\mathbf{q}^2)] + \epsilon\nabla^2 P \qquad (21.5.10)$$

where $q^2 = q_1^2 + q_2^2$, $\mathbf{q} = (q_1, q_2)$, $\nabla^2 = \frac{\partial^2}{\partial q_1^2} + \frac{\partial^2}{\partial q_2^2}$ and $\gamma = \mu - \mu_0$.

Since P will not depend on the phase of $\mathbf{q}$ its functional dependence is on the magnitude of $\mathbf{q}$ alone and Eq.(21.5.10) becomes

$$\frac{\partial P}{\partial t} = \frac{1}{q}\frac{\partial}{\partial q}[-Pq^2(1 - q^2)] + \frac{\epsilon}{q}\frac{\partial}{\partial q}\left(q\frac{\partial P}{\partial q}\right) \qquad (21.5.11)$$

where we have set $\gamma = \beta = 1$ for convenience.

The above equation needs to be supplemented by a boundary condition at $t = 0$. We set this as $P(q, t = 0) = \delta(q)$. This physically corresponds to suddenly setting the pump parameter at a value $\mu > \mu_0$ from a value below μ_0. This implies that at $t = 0$, there are no photons in the cavity (a non-equilibrium states), while the equilibrium has to correspond to a definite intensity of radiation, namely $P \propto e^{-S/\epsilon}$, where $S = -\frac{q^2}{2} + \frac{q^4}{4}$. The approach to equilibrium has three different regimes we will now consider separately.

21.5.1 Regime - I: Diffusion

The probability concentrated near $q = 0$, begins to spread out and the appropriate approximation is to use $S = -q^2/2$ for the initial growth period. The Fokker-Planck equation can be solved exactly for the quadratic potential. We illustrate this for the one-dimensional situation where $S = -x^2/2$. This leads to the following form for the operator on the left hand side of Eq.(21.3.5) (calling the operator L)

$$L = -\epsilon \frac{\partial^2}{\partial x^2} + \frac{1}{4\epsilon} x^2 + \frac{1}{2}. \tag{21.5.12}$$

The eigenvalues λ_n with the eigenfunctions ϕ_n satisfy

$$\left(-\epsilon \frac{\partial^2}{\partial x^2} + \frac{1}{4\epsilon} x^2 \right) \phi_n = \left(\lambda_n - \frac{1}{2} \right) \phi_n. \tag{21.5.13}$$

Clearly

$$\lambda_n = n + 1. \tag{21.5.14}$$

With

$$\phi_n = N_n e^{-x^2/\epsilon} H_n \left(\frac{x}{2\epsilon} \right). \tag{21.5.15}$$

Where H_n is the Hermite Polynomial of order n and N_n is the normalization constant.

Thus,

$$P(x, t) = \sum C_n e^{-(n+1)t} N_n e^{-x^2/\epsilon} H_n \left(\frac{x}{2\epsilon} \right). \tag{21.5.16}$$

Where the C_n are obtained from the fact that at $t = 0$, $P(x, t) = \delta(x)$ and hence $C_n = H_n(0)$. We now obtain

$$
\begin{aligned}
P(x, t) &= \sum e^{-(n+1)t} N_n e^{-x^2/\epsilon} H_{\bar{n}}(0) H_n \left(\frac{x}{2\epsilon} \right) \\
&= \frac{1}{[2\pi\epsilon(e^{2t} - 1)]^{1/2}} e^{-x^2/2\epsilon(e^{2t}-1)}.
\end{aligned}
\tag{21.5.17}
$$

For the D-dimensional oscillator potential x^2 becomes the r^2 where r is the radius vector. For the two-dimensional vector q of our case

$$P(q, t) = \frac{1}{[2\pi\epsilon(e^{2t} - 1)]^{1/2}} e^{-q^2/2\epsilon(e^{2t}-1)} \quad \text{(for short time.)} \tag{21.5.18}$$

The short-time approximation means that the inverted oscillator part of the potential $V(q) = -q^2/2 + q^2/4$ is being experienced. As expected for the inverted potential, the probability distribution spreads out in time and as is obvious from Eq.(21.5.18), the expectation value $\langle q^2 \rangle$ increases as $2\epsilon(e^{2t} - 1) \sim 4\epsilon t$ for short times. The validity of the approximation of dropping q^4 is going to be reasonable for times such that $\langle q^4 \rangle < 1$, which implies $2\epsilon e e^{2t}$ which is for times $t < t_0$ $O(\ln[\frac{1}{\epsilon}])$. For $t > t_0$ it is no longer permissible to drop the q^4 term of the potential. However, yet another simplification sets in.

21.5.2   Regime - II: "Scaling" or "Sliding"

When q becomes somewhat larger, we can no longer ignore the q^4 term but now the effect of noise is minimal while the system slides down the potential. In this limit the ϵ-containing terms in Eq.(21.5.11) can be dropped and we have

$$\frac{\partial P}{\partial t} = \frac{1}{q}\frac{\partial}{\partial q}\left[-Pq^2(1-q^2)\right] = -\left[q(1-q^2)\frac{\partial P}{\partial q} + 2(1-2q^2)P\right]. \quad (21.5.19)$$

Substituting $\tau = e^t$ and $q^2 = y$, we find

$$\frac{\tau}{2}\frac{\partial P}{\partial \tau} = -\left[y(1-y)\frac{\partial P}{\partial y} + (1-2y)P\right]. \quad (21.5.20)$$

This has the same form as the Callan-Symanzik equation and hence the solution can be written down as

$$P(y,\tau) = \frac{1}{y(1-y)}\Phi\left(\ln \tau^2 \cdot \frac{(1-y)}{y}\right) \quad (21.5.21)$$

where Φ is any function. The functional form of Φ is determined by matching between regimes (i) and (ii). In regime (i), the solution has the form $P \sim e^{-q^2/\epsilon(e^{2t}-1)}$, which is valid for $\langle q^2 \rangle < 1$.

To match this form with the solution given in Eq.(21.5.18), we require $\Phi[\ln e^{2t}\frac{(1-q^2)}{q^2}] = e^{-q^2/\epsilon(1-q^2)(e^{2t}-1)}$, which is of course $e^{-q^2/\epsilon e^{2t}(1-q^2)}$ for $e^{2t} \gg 1$, i.e., corresponding to the regime (ii). Taking care of the normalization, we have for the combined regimes (i) and (ii)

$$\begin{aligned} P(q,t) &= \frac{1}{q^2(1-q^2)} \\ &\times\ e^{-q^2/\epsilon(1-q^2)(e^{2t}-1)}. \end{aligned} \quad (21.5.22)$$

The time development of the probability is shown in the sequence of figures in Fig.21.2.

The solution given in Eq.(21.5.22) does show that the probability acquires peaks near $q^2 = 1$, but it does not spread beyond $q^2 = 1$ and this cannot give the

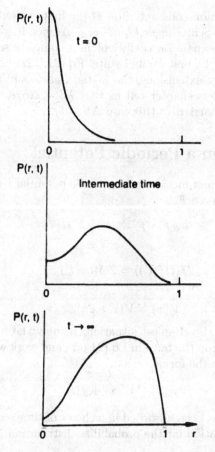

Figure 21.2: The time evolution of $P(q,t)$

true equilibrium distribution as $t \to \infty$. The limit of Eq.(21.5.22) as $t \to \infty$ is the accumulation of the probability at $q^2 = 1$. The true equilibrium is obviously proportional to $e^{\left(\frac{q^2}{2} - \frac{q^4}{4}\right)/\epsilon}$ and hence there is a third regime where the true equilibrium sets in.

21.5.3 Regime-III: Kramers

In this regime which sets in when $\langle q^2 \rangle \sim 1$, i.e., for time scales greater than that obtained from $\epsilon \langle 1 - q^2 \rangle e^{2t} \sim 1$, the noise is important once more and causes occasional transition between the two wells. The solution for P(q, t) in this final regime has the form

$$P(q,t) = N \left[e^{\frac{q^2}{2\epsilon} - \frac{q^4}{4\epsilon}} + a_1 \phi(q) e^{-\lambda_1 t} \right] \qquad (21.5.23)$$

where N is a normalization constant, $\Phi(q)$ is the first excited state wavefunctions in the equivalent Schrodinger equation corresponding to the potential $-q^2/2 + q^4/4$, a_1 is an expansion coefficient in a complete set expansion and λ_1 is the eigenvalue for the first excited state. Eq.(21.5.22) is valid for $t \to \infty$, when in the complete let expansions, the λ_1-term alone survives. The general arguments of the previous chapter tell us that $\lambda_1 = \Delta S/\epsilon$, where ΔS is the height of the potential barrier. In this case $\Delta S = 1/2$.

21.6 Diffusion in a Periodic Potential

The form of the equation of motion for a periodic potential $V(x)$ and an externally applied constant force F is

$$\ddot{x} + k\dot{x} = F - V'(x) + f(t) \tag{21.6.1}$$

where

$$\langle f(t)f(t')\rangle = 2\epsilon\delta(t - t') \tag{21.6.2}$$

and

$$V(x) = V(x + 2\pi). \tag{21.6.3}$$

We will deal vith the simplest situation, namely the high friction limit and consequently will drop the term in Eq.(21.6.1) and work with the Langevin equation which acquires the form

$$\dot{x} = F - V'(x) + f(t) \tag{21.6.4}$$

where the constant k has been absorbed in a choice of time scale. This will lead to a Fokker-Planck equation for the probability distribution $P(x,t)$:

$$\frac{\partial P}{\partial t} = -\frac{\partial}{\partial x}[(Fx - V)P] + \epsilon\frac{\partial^2 P}{\partial x^2} \tag{21.6.5}$$

where 2ϵ is the coefficient of the delta function of the fluctuating force when the re-scaled time variable is used. The stationary solution P_0 is clearly given by $\left(\frac{\partial P_0}{\partial t} = 0\right)$, i.e.,

$$(Fx - V)'P_0 - \epsilon\frac{\partial P_0}{\partial x} = \text{constant} = S \tag{21.6.6}$$

(the prime in $(...)'$ denotes derivative w.r.t x) leading to

$$P_0(x) = e^{-W(x)/\epsilon}\left[N - \frac{S}{\epsilon}\int_0^x e^{W(x')/\epsilon}dx'\right]$$

where

$$W(x) = V(x) - Fx \tag{21.6.7}$$

is the total potential.

We now show that if $P_0(x)$ is bounded for large x, then P_0 is periodic with $P_0(x) = P_0(x + 2\pi)$. The proof proceeds by constructing the integral

$$\int_0^{2\pi n+x} e^{W(x')/\epsilon} dx' = \int_0^{2\pi} e^{W(x')/\epsilon} dx' + \ldots + \int_{2\pi(n-1)}^{2\pi n} e^{W(x')/\epsilon} dx' + \int_{2\pi n}^{2\pi n+x} e^{W(x')/\epsilon} dx'$$

Now,

$$
\begin{aligned}
W(x + 2n\pi) &= V(x + 2n\pi) - (x + 2n\pi) - (x + 2n\pi)F \\
&= V(x) - Fx - 2n\pi F \\
&= W(x) - 2\pi n F
\end{aligned}
$$

and thus the above integral becomes

$$
\begin{aligned}
\int_0^{2\pi n+x} e^{W(x')/\epsilon} dx' &= I + Ie^{-2\pi F/\epsilon} + \ldots + Ie^{-2\pi(n-1)F/\epsilon} \\
&\quad + \int_0^x e^{W(x')/\epsilon} dx' e^{-2\pi n F/\epsilon} \\
&= I\left[\frac{1 - e^{2\pi n F/\epsilon}}{1 - e^{-2\pi F/\epsilon}}\right] + e^{-2\pi n F/\epsilon} \int_0^x e^{W(x')/\epsilon} dx'
\end{aligned}
$$

$$(21.6.8)$$

where

$$I = \int_0^{2\pi} e^{W(x')/\epsilon} dx'. \qquad (21.6.9)$$

Thus,

$$
\begin{aligned}
P_0(x + 2n\pi) &= e^{w(x)/\epsilon}\left[N - \frac{S}{\epsilon} \int_0^{2n\pi+x} e^{W(x')/\epsilon} dx'\right] e^{2\pi n F/\epsilon} \\
&= e^{-W(x)/\epsilon} e^{2\pi n F/\epsilon}\left[N - \frac{S}{\epsilon} I \frac{1}{1 - e^{-2\pi F/\epsilon}}\right] \\
&\quad + \frac{e^{-W(x)/\epsilon}}{1 - e^{-2\pi F/\epsilon}} \frac{SI}{\epsilon} - \frac{S}{\epsilon} e^{-W(x)/\epsilon} dx' \qquad (21.6.10)
\end{aligned}
$$

where we have used Eq.(21.6.9). If we now require that $P_0(x + 2n\pi)$ is bounded for large n ($n > 0$ if $F > 0$ and $n < 0$ if $F < 0$), then first term in Eq.(21.6.11) needs to be removed and hence we require that

$$N = \frac{S}{\epsilon} I \cdot \frac{1}{1 - e^{-2\pi F/F}}. \qquad (21.6.11)$$

Now using this in Eq.(21.6.11) in the non-vanishing part, we have (see Eq.(21.6.7))

$$P_0(x + 2n\pi) = P_0(x). \tag{21.6.12}$$

Because of the periodicity, the probability is normalized in the periodicity interval and we have

$$\int_0^{2\pi} P_0(x)dx = 1. \tag{21.6.13}$$

The mean drift velocity in the stationary state can be found from the Langevin equation given in Eq.(21.6.4) as

$$
\begin{aligned}
\langle \dot{x} \rangle &= \langle F - V'(x) + f(t) \rangle \\
&= \langle F - V'(x) \rangle \\
&= \int_0^{2\pi} (F - V'(x)) P_0(x) dx \\
&= \int_0^{2\pi} \left(S + \epsilon \frac{\partial P_0}{\partial x} \right) dx \\
&= 2\pi S
\end{aligned}
\tag{21.6.14}
$$

where we have used Eq.(21.6.6). The integration constants N and S are obtained from Eqs.(21.6.11) and (21.6.13). Elimination of N between these two equations, and taking the limit of $F \to 0$, yields the mobility μ as

$$\mu = \lim_{F \to 0} \frac{\langle \dot{x} \rangle}{F} = 2\pi \frac{2\pi}{\int_0^{2\pi} e^{V(x)/\epsilon} dx} \cdot \frac{2\pi}{\int_0^{2\pi} e^{-V(x)/\epsilon} dx} \tag{21.6.15}$$

22

Fluctuation Dissipation Theorem

22.1 Introduction

In this final chapter, we consider the response of a system to an external perturbation and show how the effect of the perturbation can be found from the fluctuation spectrum in the absence of the perturbation. Let x be an observable of a dynamical system with Hamiltonian $H_0(x)$ subject to thermal fluctuations. The observable fluctuates about its mean value $\langle x \rangle_0$ and the fluctuation is characterized by the correlation function $S(\omega)$ in frequency space. We now introduce a perturbing term in the Hamiltonian which is explicitly time dependent and the new Hamiltonian is

$$H = H_0(x) - xf(t). \tag{22.1.1}$$

The response of the system to this $f(t)$ is characterized by a linear response function χ, defined by

$$\langle x \rangle = \langle x \rangle_0 + \int_{-\infty}^{t} f(\tau)\chi(t-\tau)d\tau. \tag{22.1.2}$$

The fluctuation dissipation theorem states

$$S(\omega) = \frac{2k_BT}{\omega}\mathrm{Im}\chi(\omega) \tag{22.1.3}$$

where the right hand side is related to the energy dissipated when the system is pumped by a perturbation $f = f_0 \sin \omega t$.

Before considering the sinusoidal perturbation, we tackle the simpler situation of switching on $f(t)$ at $t = -\infty$ and switching it off at $t = 0$, so that $f(t) = f_0\theta(-t)$. The average value of $x(t)$ at any time t can be computed if we know the distribution function $W(x,t)$ at $t=0$ and the transition probability $P(x',t|x,0)$ from x at $t=0$ to x' at t. Then

$$\langle x(t) \rangle = \int dx' \; x' \int dx \; P(x',t|x,0)W(x,0). \qquad (22.1.4)$$

The normalized distribution $W(x,t)$ is

$$W = \frac{e^{-\beta H}}{\int dx' e^{-\beta H(x')}}. \qquad (22.1.5)$$

Treating $xf(t)$ as a perturbation,

$$W = W_0(1 + \beta f_0 x). \qquad (22.1.6)$$

Inserting this W in Eq.(22.1.4), we have

$$\begin{aligned} \langle x(t) \rangle &= \langle x_0 \rangle + \beta f_0 \langle x(t)x(0) \rangle_0 \\ &= \langle x_0 \rangle + \beta f_0 C(t). \end{aligned} \qquad (22.1.7)$$

Using the definition of χ from Eq.(22.1.2), we have

$$\begin{aligned} \beta f_0 C(t) &= \int_{-\infty}^{t} f(\tau)\chi(t-\tau)d\tau \\ &= f_0 \int_{-\infty}^{t} \theta(-\tau)\chi(t-\tau)d\tau \end{aligned} \qquad (22.1.8)$$

Writing $t-\tau = \tau'$, the right hand side of Eq. (22.1.8) becomes $f_0 \int_0^\infty \theta(\tau'-t)\chi(\tau')d\tau'$. Differentiating Eq. (22.1.8) with respect to t with this from of the right hand side immediately yields

$$\beta \frac{d}{dt}(C(t)) = -\chi(t). \qquad (22.1.9)$$

For a stationary process, [where $\langle x(t_1)x(t_2) \rangle$ depends only on $|t_1 - t_2|$], $S(\omega) = 2C(\omega)$, [where $C(\omega)$ is the Fourier transform of $C(t)$] and we arrive at Eq.(22.1.3) form Eq.(22.1.9).

22.2   The General Case

In this section, we present the general derivation by Callen and Welton of the fluctuation dissipation theorem and apply it to have yet another look at the Brownian motion. We begin with definitions.

A system is said to be dissipative if it is capable of absorbing energy when subjected to a time-periodic perturbation (e.g., electrical resistor absorbing energy from an impressed sinusoidal voltage). The system is linear if the power dissipated is proportional to the square of the magnitude of the perturbation.

For the electrical case, power=(voltage)$^2 \frac{R}{|Z|^2}$, Z being the complex number $Z = Z_1 + iZ_2$ with $i = \sqrt{-1}$, $Z_1 = R$ (the resistance), and Z_2 is the inductive and capacitative part. The applied perturbation will be treated quantum mechanically and the power dissipated related to the matrix element of the perturbation operator. The Hamiltonian is,

$$H = H_0 + V(t)\varrho \tag{22.2.1}$$

where ϱ is a function of coordinates and momenta and $V(t)$ is a function of time alone. For a sinusoidal variation $V(t) = V_0 \sin \omega t$. We write the eigenvalues and eigenfunctions of H_0 as $\{E_n\}$ and $\{\psi_n\}$ with

$$H_0 \psi_n = E_n \psi_n \tag{22.2.2}$$

and expand the wavefunction $\psi(t)$ at any time as

$$\psi(t) = \sum_k C_k(t)\psi_k. \tag{22.2.3}$$

The transition probability of a system initially in the state ψ_n is given by Fermi's golden rule as

$$\frac{\pi}{2} V_0^2 \hbar^{-1} [|\langle E_n + \hbar\omega|\varrho|E_n\rangle|^2 \rho(E_n + \hbar\omega + |\langle (E_n - \hbar\omega|\varrho|E_n\rangle|^2 \rho(E_n - \hbar\omega)] \tag{22.2.4}$$

where $\langle E_n + \hbar\omega|\varrho|E_n\rangle$ is the matrix element of ϱ between unperturbed states whose energies are E_n and $E_n + \hbar\omega$ and $\rho(E)$ is the density in energy of the quantum states between E and $E + dE$.

Each transition from E_n to $E_n + \hbar\omega$ involves an absorption of energy $\hbar\omega$ and each transition from E_n to $E_n - \hbar\omega$ involves a release of energy $\hbar\omega$ and hence the rate of absorption of energy by a system initially in the state n is

$$\frac{\pi}{2} V_0^2 \omega \left[|\langle E_n + \hbar\omega|\varrho|E_n\rangle|^2 \rho(E_n + \hbar\omega)|\langle E_n - \hbar\omega|\varrho|E_n\rangle|^2 \rho(E_n - \hbar\omega)\right]$$

Now the thermodynamic probability of a state with energy E_n is the Boltzmann factor $e^{-E_n/k_B T}$ (in general we will denote this probability by $f(E_n)$ and the power dissipated is then

$$
\begin{aligned}
\text{Power} &= \omega \frac{\pi}{2} V_0^2 \sum_n \left[|n\langle E_n + \hbar\omega|\varrho|E_n\rangle|^2 \rho(E_n + \hbar\omega) \right. \\
&\quad - \left. |\langle E_n - \hbar\omega|\varrho|E_n\rangle|^2 \rho(E_n - \hbar\omega)\right] f(E_n) \\
&= \frac{\pi}{2} V_0^2 \omega \int_0^\infty \rho(E)f(E)dE \left\{|\langle E + \hbar\omega|\varrho|E\rangle|^2 \rho(E + \hbar\omega) \right. \\
&\quad - \left. |\langle E - \hbar\omega|E\rangle^2 \rho(E - \hbar\omega) \right. \tag{22.2.5}
\end{aligned}
$$

where we have converted the sum to an integral by noting

$$\sum_n (\ ) \to \int \rho(E)(\ )dE.$$

We now define the impedance Z as the ratio of V to the response $\dot{\varrho}$ and thus

$$V = Z(\omega)\dot{\varrho}. \tag{22.2.6}$$

The instantaneous power is $V\dot{\varrho}R/|Z|$ and the average power absorbed at frequency ω is

$$\text{Power} = \frac{1}{2}V_0^2 R(\omega)/|Z(\omega)|^2. \tag{22.2.7}$$

Comparison with Eq.(22.2.5) leads to

$$\frac{R}{|Z|^2} = \pi\omega \int_0^\infty \rho(E)f(E)\left\{|\langle E+\hbar\omega|\varrho|E\rangle|^2\rho(E+\hbar\omega) - |\langle E-\hbar\omega|\varrho|E\rangle|^2\rho(E-\hbar\omega)\right\}. \tag{22.2.8}$$

We now remove the applied force V and leave the system in thermal equilibrium. In this isolated condition there can be spontaneous fluctuations of $\dot{\varrho}$ with a mean square value $\langle\dot{\varrho}^2\rangle$, which can be calculated as follows:

Suppose the system is in the nth eigenstate. Clearly $\langle E - n|\varrho|E_n\rangle = 0$, since

$$\dot{\varrho} = \frac{1}{i\hbar}[\varrho, H_0].$$

The mean square fluctuation of $\dot{\varrho}$ in the nth state is

$$\begin{aligned}
\langle E_n|\dot{\varrho}^2|E_n\rangle &= \sum_m \langle E_n|\dot{\varrho}|E_m\rangle\langle E_m|\dot{\varrho}|E_n\rangle \\
&= \frac{1}{\hbar^2}\sum_m \langle E_n|H_0\varrho - \varrho H_0|E_m\rangle\langle E_m|H_0\varrho - \varrho H_0|E_n\rangle \\
&= \frac{1}{\hbar^2}\sum_m (E_n - E_m)^2|\langle E_m|\varrho|E_n\rangle|^2 \tag{22.2.9}
\end{aligned}$$

Introducing a frequency $\omega = |E_n - E_m|/\hbar$, the sum over m can be split into two parts $E_n < E_m$ and $E_n > E_m$ changing the summation to an integration, we have

$$\begin{aligned}
\langle E_n|\dot{\varrho}^2|E_n\rangle &= \hbar\int_0^\infty d\omega\omega^2\left\{|\langle E_n+\hbar\omega|\varrho|E_n\rangle|^2\rho(E_n+\hbar\omega)\right. \\
&= \left.|\langle E_n+\hbar\omega|\varrho|E_n\rangle|^2\rho(E_n-\hbar\omega)\right\} \tag{22.2.10}
\end{aligned}$$

The actually observed fluctuation is obtained by multiplying the fluctuations in the n^{th}th state by the thermodynamic weight factor $f(E_n)$ and summing. We get

$$\langle \dot{\varrho}^2 \rangle = \hbar \int_0^\infty \omega^2 d\omega \int_0^\infty dE \rho(E) f(E) \left\{ |\langle E + \hbar\omega| \varrho |E\rangle|^2 \rho(E + \hbar\omega) \right.$$

$$+ \quad |\langle E - \hbar\omega| \varrho |E\rangle|^2 \rho(E - \hbar\omega) \tag{22.2.11}$$

The "voltage" fluctuation is consequently

$$\langle V^2 \rangle = \int_0^\infty |Z|^2 \omega^2 d\omega \int_0^\infty dE \rho(E) f(E) \left\{ |\langle E + \hbar\omega| \varrho |E\rangle|^2 \rho(E + \hbar\omega) \right.$$

$$+ \quad |\langle E - \hbar\omega| \varrho |E\rangle|^2 \rho(E - \hbar\omega) \} dE \tag{22.2.12}$$

Using Eqs.(22.2.8) and (22.2.12), we now need to relate $R/|Z|^2$ with $\langle V^2 \rangle$. To do so, we need to relate the two integrals $C_\pm$, where

$$C_\pm = \int_0^\infty f(E) \left\{ |\langle E - \hbar\omega| \varrho |E\rangle|^2 \rho(E - \hbar\omega) \right.$$

$$\pm \quad |\langle E - \hbar\omega| \varrho |E\rangle|^2 \rho(E - \hbar\omega) \right\} \rho(E) dE. \tag{22.2.13}$$

Working with second part of the integral in Eq.(22.2.13) we note that $\langle E - \hbar\omega |\varrho| E \rangle = 0$ if $E < \hbar\omega$ and hence one can write

$$\int_0^\infty f(E) |\langle E - \hbar\omega| \varrho |E\rangle|^2 \rho(E - \hbar\omega)\rho(E) dE$$

$$= \int_{\hbar\omega}^\infty f(E) |\langle E - \hbar\omega| \varrho |E\rangle|^2 \rho(E - \hbar\omega)\rho(E) dE$$

$$= \int_0^\infty f(E + \hbar\omega) |\langle E| \varrho |E + \hbar\omega\rangle|^2 \rho(E)\rho(E - \hbar\omega) dE.$$

Using the above result in the of Eq.(22.2.13), we get

$$C_\pm = \int_0^\infty |\langle E + \hbar\omega| \varrho |E\rangle|^2 \rho(E + \hbar\omega)\rho(E) f(E) \left(1 \pm \frac{f(E + \hbar\omega)}{f(E)} \right) dE$$

$$= \left(1 \pm e^{-\hbar\omega/k_B T} \right) \int_0^\infty |\langle E + \hbar\omega| \varrho |E\rangle|^2 \rho(E + \hbar\omega)\rho(E) f(E) dE$$

where we have used the fact that $f(E) \propto e^{-E/k_B T}$.

We return to Eq.(22.2.12) to write

$$
\begin{aligned}
\langle V^2 \rangle &= \int_0^\infty |Z|^2\, \hbar\omega^2\, d\omega\, C_+ \\
&= \int_0^\infty |Z|^2\, \hbar\omega^2\, C_- \frac{1 + e^{-\hbar\omega/k_BT}}{1 - e^{-\hbar\omega/k_BT}} \\
&= \int_0^\infty |Z|^2\, \hbar\omega^2 \frac{R}{|Z|^2} \frac{1}{\pi\omega} \frac{e^{\hbar\omega/k_BT} + 1}{e^{\hbar\omega/k_BT} - 1} \\
&= \int R(\omega) \frac{\hbar\omega}{\pi} \left(\frac{e^{\hbar\omega/k_BT} + 1}{e^{\hbar\omega/k_BT} - 1} \right) d\omega \\
&= \int R(\omega) \frac{\hbar\omega}{\pi} \left(1 + \frac{2}{e^{\hbar\omega/k_BT} - 1} \right) d\omega \\
&= \frac{2}{\pi} \int R(\omega) E(\omega)\, d\omega \qquad\qquad (22.2.14)
\end{aligned}
$$

where

$$
E(\omega) = \frac{1}{2}\hbar\omega \left(1 + \frac{2}{e^{\hbar\omega/k_BT} - 1} \right) = \frac{1}{2}\hbar\omega \coth \frac{\hbar\omega}{2k_BT}. \qquad (22.2.15)
$$

The fluctuation dissipation theorem is given by Eqs.(22.2.14) and (22.2.15). At high temperatures,

$$
E(\omega) \approx \frac{1}{2} k_B T
$$

and we get

$$
\langle V^2 \rangle \approx \frac{2k_B T}{\pi} \int R(\omega)\, d\omega. \qquad (22.2.16)
$$

Turning to the question of a particle moving through a fluid with velocity v, we know that the drag force (friction) exerted on the particle is

$$
F = -\eta v \qquad\qquad (22.2.17)
$$

where η is the viscosity. According to the fluctuation dissipation theorem for $k_B T \gg \hbar\omega$, Corresponding fluctuating force has the mean square value

$$
\langle F_x^2 \rangle = \langle F_y^2 \rangle = \langle F_z^2 \rangle = \frac{2k_B T}{\pi} \eta \int d\omega. \qquad (22.2.18)
$$

This is consistent with the fluctuating force written down in the Langevin equation of Chapter 20.

22.3 Jarzynski Equality

We return to the breakup of the Hamiltonian at the beginning of this chapter, where $H(x) = H_0(x) = V(x)$ with the form of $V(x)$ not specified. The partition function is

$$
\begin{aligned}
Z &= \int dx e^{-\beta H_0(x)} e^{-\beta V(x)} \\
&= Z_0 \int dx \frac{e^{-\beta H_0(x)}}{Z_0} e^{-\beta V(x)}
\end{aligned}
\tag{22.3.1}
$$

where $Z_0 = \int dx e^{-\beta H_0(x)}$. The unperturbed distribution function $\rho_0 = e^{-\beta H_0}/Z_0$ and hence Eq.(22.3.1) is

$$
Z = Z_0 \int dx \rho_0 e^{-\beta V(x)} = Z_0 \langle e^{-\beta V} \rangle_0.
\tag{22.3.2}
$$

The free energy is found as

$$
\begin{aligned}
F &= -k_B T \ln Z \\
&= -k_B T \ln Z_0 - k_B T \ln \langle e^{-\beta V} \rangle_0 \\
&= F_0 - k_B T \ln \langle e^{-\beta V} \rangle_0,
\end{aligned}
\tag{22.3.3}
$$

F_0 being the free energy of the unperturbed system. This is a static equilibrium result.

However, imagine that this is an equilibrium situation till $t = 0$ governed by the Hamiltonian H_0 and then a potential (time dependent) is switched on such that at time, t, the energy of the system is

$$
E_t(x) = E_0(x) + V_t(x).
$$

If $V_t \neq 0$ for $t > 0$, then for any $t > 0$,

$$
F_t = F_0 - k_B T . \ln \langle e^{-\beta V} \rangle_0.
\tag{22.3.4}
$$

If the Hamiltonian of a system has an explicit time dependence (that is what $V(t)$ does to the system), then the change in energy form time t_1 to t_2 is

$$
\Delta E = \int_{t_1}^{t_2} \frac{\partial H}{\partial t} dt.
$$

For our situation form $t = 0$ to any $t > 0$

$$
E_t - E_0 = \int_0^t \frac{\partial H}{\partial t} dt = \int_0^t \frac{dV}{dt} dt = V(t).
$$

The change in energy being the work done, W, on the system, we can write Eq.(22.3.4) as

$$\langle e^{-\beta W} \rangle = e^{-\Delta F/k_B T}. \tag{22.3.5}$$

If the perturbing potential is switched off at time 't' and the system allowed to equilibriate, then the initial and final status are both equilibrium states. However, the work done W is during a non equilibrium process and hence Eq.(22.3.5) links an equilibrium quantity with a non equilibrium one. The derivation of the connection, known as Jarzynski equality is only heuristic but gives a flavor of the advances in non equilibrium statistical physics in the late twentieth century.

References

Some of the standard texts on statistical physics are:

- L. D. Landau and E. M. Lifshitz, *Statistical Physics* – Vol. 5 of the *Course of Theoretical Physics*, Oxford: Pergamon Press (1980).

- K. Huang, *Staistical Mechanics*, John Wiley & Sons – New York (1987).

- S. K. Ma, *Staistical Mechanics*, World Scientific – Singapore (1985).

- A. Lahiri *Statistical Mechanics: An Elementary Outline*, CRC Press (2009).

- D. Chandler, *Introduction to Modern Statistical Mechanics*, Oxford University Press (1987).

- R. K. Pathria and P. D. Beale, *Statistical Mechanics*, Butterworth-Heinemann Publications, Elsevier Ltd. (2011).

- M. Kardar, *Statistical Physics of Particles*, Cambridge University Press (2007).

- F. Reif, *Fundamentals of Statistical and Thermal Physics*, McGraw Hill – Singapore (1965).

- R. Baierlin *Thermal Physics*, Cambridge University Press (1999).

- R. Kubo, *Staistical Mechanics*, North Holland – Amsterdam (1965).

- P. T. Landsberg (editor), *Problems in Thermodynamics and Statistical Physics*, Dover Publications (2014).

- D. ter Haar, *Elements of Statistical Mechanics*, Butterworth-Heinemann Ltd. – Oxford (1965).

- C. Kittel and H. Kroemer, *Thermal Physics*, W. H. Freeman and Company – USA (1980).

- J. Sethna, *Statistical Mechanics: Entropy, Order Parameters and Complexity*, Oxford University Press (2014).

- L. E. Reichl, *A Modern Course in Statistical Physics*, John Wiley & Sons – New York (1998).

- M. Plischke and B. Bergersen, *Equilibrium Statistical Physics*, John Wiley & Sons – New York (2006).

Some suggested references for certain chapters are listed below.

Chapter 1

- *Papers and reviews*
  - A. Einstein, *Kinetic theory of thermal equilibrium and of the second law of thermodynamics*, Annalen der Physik (series 4) **9** 417 (1902).
  - A. Einstein, *A theory of the foundations of thermodynamics*, Annalen der Physik (series 4) **11** 170 (1903).
  - A. Einstein, *On the general molecular theory of heat*, Annalen der Physik (series 4) **14** 354 (1904).
  - E. Wigner, *On the quantum correction for thermodynamic equilibrium*, Physical Review **40** 749 (1932).
  - D. Ter Haar, *Theory and applications of the density matrix*, Reports on Progress in Physics **24** 304 (1961).
  - J. Ford, *Stochastic behaviour in nonlinear oscillator systems*: Chapter in *Lecture Notes in Physics*, Springer (1970).
  - S. Popescu, A. J. Short and A. Winter, *Entanglement and the foundations of statistical mechanics*, Nature Physics **2** 754 (2006).

- *Book*
  - J. R. Dorfman, *An Introduction to Chaos in Non-equilibrium Statistical Mechanics*, Cambridge University Press (1999).

Chapter 4

- For powders:
  - *Papers*
    * S. F. Edwards and R. B. S. Oakeshott, *Theory of powders*, Physica-A **157** 1080 (1989);
    * A. Mehta and S. F. Edwards, *Statistical mechanics of powder mixtures*, Physica-A **157** 1091 (1989).

Chapter 5

- For linked cluster expansion:
  - *Review*
    * E. E. Salpeter, *On Mayer's theory of cluster expansions*, Annals of Physics **5** 183 (1958).
  - *Books*
    * D. L. Goodstein, *States of Matter*, Dover Publications (2014).

∗ J. E. Mayer and M. G. Mayer, *Statistical Mechanics*, John Wiley & Sons Inc. – New York (1940).

- Variants on mean-field theory:

  – *Papers and reviews*
    ∗ For Flory-Huggins solution theory: (i) P. J. Flory, *Thermodynamics of high polymer solutions*, Journal of Chemical Physics **10** 51 (1942); (ii) M. L. Huggins, *Theory of solutions of high polymers*, Journal of the American Chemical Society **64** 1712 (1942).

    ∗ B. Widom, *Some topics in the theory of fluids*, Journal of Chemical Physics **39** 2808 (1963).

    ∗ L. P. Kadanoff, *More is the same: phase transitions and mean field theories*, Journal of Statistical Physics **137** 777 (2009).

  – *Books*
    ∗ For Bragg-Williams approximation: T. Hill, *Statistical Mechanics: Principles and Selected Applications*, Dover Publications (1987).

    ∗ J. M. Yeomans, *Statistical Mechanics of Phase Transitions*, Oxford University Press – New York (1992).

    ∗ P. M. Chaikin and T. C. Lubensky, *Principles of Condensed Matter Physics*, Cambridge University Press (2000).

Chapter 7

- For tight-binding approximation in solid-state physics:

  – *Paper*
    ∗ J. C. Slater and G. F. Koster, *Simplified LCAO method for the periodic potential problem*, Physical Review **94** 1498 (1954).

  – *Books*
    ∗ N. F. Mott and H. Jones, *The Theory of the Properties of Metals and Alloys*, Dover Publications (1958).

    ∗ N. W. Ashcroft and N. D. Mermin, *Solid State Physics*, Harcourt College Publishers – Orlando (1976).

    ∗ J. Singleton, *Band Theory and Electronic Properties of Solids*, Oxford University Press (2001).

- For graphene:

  – *Paper and Review*
    ∗ P. R. Wallace, *Band theory of graphite*, Physical Review **71** 622 (1947).

∗ A. H. Castro Neto, F. Guinea, N. M. R. Peres, K. S. Novoselov and A. K. Geim, *The electronic properties of graphene*, Reviews of Modern Physics **81** 109 (2009).

Chapter 8

- *Review*

  – H. L. Stormer, *Nobel Lecture: The fractional quantum Hall effect*, Reviews of Modern Physics **71** 875 (1999).

- *Books*

  – R. E. Prange and S. M. Girvin (editors), *The Quantum Hall Effect*, Springer-Verlag (1989).

  – D. Yoshioka, *The Quantum Hall Effect*, Springer-Verlag (2002).

  – S. Das Sarma and A Pinczuk, *Perspectives in Quantum Hall Effects*, Wiley-VCH Verlag GmbH & Co. KGaA – Weinheim (2004).

  – J. K. Jain, *Composite Fermions*, Cambridge University Press (2007).

  – R. Skomski, *Simple Models of Magnetism*, Oxford University Press (2008).

Chapter 9

- For superfulidity:

  – *Book*
      ∗ J. F. Annett, *Superconductivity, Superfluids and Condensates*, Oxford University Press (2004).

- For Bose-Einstein condensation:

  – *Review*
      ∗ F. Dalfovo, S. Giorgini, L. P. Pitaevskii and S. Stringari, *Theory of Bose-Eintein condensation in trapped gases*, Reviews of Modern Physics **71** 463 (1999).

      ∗ W. Ketterle, *Nobel lecture: When atoms behave as waves: Bose-Einstein condensation and the atom laser*, Reviews of Modern Physics **74** 1131 (2002).

  – *Books*
      ∗ L. P. Pitaevskii and S. Stringari, *Bose-Eintein Condensation*, Oxford University Press – Clarendon (2003).

      ∗ C. J. Pethick and H. Smith, *Bose-Einstein Condensation in Dilute Gases*, Cambridge University Press (2008).

Chapter 10

- For conventional superconductivity and BCS theory:

  – *Books*
    * P. G. de Gennes, *Superconductivity of Metals and Alloys*, Westview Press (Perseus Books Group) – USA (1999).

    * J. R. Schrieffer, *Theory of Superconductivity*, Westview Press (Perseus Books Group) – USA (1999).

- For high-temperature superconductivity:

  – *Paper and Review*
    * D. J. Scalapino, *A common thread: the pairing interaction for unconventional superconductors*, Reviews of Modern Physics **84** 1383 (2012).

    * B. Keimer, S. A. Kivelson, M. R. Norman, S. Uchida and J. Zaanen, *From quantum matter to high-temperature superconductivity in copper oxides*, Nature **518** 179 (2015).

Chapters 11 - 14

- *Papers and Reviews*

  – K.G. Wilson and J. B. Kogut, *The renormalization group and the ϵ expansion*, Physics Reports **12** 75 (1974).

  – M. N. Barber, *An introduction to the fundamentals of the renormalization group in critical phenomena*, Physics Reports **21** 1 (1977).

  – H. J. Maris and L. P. Kadanoff, *Teaching the renormalization group*, American Journal of Physics **46** 652 (1978).

  – D. J. Wallace and R. K. P. Zia, *The renormalisation group approach to scaling in physics*, Reports on Progress in Physics, **41** 1 (1978).

  – J. B. Kogut, *An introduction to lattice gauge theory and spin systems*, Reviews of Modern Physics **51** 659 (1979).

  – E. Brezin, J. L. Gervais and G. Toulouse (editors), *Common trends in particle and condensed matter physics: Proceedings of Les Houches Winter Advanced Study Institute, February 1980*, Physics Reports **67** the entire issue (1980).

  – B. Hu, *Introduction to real-space renormalization-group methods in critical and chaotic phenomena*, Physics Reports **91** 233 (1982).

- *Books*

  – H. E. Stanley, *Introduction to Phase Transitions and Critical Phenomena*, Oxford University Press (1971).

– P. Pfeuty and G. Toulouse, *Introduction to the Renormalization group and to Critical Phenomena*, John Wiley & Sons Ltd. (1971).

– S. K. Ma, *Modern Theory of Critical Phenomena*, Westview Press (Perseus Books Group) – USA (2000).

– J. Cardy, *Scaling and Renormalization in Statistical Physics*, Cambridge University Press (1996).

– D. J. Amit and V. Martin-Mayor, *Field Theory, the Renormalization Group, and Critical Phenomena*, World Scientific – Singapore (2005).

– J. J. Binney, N. J. Dowrick, A. J. Fisher and M. E. J. Newman, *The Theory of Critical Phenomena: an Introduction to The Renormalization Group*, Oxford University Press (1992).

– N. Goldenfeld, *Lectures on Phase Transitions and the Renormalization Group*, Perseus Books – Reading, Massachusetts (1992).

– L. P. Kadanoff, *Statistical Physics: Statics, Dynamics and Renormalization*, World Scientific – Singapore (2000).

– M. Kardar, *Statistical Physics of Fields*, Cambridge University Press (2007).

– C. Domb and M. S. Green (editors), *Phase Transitions and Critical Phenomena*, Academic Press Inc. (London) Ltd.: Volume 1 (Exact Results) – articles by R. B. Griffiths, C. J. Thompson and H. N. V. Temperley (1972); Volume 2 – articles by M. J. Buckingham, B. Widom, P. G. Watson and G. S. Joyce (1972); Volume 3 (Series Expansions for Lattice Models) – articles by G. S. Rushbrooke *et al*, C. Domb (Ising Model), H. E. Stanley and D. D. Betts (1974); Volume 5a – articles by L. P. Kadanoff and M. Luban (1976); Volume 6 – articles by F. J. Wegner, D. J. Wallace, Th. Niemeijer *et al* and C. Di Castro *et al* (1976).

Chapters 15 - 18

- *Books*

  – G. M. Kremer, *An Introduction to the Boltzmann Equation and Transport Processes in Gases*, Springer-Verlag – Berlin (2010).

  – S. Harris, *An Introduction to the Theory of the Boltzmann Equation*, Dover Publications – New York (2011).

Chapters 19 - 22

- General references on non-equilibrium statistical mechanics:

  – *Papers and Lecture*
    * S. Coleman, *Fate of the false vacuum: Semiclassical theory*, Physical Review D **15** 2929.

* C. G. Callan, Jr. and S. Coleman, *Fate of the false vacuum. II. First quantum corrections*, Physical Review D **16** 1762.

* M. Bernstein and L. S. Brown, *Supersymmetry and the bistable Fokker-Planck equation*, Physical Review Letters **52** 1933 (1984).

* D. S. Ray, *Notes on Brownian motion and related phenomena*, arXiv:physics/9903033 [physics.ed-ph].

- *Books*
  * C. W. Gardiner, *Handbook of Stochastic Methods*, Springer-Verlag – Berlin (1985).

  * H. Risken, *The Fokker-Planck Equation*, Springer-Verlag – Berlin (1989).

  * K. Lindenberg and B. J. West, *The Nonequilibrium Staistical Mechanics of Open and Closed Systems*, VCH Publishers, Inc. – New York (1990).

  * R. Zwanzig, *Nonequilibrium Statistical Mechanics*, Oxford University Press – New York (2001).

  * W. T. Coffey, Y. T. Kalmykov and J. T. Waldron, *The Langevin Equation*, World Scientific – Singapore (2004).

• For glasses:
  - *Reviews*
    * S. P. Das, *Mode coupling theory and the glass transition in supercooled liquids*, Reviews of Modern Physics **76** 785 (2004).

    * L. Berthier and G. Biroli, *Theoretical perspective on the glass transition and amorphous materials*, Reviews of Modern Physics **83** 587 (2011).

  - *Books*
    * S. P. Das, *Statistical Physics of Liquids at Freezing and Beyond*, Cambridge University Press (2011).

• Further references on fluctuation-dissipation theorem:
  - *Papers and reviews*
    * H. B. Callen and T. A. Welton, *Irreversibility and generalized noise*, Physical Review **83** 34 (1951).

    * R. Kubo, *The fluctuation-dissipation theorem*, Reports on Progress in Physics **29** 255 (1966).

    * C. Jarzynski, *Nonequilibrium equality for free energy differences*, Physical Review Letters **78** 2690 (1997); *Equilibrium free-energy differences from nonequilibrium measurements: A master-equation approach*, Physical Review E **56** 5018 (1997).

∗ G. E. Crooks, *Nonequilibrium measurements of free energy differences for microscopically reversible Markovian systems*, Journal of Statistical Physics **90** 1481 (1998).

∗ H. Touchette, *The large deviation approach to statistical mechanics*, Physics Reports **478** 1 (2009).

∗ U. Seifert, *Stochastic thermodynamics, fluctuation theorems and molecular machines*, Reports on Progress in Physics **75** 126001 (2012).

Index

Printed in the United States
By Bookmasters